ECOLOGY OF NATURAL RESOURCES IN NIGERIA

Ecology of Natural Resources in Nigeria

OLUSEGUN AREOLA
Former Professor of Geography
Department of Geography
University of Ibadan, Nigeria

Empowerment For Today
Tomball Texas

Published by Empowerment For Today

Library of Congress Cataloging-in-Publication Data

Areola, Olusegun
Ecology of Natural Resources in Nigeria / Olusegun Areola. — Second Edition.
Includes bibliographical references and index.
ISBN 978-0-9968340-2-5
1. Environmental Science — Nigeria. 2. Natural Resources — Nigeria. 3. Ecology — Nigeria. I. Ecology of Natural Resources in Nigeria. Second Edition

Library of Congress Control Number: 2024903782

Printed in USA

For permission requests, write to the publisher at the address below:

Empowerment For Today
112 N. Holderrieth Blvd #614
Tomball, TX 77375
USA

Website: www.empowermentfortoday.com
Email: contact@empowermentfortoday.com

Contents

List of Figures

List of Tables

Preface

In 1991 when the first edition of *Ecology of Natural Resources in Nigeria* was published, my primary purpose was to draw attention to the ongoing disharmony between the environment and the society that was emerging due to onslaught of development activities in Nigeria. This awareness was precursor to calling for greater focus and sensitivity to contextualized planning and implementation of development to strike a sustainable balance between the environment and development. Contextual planning helps to determine strategic direction in view of the many elements that are often interwoven and composed from one locality to another. The accelerated forces of development have not abated, and they should not. The possible environmental damages from the ongoing chaotic development situation that I warned about have manifested preponderantly. The urgency and the importance of contextualized planning has greatly increased to ensure adequate planning and management of the interrelationship between development and the environment.

Consequently, the second edition of this book emphasizes building of the knowledge base needed to support contextualized planning of the interrelationships between development and the environment. It covers analysis of resource processes in Nigeria and assesses the performance of the different players in each sector, the challenges they have faced and the proffered solutions or lack of any; and the consequences on the people, the Nigerian economy, and, most important, the landscape and the natural resources themselves. Such an assessment is important to the furtherance of Nigeria's knowledge economy.

The first edition included a preponderance of examples from southern parts of the country, especially southwestern Nigeria, which was a reflection on the lopsided nature of studies and availability of data and information on resources at that time. In putting together this new edition, I had the advantage of numerous studies that had been carried out in the country since 1991 when the first edition was published. These studies are more diverse and spread over a much wider cross-section of the country than in the 1980s. This allowed a more detailed and comprehensive analysis of the ecology of natural resources in Nigeria. These studies include not only those carried out in Geography but also in the cognate fields of agriculture, agronomy, forestry, wildlife, geology, environmental engineering and environmental sciences. There are more empirical and field data, as evident in the numerous tables included in this edition, attesting to the increasing richness of the knowledge base for planning in Nigeria. There have been notable initiatives in resource mapping, biodiversity and nature conservation, environmental management, remediation of damaged landscapes, all of which point to what could be achieved in future if the initiatives could be properly contextualized and pursued with the seriousness they deserve. This edition addresses many issues, conflicts and resource management initiatives which were only nascent when the first edition was published in 1991. There is also more focus on resource users at the grassroots level both in the rural and urban settings to draw attention to the growing problem of poverty and its dire consequences for social, economic, and political stability in the country. Nigeria has long grappled with problems of drought and desertification in the northern extremities of the country. The current worldwide attention to climate change is welcome news in Nigeria. Climate change is widely addressed in the book particularly in areas of water resources, irrigation, deforestation, grazing, wildlife management and land use and occupational changes in some parts of the country.

Glossary

aba	fertile broad river valley wetlands heavily cultivated to yams and other crops in eastern Nigeria, especially in Ebonyi State
abiotic	not of biological origin
àkùrọ̀	river valley wetlands cultivated to grow dry season crops in parts of western Nigeria
Basement Complex	a highly varied group of the old rocks of igneous and metamorphic origin in Nigeria
biotic	relating to living organisms
Boko haram	a fanatical, anti-western education, terrorist group operating the Sudan-Sahel zone of Nigeria and neighbouring countries
bunkering	stealing and illegal supply of crude or refined oil to ships, oil tankers or fuel stations in Nigeria
Burtali	cattle routes
cadastral system	a survey, map and register of landed property
catenae	a repetition of a characteristic sequence of soil profiles along hillslopes in areas of undulating topography with similar underlying rocks
cation-exchange	an internal chemical phenomenon in soils whereby a positively charged ion (cation) held on the surface of a colloid (clay or humus particle) can be exchanged for another cation present in the surrounding electrolyte
Climate Vulnerable Forum (CVF)	a 43-nation group of most vulnerable countries that negotiate as a bloc at the *United Nations Framework Convention on Climate Change*
diurnal	daily recurring phenomenon
diurnal fluctuation	the changes in magnitude of any climatological parameter recorded during the solar day

ecotone	a transitional or halfway stage between two ecosystems
edaphic	soil characteristic
eluviation	the washing out of fine materials from a soil horizon or layer, for example, from the top layer
epiphytes	plants that grow attached to the surface of another plant, like moss on a tree; unlike *parasites* they do not feed on the host and are usually harmless to it.
fadama	broad river valley wetlands or flush sites commonly cultivated in the northern drylands
firki	heavy impermeable clays commonly associated with seasonally inundated depressions in the Chad basin
forbs	soft, woody shrubs
Fulani	nomadic ethnic group
harmattan	a dry, dust laden wind blowing from the Sahara Desert
Hurumi	dry season grazing land
insolation	the intensity of solar radiation on the land
interfluve	area of high ground between rivers
isohyet	a line on a map joining points that receive equal amounts of rainfall
Jiko	a locally made energy-saving wood-burning stove in Kenya
value chain	different ways in which a forest product is transformed into something of value or service to yield income and a means of livelihood to the people
Mashekari	rainy season grazing land
mesophytic	plants that inhabit an area that is neither very wet nor very dry
miombo ombrophilous	a type of dense savanna woodland in East Africa
oxidation	formation of oxides by a combination of oxygen with mineral elements
pedogenesis	soil formation
precision agriculture	high-tech, detailed, microlevel management of farms including all types and stages of farm operations to achieve high yields, high quality and profitability of harvests and sustainability of the agricultural business
Ruga	Fulani word for "human settlement", but also acronym of "***ru**ral **g**razing **a**rea*"

saprophyte	an organism that obtains its nourishment from dead or decaying organic matter
shaduf	type of irrigation practiced in Nigeria using storage ponds and shallow ditches
stratigraphy	study of rock layers or strata
taungya	a farming system whereby local farmers are licensed to cultivate land planted with commercial trees to help control weeds in the early years of the commercial tree plantation
transhumance	mass movement of cattle from north to south in Nigeria during the dry season
xeric	desert-like or drought-resistant

Acknowledgements

The production of the first edition of this book was marred in so many ways that, in putting together this new edition, close attention had to be given to minute details to avoid the pitfalls of that first edition. By divine providence, Professor Isaac Megbolugbe, my former student at the Department of Geography, University of Ibadan, who has remained a trusted friend, an encourager, and a cheerleader, took it upon himself to solicit the services of his bosom friend, Professor Dare Afolabi, a publisher of high repute, who arduously and painstakingly put together this book, from manuscript to camera ready stage. It was a monumental task that Professor Dare Afolabi executed selflessly, expertly, and efficiently. And he did it all gratis! There's no way I can adequately appreciate the immense contributions of these two friends, but I thank them from the depths of my heart.

In producing this restructured and improved edition of the book, I benefited greatly from the thorough review and suggestions for improvement made by Professor Isaac Megbolugbe and another former student and colleague Professor Boyowa Chokor, Department of Geography, University of Benin, Benin City, Nigeria. I depended on many other former students and colleagues for illustrative materials, especially, photographs from different parts of the country.

All the maps and diagrams in this new edition of Ecology of Natural Resources in Nigeria were modified or adapted and redrawn from original sources and duly acknowledged as shown in the credits beneath each illustration. The following maps were modified, redrawn, and used with the kind permission of the individuals and publishers indicated: Figures 2.1, 5.5, 5.7, 6.1, 6.2, 8.1, 8.2 and 9.1 by Macmillan Nigeria; Fig. 10.1 by Odeh; Fig. 7.4 by the FAO.

Figures and diagrams carried over from the first edition of the book were originally permitted to be used by relevant publishers: Fig. 1.3 by Wadsworth Publishing, Inc., Belmont, California; Fig. 5.1 by Edward Arnold, London; and Figs. 7.2, 7.3, 7.5 and 7.6 by Heinemann Educational Books, Ibadan.

All the maps and diagrams were painstakingly redrawn by Mr. Joseph Muyiwa Olumoyegun of the GIS Laboratory, Department of Geography, University of Ibadan.

This edition of the book contains many more pictures than the first edition and they are all new images reflecting the present Nigerian landscape across Nigeria. The pictures were generously provided by some of my former students and colleagues: Professor Mala Daura, former Vice-Chancellor, University of Maiduguri, Dr. Abubakar Fada, Department of Geography, Usumanu Dan Fodiyo, University, Sokoto; Professor Emmanuel Oriola, Department of Geography and Environmental Management, University of Ilorin; Mr. Emmanuel Attah and Jibri Edicha, Town Planners based in Abuja; Town Planner James Edide based in Calabar; Dr. Joy O. Nwaezeigwe, Department of Geography, Obafemi Awolowo University, Ile-Ife.

Dr. Adebayo Eludoyin of Obafemi Awolowo University provided valuable data on river discharge across Nigeria, while Boye Adeyemi, an attorney and manager at Shell Petroleum Company provided the photographs from the oil producing areas of the Niger Delta.

The encouragement of Prof. Niyi Gbadegesin and Dr. F. O. A. Dada, and their support in sourcing for relevant literature, maps and other illustrations for the book, are gratefully acknowledged.

1

Nature, natural resources and people

Knowledge economy

The purpose of the current edition of this book is to provide an assessment of environmental stewardship in furtherance of Nigeria's knowledge economy. The book presents an analysis of resource processes in Nigeria and assesses the performance of the different players in each sector, the challenges they have faced and the proffered solutions or lack of any; and the consequences on the people, the Nigerian economy, and, most important, the landscape and the natural resources themselves.

Knowledge economy or knowledge-based economy (1.1) development is dependent on the quantity, quality, and accessibility of the information available, rather than on the means of production. Knowledge is used to create goods and services. Knowledge and education constitute human capital, which can be conceived as a productive asset or business (educational/intellectual) product that can be traded and exported profitably. Therefore, development is largely determined by the proportion of enlightened and skilled workers in the economy of a locality, region, country, or the world. In this knowledge economy most jobs require specialized skills, which also call for access to information and information technology to move the economy forward right from the locality or grassroots level to the regional, national, and global levels.

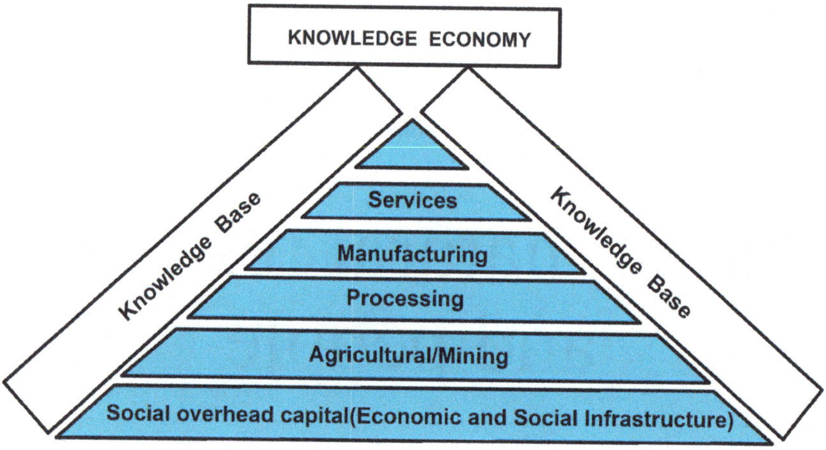

FIGURE 1.1 Knowledge economy (Source: Megbolugbe, 2013)

The global economy is propelled by access to the intellectual property of different countries or regions including trade and industrial secrets, patented products and processes and copyrighted materials. Information on resource processes, their strengths and vulnerabilities in localities, regions and the country forms part of trade information economy that influences development planning. However, significantly, the ecological perspective on the environment and its resources emphasizes knowledge of and due regard to the environmental limits of all development activities to achieve the mutually reinforcing goals of sustainable development and sustainable environments and natural resources. Natural resources are a means to an end, but the sources of these resources, the nature and the structure of the means used to deploy them and the ends to which they are put by people must be carefully thought through for the gifts of nature to be sustainably transformed to promote development and prosperity at the local, regional and national levels in Nigeria. In this book, we present a systematic analysis of natural resources in their relationships with specific sets of people (cultures, occupations), places (ecological environments), and the relevant scientific (intelligent), policy, and socio-economic platforms.

Human ecology

The field of human ecology is currently going through a process of re-examination. Popular ideas and misconceptions are being redefined or corrected. This process is a return to the core concepts and values of ecology, from which the discipline of human ecology has strayed. A most welcome recourse is the ecological concept of limits. There are limits to what people and society can do with and to the natural environment without jeopardizing its sustainability and that of its constituent natural resources. According to Fish (2020) the earth is not an empty spacious world, but an asset that does so many things that are critical to our existence and survival. The limits to the use of these assets are set, not by technology or "human capital," but by nature or the environment itself. This is not to deny the immense capabilities and achievements of human ingenuity, knowledge and technology in the exploitation and utilization of the vast potentials of the earth and its various components. But it is to emphasize that ignoring the dangers to the sustainability of the environment constitutes a threat to the survival of humanity on earth.

The developed countries are already plotting and working towards the time when this earth would no longer be able to sustain its growing population and demands on the available resources. They have lofty dreams of escaping to another planet or, in the alternative, mining the resources of other planets for use on earth. There are concerted and vigorously pursued programmes to curtail world population, which some believe could be achieved by suppressing birth rates in the less developed parts of the world. A more productive and realistic course of action would be to develop a healthy respect for the earth environment and the preservation of its ability to continue to sustain humankind within the limits set by it. This calls for deeper and more detailed knowledge of the earth environment system and its subsystems, which are manipulated by humankind to achieve its own ends. It calls for greater study and analysis of resource processes in every clime and culture to promote conservation and sustainable development. This is critical to the modern-day knowledge economy that is driving development at all levels, local, regional, national and international. It is important to provide the knowledge base for the management of the environment and ecological resources to aid planning, development and management at the local, regional and national levels in a country such as Nigeria. The national economy is a portfolio of regional economies while the regional economy itself is a portfolio of local economies comprising cities, towns and villages, farmsteads, factories and other workplaces where individuals pursue their livelihood activities and make a living. Knowledge that is tied to specific locations and areas is what is needed in the modern knowledge economy to be able to make decisions and take actions that are appropriately contextualized.

Nigeria's position in the world economy has worsened increasingly since the structural adjustment programme of the 1980s. Nigeria's economy collapsed despite the unprecedented earnings made from oil exports in the last five decades or so. The population is poorer; the gap between the rich and the poor has widened as has that between the urban and rural areas; and between the low income and high-income residential areas of cities. The natural environment has suffered from poor management and uncontrolled exploitation of natural resources, a situation that has been compounded by the impacts of climate change, such as, for example, recurrent droughts, flash floods and locust and weaver bird invasions. Nigeria's oil wealth has not been put to good use to build better the quality and productive capacity of the natural environment and resources. The social and political spheres have been bedevilled by corruption, wasteful spending by both federal and state governments, mismanagement of resources, high levels of unemployment and underemployment, low minimum wage and poor living conditions, all of which have fuelled social insecurity and political instability. Hence, in addition to natural disasters, the country's natural environment and resources have suffered considerable damage from militant activities in the oil producing areas; clashes between pastoralists and arable farmers; and insurgent activities by both home grown and foreign terrorist groups finding willing recruits in the teeming population of uneducated and unemployed youths living at the edge of society in certain parts of the country. In some other parts of the country the landscape has been destroyed by erstwhile neglected, unemployed youths turned bandits and illegal prospectors and miners of rare gems and mineral ores.

There is no better time than now to review the environmental stewardship of the government, private sector corporate operators and the millions of grassroots users of land and its resources in Nigeria. Environmental stewardship has been defined by the United States Federal Environmental Protection Agency as the responsibility for environmental quality shared by all those whose actions affect the environment. It assesses the responsible use and protection of the natural environment through conservation and sustainable resource management and utilization practices. Such an assessment represents a significant portion of the knowledge base needed for effective contextualized planning and environmental resources management.

Ecological viewpoint

This book emphasizes the ecological viewpoint to the production and utilization of knowledge essential for appraising our natural resources management.

First, it explains the concept and practice of analyzing resource processes with illustration from resource processes in Nigeria and assesses the performance of the different players in each sector, the challenges they have faced and the proffered solutions or lack of any; and the consequences on the people, the Nigerian economy, and, most important, the landscape and the natural resources themselves. Such an assessment is important to the furtherance of Nigeria's knowledge economy.

Second, it investigates the ecological basis and the environmental consequences of natural resource utilization in Nigeria. In this introductory chapter we shall also discuss some of the ecological principles pertaining to resource utilization and management.

Resource processes

A resource process (Fig. reffig1.2) was defined by Firey (1960) as the total flow of a resource from its state in nature through its period of contact with man to its disposal. Resource processes can be studied in a variety of ways, four of which were identified by Simmons (1974). The first is the economic perspective which considers the ways in which societies attempt to match the supply of resources to the demand for them. The rate and magnitude of supply is determined largely by market forces, in particular, the prices people are willing to pay for them. As demand increases so does the need to increase production. The economic perspective emphasizes the need for continued growth and production of resources in order to meet the needs of the ever-increasing population.

A second approach is that which is concerned with the ethics of resource utilization by people. It judges how people ought to use the biosphere and its resources both for particular cultures and for humankind as a whole. Every element of nature is seen to have value whether of an economic, cultural or aesthetic nature. The third approach is the behavioural one which identifies the socio-cultural traits and psychological impulses that cause different societies to make use of the land and its resources in the way they do.

Finally, there is the ecological approach which views each resource process as a set of interactions between the biotic and abiotic components of the biosphere. Human manipulation of these systems has repercussions on the environment as a whole. There is a limit beyond which people cannot go without causing serious imbalances in nature. Thus, the ecological perspective on resource processes stresses the environmental limits of human activities and human exploitation of natural resources.

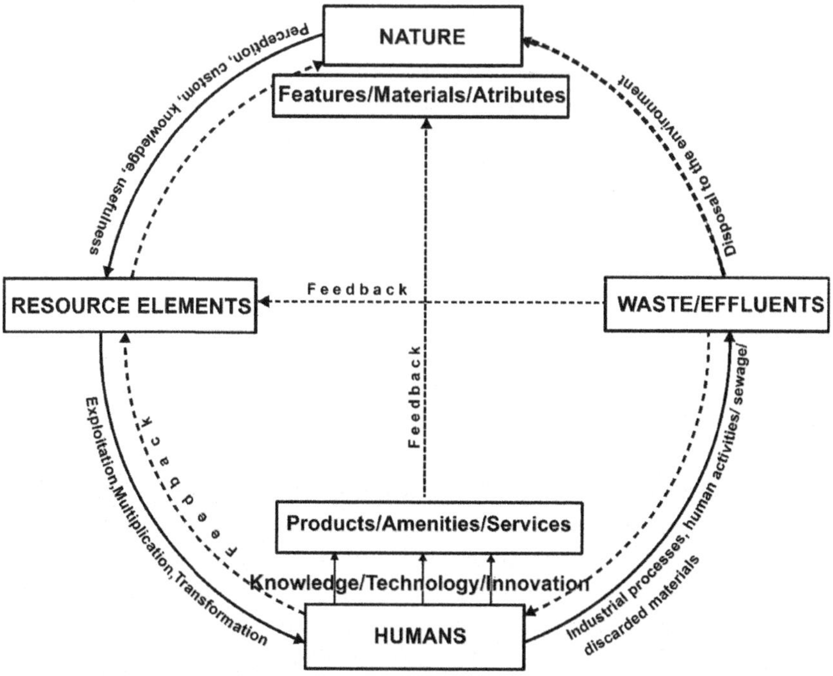

FIGURE 1.2 Schematic diagram of a resource process

Ecological viewpoint

This book emphasizes the ecological viewpoint to the production and utilization of knowledge essential for appraising our natural resources management. First, it explains the concept and practice of analyzing resource processes and then, it illustrates by analyzing various resource processes in Nigeria and assesses the performance of the different players in each sector.

Principles of ecology and resource utilization

The ultimate concern of ecology is humankind in the physical environment. However, humans cannot be studied in isolation from other forms of life such as plants, animals and micro-organisms which form part of their environment and the basis of their existence on earth. Therefore, ecology is the study of that portion of the earth called the biosphere, which contains life. This biologically active portion of the earth occupies the interfaces between the atmosphere, the lithosphere and the hydrosphere (Fig. 1.3), which are the major components of the physical environment. Living things interact with one another and with the physical environment in a constant exchange of matter and energy which we call an ecosystem. Odum (1959) defined an ecosystem as any area of nature that includes living organisms and non-living substances interacting to produce an exchange of materials between the living and non-living parts. People are the dominant factor because they have acquired skills which enable them to manipulate the ecosystem to a considerable extent. People extract matter and energy from the ecosystem for their own use, and term them resources.

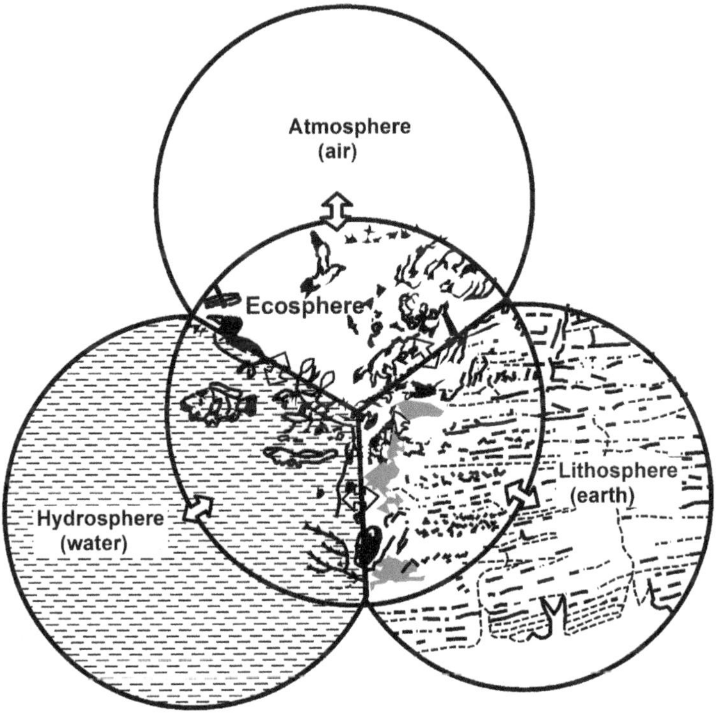

FIGURE 1.3 The ecosphere (or biosphere) Source: G. T. Miller, 1975

Resources may be classified into two broad groups: *products* and *amenities*. Products include tangible, consumable or directly utilizable materials such as minerals, timber, wild game, water and soil. Amenities involve the transformation of natural features into consumable forms such as when a waterfall, wind speed or solar radiation is used to generate electricity or when a sand beach is turned into a recreation centre. A natural resource can also be an attribute of nature which can be exploited by people to their own advantage. For example, traditional tropical agriculture depends on the natural regenerative power of vegetation for the restoration of soil fertility during the fallow period; people also manipulate the hydrologic cycle to obtain more water and take advantage of the interdependence of the elements to control certain crop or animal pests.

Nature provides opportunities but it is the people who determine which of the elements and attributes of nature constitute resources. Culture is the fusion of the elements of human knowledge, way of life and technological capabilities. Besides influencing their selection of resources, it determines their ability to harness those resources.

While people extract matter and energy from the environment, they also give something in return. Thus, human impact on the environment in any area depends on (1) the magnitude, rate and manner of extraction of resources by people; and (2) the magnitude, rate and quality of the feedback. A crucial element in either case is the production and management system adopted, in other words, the skills and techniques applied in land use activities. For instance, in assessing the possible impact of agriculture on the landscape, the crucial factors are likely to be the farming system; the management inputs in terms of implements, fertilizers, manures, pesticides, and herbicides; the intensity of cultivation; and the soil and land conservation measures being applied.

The management input is very important in that it largely determines the extent to which people can capitalize on the ability of many natural resources to replenish and maintain themselves. As Owen (1975) noted, the best type of management for a given resource depends upon the quantity, mutability, renewability and re-usability of the resource.

Some resources exist in such large quantities or are so continuously available that they may be regarded as *inexhaustible.* Some of these are immutable, in that they cannot be significantly altered or degraded by people, for example, energy from fissionable materials, wind power, precipitation .and the water power of tides. But some are capable of being misused and their quality impaired by people, for example, solar power, the atmosphere and water bodies which may suffer from the effects of environmental pollution.

On the other hand, many resources are *exhaustible*. Fortunately, a good number of these are also *renewable* and can be maintained more or less permanently depending on the way they are used. Such resources include soil, water and forest products. Others, such as certain species of wildlife or specimen wildernesses, are *non-renewable* in that once they disappear there is no hope of recreating them.

Minerals cannot be maintained indefinitely since the total quantities available are finite, and can be described as *wasting assets*. However, some minerals such as gems, precious metals, some types of iron, copper and aluminium, are *re-usable* in that, however much they are used, they can still be reworked into new products. By contrast, fossil fuels and most non-metallic minerals cannot be recycled because they are totally consumed in use (Owen, 1975, pp. 13-14).

These concepts should provide useful guides in devising rational methods of managing the resources of a country.

Some fundamental principles of resource management

The ecosystem concept provides important guiding principles for resource management. One such principle is that of *reciprocity* between the living and non-living parts of the ecosystem. When people make use of any component of the ecosystem, they may bring about changes in other components. For example, the cropping of a piece of land will bring about changes in the nutrient status of the soil.

Furthermore, interactions within the ecosystem are very complex indeed and this is reflected in the interlocking pattern of the *food chain* which is better described as *food web*. One implication of this is that most components and, therefore, most resource elements, have multiple functions in the natural environment, and most resources are therefore capable of *multiple use*. Delicate and knowledgeable management is required in order to take advantage of the multiple functions of a resource without damaging its ecosystem. For example, careful management may make it possible to use forest lands for lumbering and wildlife conservation without impairing its natural ability to protect natural water supplies.

Another consequence of the complexity of ecosystems is that they are normally capable of accommodating minor disruptions by people. For example, if a species of animal is removed from an ecosystem, others capable of performing its functions will move into the vacant niche. However, major disruptions usually set up chain reactions within the ecosystem which could not be predicted because, until recently, they had been little understood. For example, the human influence on the ecosystem is very often to simplify the food chain, reduce the lines of interaction and deplete its energy resources by removing energy-rich matter. This reduction in the diversity and vitality of the ecosystem greatly disrupts its stability. Overgrazing, over-exploitation of timber in the forest and over-cultivation of the land are examples of the destruction of the vitality and, so, the stability of ecosystems by people.

The concept of *energy flow* through the ecosystem is very important. The energy input from the sun controls life and thus everything that makes up natural resources. Energy enters the ecosystem as free solar energy and is converted by green plants into chemical energy in the process known as *photosynthesis*. Some of this energy is dissipated as heat during plant respiration. What is left is converted into organic matter and forms part of the living weight or *biomass* of plants.

The whole process by which solar energy is converted to organic matter is known as *biological production*. In agriculture most of human effort is geared towards increasing the productivity, that is, the yield, of cultivated crops. In doing this, all other elements of the ecosystem are subordinated to crop production. Weeds are plants which people do not make use of and which may compete with cultivated crops for the available ecological resources and which, therefore, should be removed. In the same way, animals, insects and birds, which, in any way, inhibit the production of the plants or animals favoured by humans, are termed pests and are destroyed.

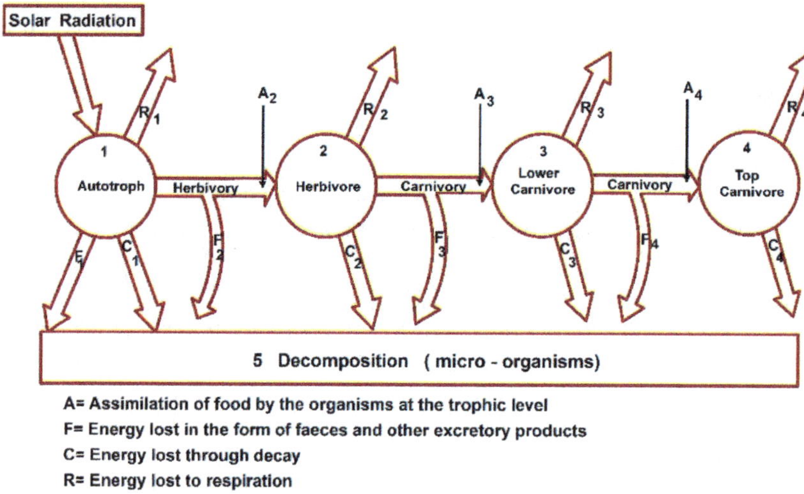

FIGURE 1.4 Energy flow and the food chain (Source: Clapham, 1973)

The energy fixed by plants and locked up in organic matter is passed through the ecosystem by means of the food chain (Fig. 1.4). At each trophic level, the potential energy in organic matter is broken down and reconstituted for the use of the organism. Some of the energy is again dissipated through respiration. This dissipation of energy at each stage is known as entropy. As a result of entropy, the energy available to humans decreases down the food chain.

The circulation of nutrient elements within the ecosystem is very important. These nutrients include oxygen, carbon and nitrogen and inorganic elements such as calcium, magnesium, potassium, sodium and phosphorus. These circulate between the biosphere and the physical environment following fixed pathways (referred to as *nutrient* cycles) but the rate of turnover is rather slow. Furthermore, some elements are locked up in such a way that they are not readily accessible for use by living organisms. In modern times, people increasingly interfere with nature to increase the efficiency of these cycles or to compensate for the short supply of some nutrients by adding artificial fertilizers. The overall result of these actions has been to speed up the natural cycles while introducing larger quantities of materials than they can cope with. The consequences have been noticeable in the pollution of the environment, the eutrophication of water bodies by excess nutrients from the crop fields and the net loss of mineral elements from the land to the sea through leaching, runoff and erosion.

Finally, biological functions in the ecosystem, especially biological productivity, have limits dictated mainly by the amount of solar energy received but also by abiotic factors, like climate and mineral nutrients. The *concept of limiting* factors, which is the basis of the need to restrict human exploitation of natural resources, is of paramount importance since the rate of replacement of the renewable resources is itself limited.

The *growth rate* of a given species population of plants or animals in the ecosystem is governed by the *birth* and *death rates* of the species, as well as the movement of individuals in and out of the population. The success with which the species competes with similar groups in the ecosystem is also crucial to its growth and survival. The maximum growth rate that a species population can ever attain under ideal environmental conditions is referred to as its *biotic potential*. But this is rarely achieved because as the population density increases *environmental resistance* sets in to check further growth. Environmental resistance refers to all environmental, biotic and physiological stresses to which a species population is susceptible and which limit its growth. The population size, at the stage where environmental resistance effectively checks further growth is referred to as the *carrying capacity* of that habitat. This is the point at which the resources available are just sufficient for the size of the population. This concept of the carrying capacity of land is very important, for instance, in animal grazing. It depends not only on the nature and quality of the pasture and the climatic conditions but also on the quality of the vegetation and livestock management techniques.

Integrating the ecological viewpoint and other perspectives on resource processes

The ecological perspective on resources is quite useful in examining the relationship between people and natural resources in Nigeria. The essentially rural nature of the country and the economy is, perhaps, an obvious reason for this. Most of the people still live close to nature and engage in primary production activities which, to a degree, are influenced by natural forces and processes. Secondly, the ecological approach stresses the interrelationship between all the components of the natural environment. It is important for planning purposes to look at a given environment as an integrated unit: no one resource element should be developed without paying due attention to the possible consequences on the other elements. This is particularly important now that Nigeria is rapidly developing her economy and is making more demands on the land and its resources. This is particularly essential because the use of nation-state and society construct in national development discourse has land and people as the two core targets that require mobilization nationwide and throughout society to trigger authentic national development processes. Consequently, the ecological perspective complement efforts that seek to activate factors of production: land, labor or people, capital, and entrepreneurship.

The ecological perspective on natural resource utilization has been much misunderstood especially by the advocates of economic growth who seem to believe that an ecological view is necessarily a sterile one involving the setting aside of large areas of land as unproductive nature reserves. The truth is that the ecologist emphasizes the need for conservation through scientific management and development to ensure the sustained yield of natural resources. The more advanced the system of production and the more sophisticated the technology applied in the exploitation of natural resources, the more crucial to the environment are those potential limiting factors. Thus, adherence to the ecological perspective on resource processes becomes more and more critical as Nigeria develops technologically and economically.

Since the last two or three decades of the 20th century, attempts have been made to integrate the ecological and economic perspectives on environment and natural resource management beyond the concept of sustained yield. The emphasis shifted to maintaining sustainable environments. A sustainable environment is one that is able to meet the demands placed on it in a given economy without diminishing its capacity to allow the people to live well in the present and in the future. This means that the prevailing rates and intensities of exploitation and use of renewable natural resources, the depletion of non-renewable resources, and the pollution and manipulation of the supporting ecosystems or environment must be sustained more or less indefinitely. This means first that there is sustainable yield of renewable resources such that the rate of exploitation does not exceed the rate of regeneration or renewal. Secondly, the rate of depletion of non-renewable resources is matched by the development of alternative renewable substitutes such as, for example, solar, wind or water energy in place of fossil fuels. Thirdly, the rate of waste (pollutant) generation does not exceed the absorptive capacity of the environment at any given time. Finally, the ecosystems that support all these continue to provide the necessary ecosystem services that promote environmental sustainability (Daly, 1990, Goodland and Daly, 1996).

However, a sustainable environment is not attainable if the economic and socio-political conditions in the country are not conducive. It is extremely difficult to advocate and promote a sustainable environment where the vast majority of the people are still grappling with meeting basic necessities, such as, food, shelter and clothing or where there is social unrest, political instability or, worse still, civil war. Hence, environmental sustainability must be accompanied by economic sustainability and social sustainability. The problem of promoting sound environmental and natural resources management in Nigeria, as in other African countries, is inextricably linked to the thorny issue of poverty alleviation at the rural grassroots level. As noted by Areola (2001), in the case of deforestation and its consequences, the problems facing environmental and natural resources management in Nigeria require that policy makers and development agencies focus more on what is happening at the grassroots level than at the national or global level. The need for contextualization is so important because the activities of the millions of people at the grassroots level are having the greatest impact on the natural environment and natural resources in Nigeria. Their activities are mostly informal and, therefore, much more difficult to organize and control, but the essence of their practices is not impossible to discern. These practices could be radicalized even before formalization could be achieved. The operators are moved less by appeals to preserve the productive capacity and the aesthetic quality of the environment as by the need to meet the necessities of life and survive in an increasingly hostile urban-dominated social and economic system. Indeed, it would make great sense to weave strategies for achieving sustainable natural environments and natural resources into comprehensive community-based participatory poverty alleviation programmes at the grassroots level directed ultimately to progressive advancement of the welfare of the society.

Policy makers and development agencies will have to work closely with local institutions including hometown associations or development unions, farmers' unions and cooperatives, various occupational guilds, thrift and credit societies, women societies to harness the underlying principles and meaningful practices that could be transformed into appropriate technologies for achieving sustainable environments. It is inconceivable that poverty alleviation programmes in the rural areas will not be based on the available local land and natural resources. Therefore, the philosophy of natural resources development programmes must be that the local people will not only be involved in project formulation, planning and implementation but must also be the first beneficiaries of the proceeds of such programmes. Although, there is increasing interest and participation of educated or literate young people in agricultural enterprise in the rural areas, they are still too few to replace the millions of traditional rural farmers, who are still likely to continue to produce a large proportion of the food and raw materials required by the urban-industrial sector of the Nigerian economy

It is now a firmly accepted wisdom since the days of the defunct Natural Resources Conservation Council of Nigeria, to involve the local people in environment and natural resource development projects (Ajayi, 1991). This goes beyond the practice of seeking the views and preferences of stakeholders entrenched in the environmental impact assessment process. It involves a *community-based participatory approach* to the formulation, design and implementation of development projects that pertain to the livelihoods of the people. It is accepted that the first beneficiaries of any such projects will be the local communities themselves. This approach is to ensure the cooperation of the local people that is so vital to the success of development initiatives particularly in the rural areas. The rural communities must take ownership of resource development projects in their neighbourhoods for such projects to have any reasonable prospects of sustainability.

A participatory approach should also recognise the importance of the accumulated *indigenous knowledge* of local peoples on the environment, natural resources and resource use and management practices. Policy makers and development planners on the environment and natural resource use and management may discover that adaptation of local people's management practices may yield greater results than wholesale introduction of technologies from the advanced countries. This was a lesson the colonial administration of Nigeria learnt a long time ago when their mechanised groundnut development schemes in the savanna lands failed woefully. Modern-day development agencies must not make the same mistakes; it is important to seek to understand the crop and animal husbandry practices of the people as they seek to increase the efficiency and productivity of resource processes in the country. It is very instructive that in the advanced regions of the world African-style organic farming has become a major resource process as people have turned against the massive, unbridled use of chemical fertilisers in western primary production systems and its nutritional, health and environmental implications.

Chokor (2019) has suggested also that, recent developments in cognitively-derived measures provide opportunities for engaging ordinary people in the task of environmental assessment with wider benefits to environmental planning and sustainable environmental management. He believes that contemporary developments in environmental economics offer alternative frameworks for putting economic and social values on ecological assets, which hitherto had no market or were grossly undervalued. Economists are now reconciled to the fact that group consensual discussion could be employed to assign value to ecological resources or environmental goods rather than reliance on individual or professional valuation. This represents some methodological meeting point and opportunities for cooperation between ecologists and environmental economists.

There is also the need to gain a better understanding of how African traditional practices could be strengthened and integrated into natural resources management and conservation strategies by exploring the use of space in communities. This is because the use of space conveys a substantial aspect of the various values placed on the environment and the kinds of knowledge gained or manipulated in the process of natural resources use (Chokor, 2018).

For instance, in the field of soil and land evaluation, experts have learnt that indigenous peoples have their own criteria and systems of classifying soil and land, which have proved useful in community-based participatory land capability assessment for various uses. Unfortunately, Chokor (2018) rightly noted that values and culturally filtered regulations are changing and traditional rules of land and resource use are less stringently adhered to or enforced in post-colonial societies. This is because development, modernization, and market processes under capitalism have tended to erode community-held beliefs and encouraged movement away from informal group beliefs and norms to more formal ones. It is important to revert to a more structured study of traditional practices and the sociocultural values, beliefs, and norms that mediate in the use of space, which in turn mediates the management and utilization of ecological resources.

Climate change

Climate change is a fact of life in the 21st century world, protestations to the contrary in some parts of the capitalist western world notwithstanding. The impact of climate change is evident in the land use/land cover changes that have occurred in the Nigerian landscape. The impact of desertification in the Sahel ecological zone is evident in the drying up of Lake Chad while the mobilization and southward advance of sand dunes, which had remained stabilized for many centuries, is the reason that the Green Wall is being planted in the region to stem the advance of the Sahara Desert. All the ecological zones in the country, including the marine, are susceptible to the impacts of climate change. Therefore, climate change must be factored into both current and future policy formulations and natural resource development programmes in the country.

Ahmed (2019) has documented the devastating effects of natural resources degradation in the drylands of Nigeria as a result of global warming, drought and desertification, which exacerbated the impacts of overgrazing and deforestation in the region. In the coastal belt of the country also climate change has become an important consideration because of the vulnerability of millions of people at the grassroots levels whose livelihoods might be affected by ocean surge and flooding brought about by global warming and rise in ocean levels. The fact that small holder operators at the grassroots level dominate the various resource processes in Nigeria makes the task of combating the impacts of climate change that much more challenging. Strategies adopted for mitigating the impacts of climate change would be critical in efforts aimed at ensuring sufficiency, security and sustainability in all resource processes in the county. In other words, climate change poses a challenge not only in crop and animal husbandry which, for obvious reasons, often attract greater attention, it is also a major threat in the management of the country's forests, water resources, wildlife and biodiversity in general. Moreover, concern about land degradation in Nigeria occasioned by uncontrolled mining or quarrying of precious metals, rare ores and earth materials in certain parts of the country is partly because climate change will surely compound the problem of land devastation in those parts of the country. Climate change, and strategies for promoting rural livelihoods and alleviating poverty at the grassroots level have become recurring issues in environment and natural resources management in the 21st century in Nigeria as in all other African countries. This is the more reason why ecological perspectives on the management of the natural environment and resources are increasingly being mediated by economic and socio-cultural principles as stated above. It is important to create synergies between these different perspectives to build the knowledge and intelligence base required to create sustainable environments and livelihoods at the local grassroots level.

Nigeria's Major Ecological Zones

Details of the different elements of the Nigerian natural environment are given in the relevant chapters that deal with the different resource processes. Suffice it here to give a brief preview of the natural ecology of the country. There are distinct ecological belts in Nigeria from the Atlantic coast to the Sahel in the margins of the Sahara Desert. They include (i) the marine; (ii) the wetland; (iii) the forest; and (iv) the savanna ecological regions (fig.1.5). There are spatial variations in environmental conditions within each ecological region, hence each of them can be divided into smaller more homogeneous sub classes as we shall see in subsequent chapters of this book. Table 1.1 summarises the major climatic conditions in each of the zones and their major subdivisions.

Marine Zone

Nigeria has an appreciable area of marine ecosystems both offshore and onshore along the 800km. Atlantic Ocean coastline. The shoreline is low and lined with barrier sand ridges that created the lagoons running parallel to the shoreline. There are many rivers and creeks emptying into these lagoons thereby diluting the salt waters. The Niger delta region is a maze of creeks, rivers and distributaries that empty into the Atlantic Ocean. The tidal lagoons, creeks, distributaries and estuaries constitute a distinct aquatic environment that is very rich in biodiversity. But, at the same time, these waterways carry the bulk of shipping, local boat transport, log transportation, boat building, artisanal fishing, and oil and gas exploration activities in Nigeria. Both the biodiversity and economic activities on these coastal water bodies and environs have suffered from (i) invasion by exotic aquatic plants, particularly the water hyacinth, *Eichhornia crassipes*; (ii) oil pollution from pipe bursts and oil well blowouts; (iii) heat and air pollution from gas flares; (iv) waste pollution from open defecation around human settlements; and (v) sand quarrying along the lagoons, rivers and creeks.

Offshore, the Nigerian maritime environment extends from the coastal beaches to the continental shelf, the continental slope and the deep ocean sea bed. It is an area rich in biodiversity, earth materials (sands, gravel, red clay) and mineral resources (heavy metals, calcareous shells, glauconite, phosphorite). The bulk of the oil and gas prospecting and exploitation activities in the country is now carried

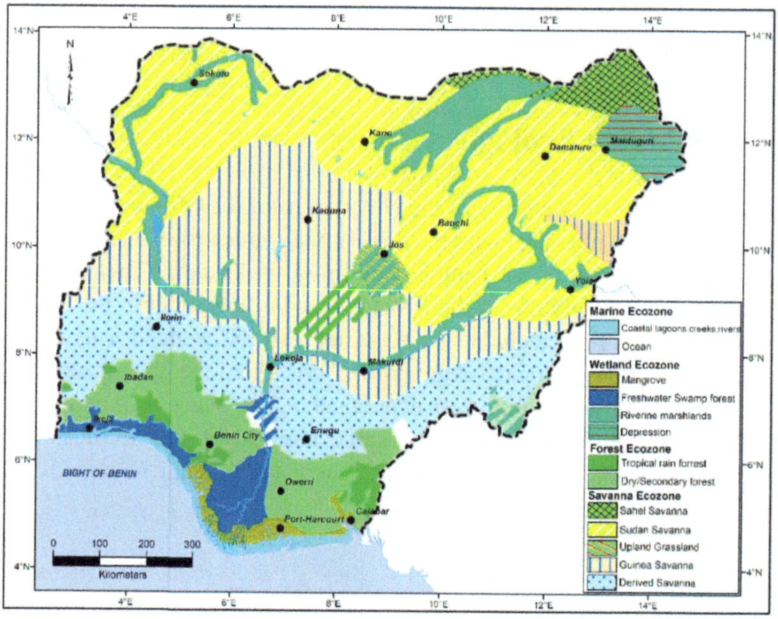

FIGURE 1.5 Major ecological zones of Nigeria

out offshore in the country's 256,000 km^2 Exclusive Economic Zone. The area also supports many livelihood economic activities focused on the gathering of sea weeds, marine snails, turtles, clams, crabs, shrimps, prawns, lobsters, and a wide range of fish species.

Wetlands

The coastal and deltaic wetlands comprise the mangrove swamps found in areas of brackish water and the freshwater swamp forests, which cover most of the Niger River Delta and coastal river valleys. The underlying marine and alluvial deposits are mostly sand and silt and mud. The barrier ridges have poorly developed sandy soils (regosols) while the lower slopes and valley have acidic hydromorphic soils (fluvisols, gleysols, nitisols). The mangrove swamps are found along the river estuaries and lagoons where brackish water stagnates. These mangrove swamps are important as habitat and/or breeding grounds for fish and other terrestrial and aquatic life forms. They also provide refuge for many lifeforms whose survival in the coastal region has been greatly threatened by the advance of human activities and

TABLE 1.1 Climatic data of major ecozones

Major ecozones	Annual rainfall (mm)	Rainy days per year	Min temp.	Max temp.	Mean annual temp.	Altitude (m)
Fresh water swamps and coastal vegetation	3,000–4,400	290	21°C	32°C	25°C	0–15
Lowland Rainforest	2,000–3,000	250	20°C	36°C	25°C	15–150
Derived Savannah	1,500–2,000	220	15°C	36°C	25°C	150–600
Jos Plateau (Montane)	1,300–1,500	185	13°C	31°C	22°C	1,200
Mambila Plateau (Montane)	1,500–1,800	200	10°C	30°C	20°C	1,500–1,850
Sudan Savannah	800–1,000	90	14°C	38°C	27°C	300–700
Southern Guinea Savannah	900–1,200	200	18°C	36°C	27°C	150–200
Northern Guinea Savannah	1,000–1,300	180	14°C	35°C	25°C	600–700
Sahel Savannah	200–400	80	13°C	40°C	27°C	300–700

Source: NAERLS 1992

the attendant deforestation. The vegetation and rich biodiversity of the mangrove forests have suffered also from invasion by the exotic Nypa palm, *Nypa fruticans,* which has spread rapidly to dominate many mangrove swamps

The deltaic freshwater swamp forests are among the few remaining areas of mature tropical rain forests in the country and are the target of efforts by both local and international organizations to create nature reserves and wildlife sanctuaries in the costal belt of Nigeria. The forests are under pressure from agriculture (especially, plantain cultivation); road construction, quarrying of sand and other earth materials and expansion of human settlements and industrial establishments.

Inland the riverine and lake shore marshlands and woodlands constitute the major wetland ecosystems. While they are important as habitat and breeding grounds for fish, crabs, birds (indigenous and migratory) and reptiles they are under pressure from flood land agriculture, modern irrigation and municipal water supply schemes.

Forest

The forests comprise (i) the tropical rainforest, and (ii) the secondary forest. The tropical rainforests now occur in patches along the sedimentary rock belt that runs west-east from the border with Benin Republic to the Cameroon border hills and mountains. The sedimentary rocks include the Coastal Plains Sands, and Tertiary and Cretaceous sediments (sandstones, shales). The soils are highly weathered, leached and acidic *ferralsols* (*Ultisols, Oxisols, lixisols*). The tropical rainforests are the major timber yielding forests remaining in the country and they are dwindling in area very fast. The forest region has been taken over by vast plantations of exotic and indigenous tree species; and tree crops such as cacao, oil palm, rubber, and fruit trees. The more extensive and dense forests are restricted to the major forest reserves that abound in the belt. Much of the former rainforests has been degraded to secondary forest. Inland, the equatorial climate of the tropical rainforest belt gives way to the tropical wet and dry climate with highly seasonal rainfall. Therefore, the rainforest has given way to "dry forest" which is rapidly being turned into "derived savanna" by human activities. Crop cultivation (root, tubers, maize, rice), charcoal burning, grazing, bush burning, urbanization and infrastructural developments are the main drivers of deforestation and conversion to derived savanna.

Savanna

The savanna ecological region covers more than three-quarters of Nigeria's land area, characterized by highly seasonal rainfall and drier, more compact, ferruginous soils, including ferrisols and luvisols, developed mostly over crystalline Pre-Cambrian Basement Complex rocks and metasediments as well as Tertiary and Cretaceous sedimentary rocks, primarily sandstones. There are five main subtypes of savanna: (i) Derived savanna; (ii) Guinea savanna; (iii) Sudan savanna; (iv) upland grasslands (Jos, Mambila upland areas); and (v) Sahel savanna, a semi-arid scrub land. The savanna region encompasses major game reserves and national parks in the country, including Yankari, Kainji, and Old Oyo National Parks. The savanna zone is noted for (i) cultivation of grain crops (mainly maize, guinea corn, millet, rice) and legumes (groundnuts, benniseed, soya beans, potatoes), and (iii) nomadic livestock grazing and its attendant incessant and widespread annual bush burning. The upland grasslands provide a more equable, tsetse fly free environment for livestock grazing. The savanna lands have been the major focus of the modern irrigation schemes and agricultural development projects, which have been the major preoccupation of the River Basin and Rural Development Authorities established since the 1970s. The region has witnessed serious conflicts between farmers and herdsmen.

Land use/land cover changes, 1975–2013

The launching of NASA's Earth Resources Technology Satellite (ERTS-1, later renamed LANDSAT-1) in July 1972, marked the advent of space-borne earth resources imaging systems that make it possible to obtain, store and analyse real time data on the earth environment on a continuous basis in order to monitor and digitally analyse the changing character of the earth's surface through time. For example, the United States Geological Surveys (USGS) processed sequential time series LANDSAT images to analyse land use/land over changes in West Africa between 1975 and 2013 using scenes for the years, 1975, 2000 and 2013 (CILSS, 2016). The analysis for Nigeria indicated that the country's landuse/landcover had witnessed rapid and significant changes in the period covering the last two-and-a-half decades of the 20th century and the first one-and-a-half decades of the 21st century. Rainfed agriculture doubled its area coverage over the period from 20 per cent to 40 per cent (that is, 380,000km^2) of Nigeria's territorial expanse. Thus, Nigeria had by far the largest area under cultivation of all the countries of West Africa. Nigeria also was the only country in the region where agriculture had traded places with savanna and doubled its area.

The CILSS (2016) noted further:

> The expansion of agriculture was observed across ecoregions from the forest zone of southern Nigeria, where root and tree crops dominate, to the forest-savanna transition of the centre of the country, where many root crops are found, to the grain belt of semi-arid northern Nigeria. The Niger and Benue Basins, promoted as a prime agricultural development area and future bread basket since the 1970s, have seen the most prominent encroachment of agriculture into the savanna, sharpening the outlines of the remaining protected areas. Not all protected savanna areas have been spared from the fast agricultural expansion.
>
> While the transition from savannah to agriculture constituted the largest land cover change in terms of area, some changes in the smaller land cover categories also stand out as important. High rates of change were observed for settlements, irrigated agriculture, plantation, and open mines, with gains accelerating from 1–2 per cent per year between

> 1975 and 2000 to 2–4 per cent per year in the 2000–2013 period. Under the pressures of a rapidly growing population and economy, forests, gallery forests and woodlands, in addition to the savanna land cover types, were all being diminished, with loss rates increasing to over 2 per cent per year during 2000–2013 period. Forest area decreased by 45 per cent from 1975 to 2013.
>
> While the area of mangroves and swamp forests along the coast and in particular in the Niger Delta–important hotspots of biodiversity – has decreased less than that of some other land cover types, the health of these ecosystems has been severely damaged by recurrent oil spills caused by accidents, poor maintenance, and sabotage of the large-scale oil extraction infrastructure in the Delta. Environmental regulations are weak and rarely enforced, and there are no effectively protected areas in the Delta, whose forest and animal populations are considered under severe threat (World Wildlife Fund, 2016). CILSS, 2016 p.167

Abbas (2009) also analysed the pattern of land cover change in Nigeria between 1975 and 2005 using 1975 Landsat images and 2005 SPOT XS images. The findings in Table 1.2 are broadly similar to those of CILSS but Abbas was concerned more about the environmental processes that triggered or resulted from the land cover changes. For example, Abbas inferred that the noticeable expansion of agriculture to new areas, also highlighted in the CILSS study, was primarily as a result of degradation of prime arable lands in other parts of the country particularly the northern Sudan savanna areas. Thus, one can infer that the expansion of agriculture into new territories was no more than shifting cultivation on a grand scale.

Abbas noted further that the marked increase in the areas of disturbed forest, extensive small holder agriculture and rainfed arable crop plantation suggested an intensification of the southward movement of the grazing zone. The marked increase in flood plain agriculture and intensification of cultivation in the *fadama* was due to the fact that the adjoining dry lands were already close to the climatic limits of cultivation. There was over 50 per cent decrease in the area of undisturbed forest.

TABLE 1.2 Changes in some land use-land cover in Nigeria between 1975 and 2005

		Area (km^2) 1975	Area (km^2) 2005	Change 1975–2005	%change
1	Agricultural Tree Crop plantation	824.15	1656.88	832.73	101.0
2	Alluvial	523.61	282.38	−241.23	−46.1
3	Discontinuous grassland dominated by grasses and bare surfaces	7614.72	12517.23	4902.51	64.4
4	**Disturbed Forest**	**14677.70**	**19491.29**	**4813.59**	**32.8**
5	Dominantly grasses with discontinuous Shrubs and scattered trees	13053.77	12487.62	−566.15	−4.3
6	Dominantly shrubs and dense grasses with a minor tree component	118529.55	85020.98	−33508.57	−28.3
7	Dominantly trees/woodlands/shrubs with a subdominant grass component	154933.40	83281.15	−71652.25	−46.2
8	**Extensive small holder rainfed agriculture.**	**170837.55**	**192892.33**	**22054.77**	**12.9**
9	Extensive Small Holder Rainfed Agriculture with Denuded Areas	447.88	10118.47	5700.58	129.0
10	**Floodplain Agriculture**	**9671.81**	**21576.03**	**11904.21**	**123.1**
11	**Forest Plantation**	**1000.85**	**1581.24**	**580.39**	**58.0**
12	Forested Freshwater Swamp	18564.71	16696.51	−1868.20	−10.1
13	Graminoid/sedge Fresh Water Marsh	5882.74	1136.51	−4746.22	−80.7
14	Grass Land	1196.74	8146.74	6950.00	580.7
15	Gullies	125.35	19070.48	18945.13	15113.2

Continued on next page

Table 1.2 – continued from previous page

		Area (km²) 1975	**Area (km²) 2005**	**Change 1975–2005**	**%change**
16	Intensive row crops	3292275.97	373481.34	44253.37	13.4
17	Irrigation Project	148.85	1008.86	860.01	577.8
18	Livestock Project	51.02	139.65	88.63	173.7
19	Major Urban	1102.58	1362.37	259.79	23.6
20	Mangrove Forest	10157.12	10067.31	−89.81	−0.9
21	Minor Urban	958.69	4022.98	3064.29	319.6
22	Montane Forest	7900.02	8053.76	153.74	1.9
23	Montane Grassland	2502.27	3898.15	1395.88	55.8
24	Natural Water bodies / Ocean	6766.53	15588.36	8821.83	130.4
25	**Rainfed Arable Crop Plantation**	**15.92**	**521.38**	**505.46**	**3175**
26	Reservoir	1331.41	2901.16	1569.75	117.9
27	Riparian Forest	7506.46	5330.46	−2176.01	−29.0
28	Rock Outcrop	1445.15	2647.96	1202.81	83.2
29	Salt marsh /Tidal Flat	18.84	596.92	578.08	3068.37
30	**Sand dunes/ Aeolian**	**1032.77**	**5428.30**	**4395.53**	**425.6**
31	Shrub/Sedge Graminoid freshwater Marsh/Swamp	17749.63	10251.68	−7497.95	−42.24
32	Teak/Gmelina plantation	624.44	1156.43	531.99	85.19
33	**Undisturbed Forest**	**28022.42**	**13477.90**	**−14544.52**	**−51.90**
34	Canal		30.76	30.76	
35	Mining Areas		61.15	61.15	

Source: Department of Geography, ABU, Zaria (Abbas, 2009)

Table 1.2 – concluded

Forest plantation increased by 58% but this was due to afforestation and the shelter breaks project in some northern states to counter the encroachment of desert conditions on the Sudano-Sahelian zone. The area of sand dunes increased by 425 per cent during the period, 1975–2005. In essence, the land cover changes over Nigeria in the last 50 years clearly indicated a degradation of the quality and productive capacity of the land and natural resources in most ecoregions from the coastal zone to the Sahel at the desert margins.

Overview

After this introductory discussion of the relationships between people and nature, the ecology of each of the major resource processes in Nigeria is discussed in Chapters 2 to 9. In dealing with the major resource elements the progression is from those resource processes that involve the least human inputs to those that involve more human intervention and input and, which result in drastic alterations of the natural ecosystem. Focus is not only on the large-scale commercial or export-oriented operators but also, significantly, on the myriads of grassroots and informal sector operators. Chapter 10 gives some perspectives on future environmental and natural resource management in Nigeria.

2

Forests and forest resource processes

In the analysis of land use/land cover changes discussed in Chapter 1, it was noted that the forest cover declined by 45% between 1975 and 2013. While, by 2005 the coverage of 'undisturbed forests' had dwindled by over 50%, the area of 'disturbed forests' had increased by over 32% and that of forest plantations by over 120%. This chapter focuses on the use made of forests outside the protected zones, that is, outside the nature reserves (locally called forest reserves) where plantation forestry is predominant. While the timber or wood resources have attracted more attention, the forests also provide a wide variety of non-timber products that provide foods, animal feed, medicines, stakes, pole wood, fibres, leaves and other materials for craft and household use. Thus, the resource processes under consideration in this chapter are (i) the harvesting of non-wood or non-timber forest products; and (ii) the extraction of roundwood from the forests, including the savanna woodlands, primarily for timber, pole wood, firewood and charcoal making. While in the protected zones the utilization of forest resources is planned and, at the least, some effort is made to enhance quality and sustain yield, the forests outside the protected zones have suffered much abuse. This is especially because the forests attract millions of grassroots rural operators that depend so much on them for their means of livelihood. Nigeria's natural forest and savanna communities, like those in other tropical regions, are among the most complex and delicately balanced of natural ecosystems (Fig. 2.1). It is therefore not surprising that any form of human interference leaves scars on the landscape and the vegetation that take long to remedy.

Aspects of tropical forest ecology

A broad review of the tropical forest environment and ecology is necessary in order to place the impact of man's exploitation in the proper perspective. Of all terrestrial ecosystems the tropical rain forest probably exercises the greatest influence on the immediate atmosphere and the underlying soil. It creates its own internal climate which is most crucial to the sustenance of its diverse life forms and the stability of the whole ecosystem. Much of the negative impact of man on forests results from the modification of this unique internal climate.

Longman and Jenik (1974) summarized existing data and information on the tropical forest environment with particular reference to West Africa. The elements of the forest environment include light, temperature, water supply, soils and the nutrient cycles, and the biotic factors.

Light

Different intensities of light and variation in its spectral composition affect plant growth, reproduction and biological production and thus, indirectly, the structure of the forest. The greatest light intensity is received at the tops of the tallest trees, which are referred to as the emergents. The twigs of these trees are often covered by such epiphytes as ferns, orchids and lichens which can tolerate full light. The interior of the emergent crowns is still well illuminated but the light intensity depends on the density of the foliage. Relative light intensity in crowns with dense foliage may be only 24 per cent of full light. At the closed canopy of the middle and lower tree layers the relative light intensity drops to about three per cent of full light. This zone is characterized by intense competition for light among plants. The forest floor is the dim layer where relative light intensities of less than one per cent have frequently been recorded.

Light intensity, especially at the forest floor, changes with the angle of incidence of the sun's rays. It also changes with the seasonal variations in atmospheric cloudiness or haziness. Horizontal or spatial variation in light intensity is especially important because it determines plant distribution (the undergrowth) on the forest

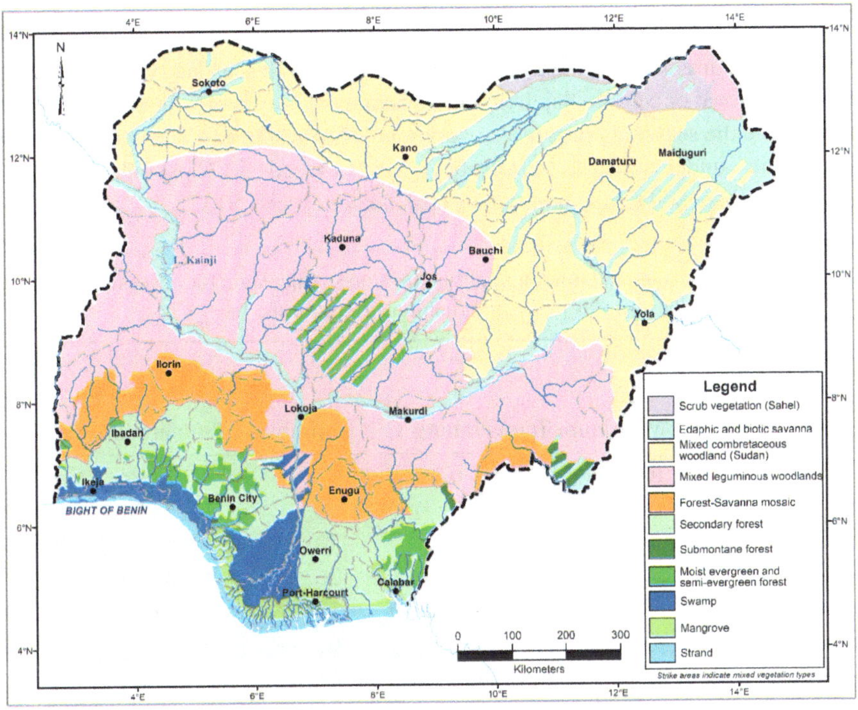

FIGURE 2.1 Major Types of Vegetation in Nigeria (Macmillan Nigeria, 2010)

floor.

Forest is composed of plants with varying degrees of tolerance to light. The shade provided by the upper tree layers is crucial to the survival of many plants of the undergrowth among them seedlings of small-size representatives of the upper layer trees prevented from growing up by the low light intensity inside the forest.

Temperature

According to Longman and Jenik (1974), mean monthly temperatures for the meteorological stations in the tropical forest zone give the impression of uniformity from month to month. However, these monthly mean temperature values disguise diurnal and periodical fluctuations. Maximum day time and minimum night temperatures can vary greatly (Table 2.1) and considerable temperature differences occur between the various layers of the forest (Table 2.2).

Soil temperature

Soil temperature fluctuates less than air temperature both seasonally and diurnally. Furthermore, at or below a depth of about 75cm there may be no diurnal fluctuation at all while the seasonal range is also very small (Schulz, 1960). The maximum soil temperature under a closed forest probably never exceeds 30°C in contrast to open clearings where soil temperature may occasionally exceed 50°C.

TABLE 2.1 Maximum and minimum air temperatures (calculated for a weekly period) at various heights in the ombrophilous forest of the Ivory Coast

Height above ground (m)	Maximum Temperature °C		Minimum Temperature °C	
	July	February		July
46	25.8	32.3	46	25.8
33	–	31.1	33	–
26	24.1	30.3	26	24.1
12	23.8	30.0	12	23.8
6	–	28.5	6	–
1	23.8	28.2	1	23.8

Source: Longman, K. A. and Jenik, J. (1974)

TABLE 2.2 Diurnal range of air temperature (calculated for a week period) at various heights in the closed ombrophilous forest of the Ivory Coast

Height above ground (m)	Diurnal Range of Temperature	
	December	June
46	10.8	4.0
33	10.0	3.8
26	9.9	3.4
12	6.6	2.8
6	5.2	2.2
1	4.4	1.7

Source: Longman, K. A. and Jenik, J. (1974)

Water supply

Although tropical forests are found in high rainfall regions, they thrive best on freely-drained rather than swampy soils. Thus, from the ecological point of view, most tropical forests are mesophytic. Usually, there is variation of soil and plant communities along relief catenae. Soil moisture is a crucial factor in determining such local patterns of vegetation.

Rainfall is the chief source of water supply in tropical forests; but dew, fog, and low clouds are also important sources. In Nigeria forest occurs in areas with at least 1143mm annual rainfall. Apart from the actual water supply, the relative humidity of the air is an important ecological factor in tropical forests. Moist atmospheric conditions both around and within the forest are crucial to the survival of its varied life forms. However, prolonged saturation of the air is undesirable because it interferes with the process of evapotranspiration. This is an important process which affects the transportation or circulation of mineral elements within the plants.

Day-time relative humidity varies widely from the upper tree layer (70 per cent) to the forest floor (90 per cent). In forest margins affected by seasonal rainfall or the harmattan, day-time relative humidity values may fall, thereby temporarily creating dry weather conditions. For example, relative humidity on the forest floor can drop to about 70 per cent which may be too dry for certain plant species in the undergrowth.

Nutrient cycling

Nutrient cycling in the forest is restricted largely to the top few centimetres of the soil that is rich in humus. The reason for this can be found in the nature of forest soils. The soils are predominantly deeply weathered and intensely leached ferralitic soils with low base saturation (see Chapter 6). The clay fraction consists mainly of kaolinite with abundant iron and aluminium oxides; therefore, the cation-exchange capacity of the soils is very low except for the humus-rich top layer. In addition, the reserves of weatherable minerals are low. The soils are acidic in reaction with some having pH values as low as 4.0. The C/N ratio may be as high as 10.

Because of the poor nutrient holding capacity of the soils the greater proportion of nutrient capital of a mature tropical forest ecosystem is contained in the vegetation particularly in the tree trunks, leaves and twigs (Table 2.3). Nutrients are immobilized for long periods in the tree trunks. Soil nutrient status is maintained through the supply of litter by the vegetation, and through root exudates. The soil nutrients are protected against leaching and erosion by the dense vegetation and root systems which capture the nutrients washed down by drainage water and return them to the top layers of the soil. Indeed, in a mature forest, nutrients are kept more or less in a closed cycle between the vegetation and the topsoil.

Thus, the tropical forest lives precariously on the products of its own decay, and once a forest area is destroyed, regeneration is very slow. With the removal of the vegetation, the ecosystem loses a great deal of energy and nutrient-rich material; the nutrient cycle is greatly disturbed and soil humus content falls. The change in microclimatic conditions, especially the drier air and the more intense insolation on the ground surface, will promote faster rates of organic matter decomposition.

TABLE 2.3 Nutrient elements (in Kg/ha) stored in the upper 30 cm layer of the soil, together with those stored in the biomass plus litter, in a mature secondary stand within a tropical rain forest

	N	P	K	Ca	Mg
Soil	1,830	125	820	2,250	345
Stand	4,580	12	650	2,580	370

Note: After Nye, P. H. and Greenland, D. S. (1960)

Biotic factors

The tropical forest is characterized by its diversity of life and complex interactions between organisms. The diversity of life is thought to be due to the multiplicity of microhabitats. Richards (1952) classified the plants into two types: the autotrophic plants which have chlorophyll and the heterotrophic plants which do not. The *autotrophs* include mechanically independent plants such as trees, shrubs, forbs and graminoids; and the mechanically dependent plants such as the climbers, stranglers and epiphytes. The *heterotrophs* consist of saprophytes such as fungi, bacteria and vascular saprophytes; and vascular and non-vascular parasites.

Insects are the most numerous and prolific of forest fauna. They inhabit particularly the litter layer and the forest canopy. They thus constitute a major link in the nutrient cycles within the forest. Other animals include many tree frogs, tree snakes and lizards, mammals and arboreal birds. Like the plants, there is vertical stratification amongst the animals with particular species feeding, moving and resting at specific layers in the forest. However, the upper tree layers where food is most abundant carry the largest animal populations.

The forests of Nigeria

In order to appreciate the extent of the timber resources of Nigeria a brief survey of the vegetation is necessary. There are two major types of vegetation: forest and savanna. The forests comprise of swamp forests, tropical rain forests and secondary forest regrowth. The swamp forests are found in the coastal and deltaic regions and may be subdivided into two broad categories: the mangrove and the freshwater swamp forests. Undisturbed mangrove forests are found mostly along the coastal islands, the lagoon littorals, the creeks and estuarine zones. Studies of the mangroves near Lagos have shown that the plants lack many of the adaptive features normally associated with species in a saltwater environment. Hence, the mangroves constitute more or less an ecotone between pure mangrove vegetation and freshwater communities. It is probable that the dilution of the saltwater by the numerous streams which flow into the lagoons is responsible for this condition.

The mangroves are dominated by varieties of the same tree species *Rhizophora* which include *R. mangle*, *R. racemosa* (the red mangrove) and *R. harrisonii*. The tree species *Avicennia gaminans nitida* (white mangrove) and the shrub *Laguncularia racemosa* (white button wood, false mangrove) are much less abundant. The most important of the tree species is *R. racemosa,* which, if left undisturbed at the outer margins of the mangrove swamp, may reach up to a height of 45 metres and attain a girth of 2.7 metres. However, frequent cutting for firewood has prevented many of the trees from reaching maturity.

The freshwater swamp forests occur extensively in the Niger delta, on the flood plains of the large coastal rivers, for example, Ogun Forest Reserve; the Osun, Siluko and Osse valleys; and in the smaller valleys dissecting the coastal sandy plains. In many of these forests the ground is waterlogged and water stagnates on or near the surface resulting in the development of a highly organic, or peaty, topsoil. The undisturbed forest consists of tall, slender trees some 30-50m high many of which do not appear to reach maturity before rotting away. This is attributable to the extremely poor anaerobic soil conditions.

Nevertheless, widely scattered through the forests, are substantial numbers of healthy merchantable trees of varying timber value. These include *Symphonia glob uliferia* (hog gum tree), *Zanthoxylum zanthoxyloides* (candlewood), *Piptadeniastrum africanum* (African greenheart), *Ficus spp.* (fig), *Alstonia congensis (boomei)* (alstonia), *Spondianthus preusii, Mitragyna ciliata, Albizia zygia* (West African Albizia), *Napoleona vogelii*, *Lophira alata* (red ironwood), *Nauclea gillettii* (West African boxwood), *Diospyros monbuttensis* (Yoruba ebony) and *Oxystigma manii*. The disturbance of the natural forest by humans seems to give the *Raphia* palm a competitive advantage; it spreads quickly to dominate entire swamps. In general, the main swamp forest canopy is open and irregular. In the gaps, dense, almost impenetrable tangles of shrubs and lianas develop, a factor which, together with the poor ground conditions, makes logging particularly difficult in these forests.

The lowland tropical rain forest was described by Keay (1952), as 'a complicated mosaic of communities of different status and floristic composition'. Structurally, the forest consists of three tree layers, a shrub layer and the undergrowth. Many of the species of forest emergents in the top layer are important economic trees. It used to be assumed that these emergents grew rather slowly and required a fair amount of overhead shade in their early life in order to survive and thrive. But it is now recognized that many of them grow very rapidly in full light conditions. In fact, in the forest they exist for long periods in a state of suppressed growth, under the shade of the canopy, until a gap occurs which allows them to grow up. The gaps are created by the falling down of dead trees. Such gaps are infrequent and widely scattered, consequently only a few of the emergent tree seedlings actually succeed in growing up to pole-size from which threshold they can successfully compete and survive in the forest. Thus, there is a high rate of mortality among the seedlings of emergent species. In contrast many of the middle and lower storey trees have shade-tolerant seedlings which grow up under the dense overhead shade of their parents. Generally speaking, competition is intense among the trees in the forest that only a small proportion of the pole-size seedlings under the shade actually grow into big trees. These are some of the problems which have led to a gradual change in Nigeria's forestry policy and the trend towards growing trees in commercial plantations.

Floristically, Nigerian forests are less rich than similar forests in other parts of the humid tropics. Even so, there is a wide variety of species which increase in number from the drier to the wetter parts of the forest region. The drier forests are dominated by tree species belonging to the family *Sterculiaceae* and to a lesser extent by species of the families *Ulmaceae* and *Moraceae*. Examples of the *Sterculiaceae* include *Triplochiton scleroxylon* (African maple, obeche), *Mansonia altissima* (mansonia), *Sterculia tragacantha* (African tragacanth, Kanaya), *S. rhinopetala* (brown sterculia), *S. oblongata* (now, *Eriobroma oblongata* (gum tree), *Cola gigantea, C. millenii, Pterygota macrocarpa* (African pterygota) and *Hildegardia barteri.* Examples of the *Ulmaceae* and *Moraceae* include *Celtis brownll, C. mildbraedii, C. zenkeri, Holoptelea grandis, Trema guineensis* (syn. *Orientalis*), *Morus mesozygia* (West African mulberry), *Melicia excelsa* (iroko), *Antiaris toxicana (antiaris), Ficus spp.* (fig) and *Trilepisium madagascariense. Terminalia superba* (afara) and *Khaya grandifoliola* (broad leaved mahogany) are also present in the drier forests.

The more humid forests of the Benin region contain species of the families *Meliaceae* and *Leguminosae* such as *Khaya ivorensis* (Lagos mahogany), *Entandrophragma spp.* (African mahoganies), *Lovoa trichiloides* (African walnut), *Guarea cedrata* (pink mahogany), *G. thompsonii* (scented mahogany) *Cylicodiscus gabunensis* (denya), *Distemonanthus benthamianus* (satin wood) and *Piptadeniastrum africanum* (African greenheart). The still wetter forests of south-eastern Nigeria contain such important species as *Lophira alata* (red ironwood), *Nauclea diderrichii* (West African boxwood), *Mimusops spp.* and *Omphalocarpum spp.*

Redhead (1971) classified Nigerian forest tree species into utilization grades on the basis of their timber value. The economic value of the trees is normally enhanced by their long, straight un-branched stems although, in the damp conditions that exist inside the forest, damage to the bole by insects and parasitic growths is not uncommon. One of the problems with timber species is that they do not grow in pure stands.

Secondary forest zone

In its northern fringes the forest has been degraded to secondary forest. This now covers much of the ecoregion that used to be referred to as the 'dry forest' zone (Keay, 1952). In south-eastern Nigeria this zone is represented by the oil palm bush which covers hundreds of square kilometres in Abia, Imo and parts of Anambra and Akwa Ibom States. Once the forest is cleared the soils tend to dry out progressively and some may develop clay pans and lateritic crusts. Tree regeneration is thus made more difficult. Pressure on the land due to high rural population densities, the practice of shifting cultivation and annual bush burning all combine to degrade the forest into derived savanna, in which susceptible forest trees are progressively eliminated leaving only fire tolerant savanna species such as *Parkia clappertoniana* (West African locust bean), *Daniellia oliveri* (West African copal), *Afzelia africana* (mahogany bean tree), and *Butyrospermum paradoxun* (shea butter tree), some of which may yield valuable and locally useful products. Isolated occurrences of such emergents as *Bombax costatum* (Kapok) and *Adansonia digitata* (baobab) may be found.

Savanna vegetation covers nearly three-quarters of the land area of Nigeria but it contains very few timber species. In fact, most of the savannas have a crop fit only for firewood. But forest outliers, especially along rivers and on rocky ground, occur deep into the savanna as far north as southern Zaria and the Adamawa Highlands. In addition, there are gallery forests particularly in the Guinea savanna zone. These forests contain valuable timber trees, many of which have a close affinity with those in the drier parts of the forest region. They include *Lophira lanceolata, Terminalia glaucescens, Afzelia africana* (mahogany bean tree), *Terminalia macroptera, Mitragyna inermis and Daniellia oliveri* (West African copal).

Non-Timber Forest Products

Many people including government officials and forestry experts in Nigeria had tended to see forests as sources of timber or wood with little regard for the non-timber products. This attitude dates to the colonial period when the economy emphasized the export of raw materials, including timber. But the situation has changed considerably in the 21st century as more and more researchers are focusing on the roles of the forests in rural development.

Arnold (2004) highlighted the reasons for the rising interest in non-timber or non-wood forest products in the developing countries. The *non-timber forest products* (NTFPs) have always constituted a large part of the forest economy in developing countries as individual products provide inputs and income directly to huge numbers of rural and urban households. Although the aggregate of non-timber forest products contributes as much, if not more, to national product as industrial roundwood, they have suffered from neglect because they have been regarded as minor forest products produced and consumed largely by the informal sector outside the monetary economy. Until recently, they had attracted limited interest and attention by researchers; therefore, the available data are still quite limited. Arnold (2004) attributed the recent increase in interest in the NTFPs to "shifts in developmental focus" that have made policy and development planners to recognize the importance of rural development and poverty alleviation at the local grassroots level. Forest resource processes that focus on the NTFPs have attracted attention as the non-farm component of the rural economy. In fact, Etowa *et al.* (2015) noted that some economists now view NTFPs extraction and marketing as a better alternative to timber exploitation as a rural livelihood strategy, which could also satisfy the goal of forest resource conservation touted by ecologists and forestry experts. According to Etowa *et al.* (2015), there are indications that the NTFP sector is growing faster than the timber industry (which is in decline due to increasing scarcity of timber species in the forests) due to the fact that the products are important items of cross-border trade in the humid forest zone of West and Central Africa.

FAO (1999) defined "*non-wood forest products* (NWFP)" as forest materials derived from soil mineral, water, fauna and flora resources other than round wood (sawn wood). The NWFP are classified into non-wood and non-timber products. The non-wood products are derived from wild animals, herbs, leaves, latex, gum, resins, ropes, fruits, seeds, fungi, fodder, forage, gravel, clay, limestone and natural salt. Whereas the woody but non-timber products include poles, fuel-wood, charcoal, rattan canes, sponge, chewsticks, bamboo and others. Researchers do not usually make this distinction between NTFPs.

FAO (1999) classified the NWFPs in Nigeria into

(i) forest foods (floral resources) made up of species of fruit trees and shrubs, vegetable plants; spices; mushrooms, seed bearing plants, oil producing plants, and drink producing plants.

(ii) forest foods (faunal resources) e.g. worms, frogs, insects, snails, reptiles, mammals, birds and fish.

(iii) flora products of medicinal value: All plant forms and parts are used for traditional medicine, including leaves, flowers, fruits, seeds, nuts, tubers, roots, seedlings, latex and forest litter.

(iv) fauna products of medicinal value: A lot of faunal resources not accepted for food and those not socially accepted because of taboos are used for traditional medicine. Utilization of faunal products for medicine is clearly based on small part utilization-skins, claws, feathers, bones, faeces, scales, fur and others. Typical wild animal used for food and traditional medicine include termites, rodents, antelopes, duiker, grasshopper, periwinkle, caterpillar of beetle, pangolin, crustaceans, crocodile, molluscs, fish, pangolin, monkeys, grasscutter, gazelle, lizards, snail, tortoise, warthog, ground squirrel, oysters, crabs, snakes, pakara, mudskipper.

(v) Fodder: trees, shrubs, and grasses primarily in the savanna ecological zones

(vi) other non-wood products: natural dyes, leaves and barks, wrapping leaves, palm wine, gum, tannin, toxins, materials/fibres for rope and mat weaving.

Enhancement of rural livelihoods through NTFPs

In terms of marketing, the NTFPs may be categorized as follows: (i) products consumed domestically by producers (Alao *et al.*, 2016); (ii) products consumed and/or marketed at local markets; (iii) products sold at urban and interstate markets; and (iv) products involved in cross-border trade. A number of studies have dealt with the role of NTFPs in enhancing rural livelihoods. There are so many operators in the production and marketing of NTFPs and this makes it difficult to organize and control. Further, many participate in the resource process on part-time basis but those operating across states and cross border invest more in the process. For some products, there are wholesalers and retail sellers (chew stick, snails, bushmeat, etc.). Wholesalers operate across states and cross-border.

According to Aiyeloja *et al.* (2015), in their study in Osun State, Nigeria the seasonality of the availability of many products accounts for seasonal variations in marketing and income generation from the enterprise. Apart from seasonal variation, another disincentive is the availability of substitutes for almost any NTFP one can think of, hence demand is elastic. Etowa *et al.* (2015) noted that the seasonality of products often creates an economic gap in the livelihood system for many dependent communities and have substantial effects on household welfare. Other problems identified include high transport cost and poor transportation, which is tough on those brought from distant places. Loans and credit facilities available to operators in the market economy are not yet available to producers and marketers of NTFPs. Furthermore, the forest area has been dwindling rapidly in many parts of the country.

In spite of the constraints highlighted above, researchers have drawn attention to the fact that NTFPs can be important in diversifying household farm economy and provide counter sources of food, fodder and income. They can provide reserves that can be used for subsistence and income generation in hard times, such as, periods of crop failure, drought, and shortage of wage employment. Finally, some NTFPs feature significantly in the export trade, a typical example being the kernels of *Irvingia spp* traded from the forested south to the northern savanna from where they are exported across borders (Ladipo and Boland, 1994). Omoluabi and Abang (1994) estimated a combined export trade volume of US$1.6822 million (2015 exchange rate) for nine NTFPs from Cross River State, namely, gnetum, irvingia, garcinia, randia, bushmeat, pole, carpillobia and flitches. Tables 2.4 and 2.5 give assessment respectively of the sustainability of some NTFPs and the effects of seasonality.

NTFPs and forest conservation

The resource base for NTFPs is rapidly deteriorating in sub-Saharan Africa and especially in Nigeria. As was discussed in Chapter 1, land use/land cover analysis clearly indicated the rapid disappearance of forest and savanna in relation to land occupied or degraded by farming, overgrazing, mining, urbanization and physical developments. It has also been observed that the harvest/trade in NTFPs does not satisfy the concept of good *extractivism*, which is an approach of NTFPs collection that preserves the natural forest while enhancing income (Etowa *et al.*, 2015). The resource process suffers from a pricing system that does not obey the usual

TABLE 2.4 Sustainability of selected NTFPs in Nigeria

Resource	Part Harvested	Impact of harvesting on sustainability	Level of sustainability
Bush mango	Fruit	Low	Relatively sustainable, good regeneration and community-level cultivation
Erú	Leaves	Low to medium depending on technique	Relatively sustainable if leaves are plucked, and the stem is not cut, but destructive unsustainable harvesting is often undertaken
Chewing stick	Bole	High	Highly unsustainable; species has limited geographical range and is in danger of extinction over the long term
Bush pepper	Leaves & fruits	Moderate	Relatively sustainable if leaves and fruits are plucked and the stem is not cut
Hausa stick	Stem	High	Highly unsustainable because of removal of whole stem, including root collar
Rattan canes	Mature stems	Low to medium	Relatively sustainable at current levels of harvesting

Source: Sunderland, Besong and Ayeni, 2003 (quoted by Etowa *et al.*, 2015)

laws of supply and demand. High prices have not led to increases in the absolute number and variety of products. In fact, with increasing trade, large, exploitation of exportable superior non-tradeable goods will intensify and their abundance will decline. Large slow-growing, and slow-reproducing species will be more prone to extinction in certain localities.

Intervention by forest departments and experts maybe needed to ensure forest conservation and preservation of the NTFPs. The Natural Resources Conservation Action Plan (Ajayi, 1991) recommended the adoption of community-based participatory approach in the conservation and management of Nigeria's natural resources as stated in Chapter 1. The same strategy is also proposed in the Nigerian Economic Empowerment Development Strategies (NEEDS) in its national biodiversity conservation strategy and action plan. The community-based participatory approach that involves forestry departments partnering with local communities involved in

TABLE 2.5 Impact of seasonality on availability of selected NTFPs

Resource	Impact of seasonality	Availability
Bush mango	High	Rainy season type, *I. gabonensis*, available June to September; dry season type, *I. wombolu*, available February to April
Erú	Moderate	All year, although there is less plucking and reduction in supply during early rains as people are occupied with farming
Carpolobia cattle stick	Low	All year, although transportation problems in rainy season restrict supply to markets
Randia chewing stick	Low	All year, although transportation problems in rainy season restrict supply to markets
Garcinia chewing stick	Moderate	All year, although increased availability in rainy season because of better boat access to remote creeks in forest
Bush pepper	Moderate	Fruits produced in dry season; leaves can be harvested all year
Bush onion	Moderate	Fruits produced during rainy season, but after drying, they can be stored for some time.

Source: Sunderland, Besong and Ayeni, 2003 (quoted by Etowa *et al.*, 2015)

NTFPs production and sales, is a viable management option precisely because it promotes local roles and sidesteps forest tenure issues.

But there are many key issues to tackle as identified by NEEDS (Etowa *et al.*, 2015), namely, (i) environmental degradation such as deforestation and desertification,; (ii) weak enforcement of environmental laws including forest conservation policy; (iii) loss of biodiversity; (iv) inadequate environmental data on which to plan, including data on the forest resources; (v) absence of a system of national accounting that captures the contribution of the environment including forest resources (including NTFPs) to development; (vi) most NTFPs are prone to market failure and, so, are not accounted for in GDP calculation; (vii) management of NTFPs has been difficult to coordinate and harmonize at the state (regional) and national levels in the country; (vii) inadequate knowledge of species in terms of biology, ecology and utilization makes it difficult to capture the NTFPs at the state and national levels; and (viii) most NTFPs originate from conservation areas (forest reserves) which are under *in situ* management programmes within the context of common property resource utilization while some are subject to taboos and religious strictures to their harvest and utilization. Arnold (2004) threw more light on the cause of the rudimentary nature of available knowledge on the NTFPs, which relates to "the complexities of researching and understanding such a highly diverse group of products, produced in such a wide range of different ecological and socio-economic situations."

In spite of these constraints, experts still believe that concerns that development activities be consistent with environmental integrity, and not prejudice the future potential of forest and land resources have reinforced arguments that managing the forests for NTFPs might be less environmentally damaging than alternative uses of forests, especially, lumbering (Arnold, 2004). The devolution of control and management away from central governments and institutions to local institutions has made NTFPs a potential incentive to local forest management especially through the community-based participatory approach.

Based on studies of several NTFP value chains in the Cameroon Republic, Ingram (2014) showed that success or failure depended on governance. She showed that the ecological, socio-economic and governance contexts in which the NTFP value chains were located strongly influenced how the chains had emerged, evolved and were currently configured. Cultivation of identified NTFPs in the face of unrelenting deforestation and dwindling reserves of NTFPs in the wild seemed to have produced positive outcomes in many of the value chains because the people responded appropriately to the challenges they were facing. The governance context is characterised by a shift from largely separate customary and formal systems in operation since colonial times, governing respectively access to resources and access to markets, to an increasingly comprehensive regulatory framework. Thus, according to Ingram:

> "Harvesters and new entrants to the chains (farmers, agro-foresters and different community, public and private sector organisations) have found ways to counter scarcity in the wild and vulnerability, with all the NTFPs studied being cultivated in varying degrees. Cultivation rates have been maintained or increased over time. *Cola spp.*, *Raphia spp.* and *Apis mellifera adansonii* have all been cultivated between at least the last 50 to 100 years, such that the majority of products now originate from cultivated sources. Beekeeping has replaced wild honey hunting in the Northwest and also largely in Adamaoua in the last twenty years. Products such as *Prunus africana* have been cultivated more recently as demand and their values increase, and wild sources diminish. *Irvingia spp.* has been semi-cultivated in the last forty years, and bamboos, *Acacia spp.* and *Gnetum* in the last twenty-five years. This suggests a general trend ... of increasing cultivation as demand rises and availability in natural forests decreases, mirroring global trends of declining forest quality through deforestation and degradation with expanding small and largescale tree planting in tropical countries...
>
> The livelihoods context has shaped how the products are valued, used and traded. One of the most defining socio-economic contexts, prevailing over time, is poverty. The majority of the most numerous groups of actors in the chains, harvesters and retailers, live on or under the poverty level... Given this context, the majority have engaged in NTFP harvesting, seeking to diversify and add sources of cash household income to meet their most basic needs for food, schooling, healthcare and housing. For the species with multiple, largely food, medicinal and fuel uses, these products also contribute to livelihoods, although with lesser importance than their cash contribution." (Ingram, 2014, p. 256–257).

But Ingram also noted further that "the fairly long periods of time over which harvesters have engaged in trade, around eight years, suggest both a continued reliance on NTFPs but also entrapment in poverty, particularly for those chains where incomes are lower than the national average"(p.257) This may be so, but it does not mean that the NTFP system has failed; rather it is a stack reminder of the continued failure of governance at regional and national levels to provide social security and cater for the welfare of the citizens in general. In fact, it is noted that the governance context, that is the regulatory framework within which the NTFPs operate, is not always well integrated to enable things run smoothly.

All is not lost, because Ingram also noted that whilst some species and products have remained in subsistence use, overall a rising pattern of commodification has emerged, which means the prospects of rising above the poverty level is increasing among producers in the more lucrative NTFPs, especially the 'global niche NTFPs' that is, highly profitable products that have specific uses and increasingly global chains. Furthermore, activities in traditional labour intensive NTFPs are being transformed by labour saving technologies and knowledge, for example, harvesting, storage and particularly processing. Moreover, "an increase in collective action and its formalisation, and in market-based arrangements are evident in recent decades, whereby actors have sought to control and manage access to NTFP species and markets, particularly in the face of scarcity and/or increasing values, mainly economic." (p. 258).

The inference that can be made from the Cameroonian experience is that proposed move towards community-based management of NTFP value chains is a step in the right direction. We learn from the Cameroonian experience that the governance context is still the weakest link in the whole system. One can only hope that things would improve over time.

Forestry in the National Economy

Table 2.6 gives the value and proportions of timber exports from 1964 to 1985, while Table 2.7 gives the comparative status of wood exports in the first two decades of the 21st century. Nigeria's wood exports have declined sharply since independence in 1960. However, as shown by the average unit values of the exported timber and plywood, earnings from the export trade have remained high due largely to the high demand for Nigeria's timber products. The timber export trade was also the most broadly based of all Nigerian export products in terms of the international currencies it attracted.

In 1970, Nigerian timber was bought by 34 countries compared to 21 for petroleum, 19 for cocoa and groundnuts and 7 for palm kernels (Adeyoju, 1975). Unfortunately, as Table 2.6 shows, Nigeria's timber and plywood export trade had become virtually non-existent by the 1980s. However, the export trade picked up in the late 1990s partly as a result of renewed drive to reduce dependence on crude oil exports. Table 2.7 shows that earnings from wood exports fluctuated widely because of the vagaries of the supply. Indeed, since about 2011, there has been a decline in the value of exports except for the year 2013. Data by the Bureau of Statistics, however, indicate that Nigerian wood has continued to attract demand from about 49 countries.

TABLE 2.6 Nigeria timber and plywood exports, 1965–1985

Year	Quantity ($\times 1{,}000m^3$)	Value (₦ million)	Average unit value (₦)	% of total export
1964	692.6	18.18	28.3	4.3
1965	569.0	15.30	28.3	2.9
1966	534.8	13.60	28.3	2.4
1967	328.2	8.60	28.3	1.8
1968	321.3	8.60	28.3	2.1
1969	343.4	10.40	28.3	1.7
1970	239.9	8.00	35.3	0.9
1971	223.0	7.00	31.4	0.5
1972	232.0	8.10	34.9	0.6
1973	370.0	14.30	38.6	0.6
1974	305.0	14.00	45.9	0.2
1975	107.0	4.80	44.9	0.1
1976	8.0	1.40	109.5	0.02
1977	12.8	0.50	39.1	–
1980	52.0	0.4	1.9	*
1981	–	–	–	*
1982	1.0	0.1	100.0	*
1983–85	–	–	–	*

Note: * denotes Negligible quantity
Source: Federal Office of Statistics: *Review of External Trade, Nigeria* 1965, 1970, 1975, 1985

The fall in the volume of timber exports over time was partly due to over-exploitation, resulting in a dwindling in the number of trees of timber value in the forests. Another reason was the failure of the timber firms and the forestry departments of the various states to develop the so-called 'secondary species' (mostly the middle and lower tree layers) for export purposes. These were still quite abundant in the forest areas of Nigeria due to the selective logging of the so-called merchantable timber species.

TABLE 2.7 Nigeria Wood Exports by Regions, 2006–2017

REGION	Nigeria Wood Export (US$'000)										
	2006	**2007**	**2008**	**2009**	**2010**	**2011**	**2012**	**2013**	**2014**	**2016**	**2017**
WORLD	7,815.24	352,956.53	268,376.70	111,886.82	254,082.78	53,354.11	34,051.80	307,918.95	109,104.81	19,178.38	39,491.78
Europe & Central Asia	1,840.95	317,228.57	222,414.48	31,008.51	5,839.37	34,383.03	25,409.60	128,264.43	85,385.54	528.30	6,681.63
East Asia & Pacific	144.41	10,346.84	10,563.56	13,052.64	41,158.50	5,696.50	2,392.73	38,521.70	3,891.30	10,143.78	25,417.48
South Asia	0.00	5,016.95	5,477.23	5,816.82	135,717.31	861.85	590.22	7,676.50	6,955.74	12.25	49.21
Middle East & North Africa	101.05	11,873.51	17,836.71	31,523.24	29,079.95	1,195.41	166.51	19,730.50	1,665.13		3,186.84
Sub-Saharan Africa	670.01	5,552.01	7,579.64	2,914.79	3,970.12	4,777.18	291.37	21,832.24	5,736.24	8,494.05	4,125.16
Latin America & Caribbean	3.94	689.96	2,652.99	21,526.28	5,442.18	5,273.44	3,350.63	129.16	2,628.88		13.74
North America	4,785.98	1,717.64	1,752.54	128.81	26,677.27	709.95	1,771.94	91,377.86	2,407.12		-

Source: Nigerian Trade Statistics WITS (`wits.worldbank.org`)

Perhaps, the most important factor responsible for the fall in timber exports up till now has been the rapid increase in the internal use and trade in wood as evidenced by the numerous sawmills that have been established in the southern parts of Nigeria over time (Tables 2.8–2.10a, 2.10b & 2.10c). The demand for sawn wood in the country has continued to increase tremendously because of the boom in the building and furniture industries. The Federal Department of Forestry found that by 1981 there were as many as 1,030 sawmills in the country employing 12,460 workers and producing 2,929,000m^3 of sawn wood (Table 2.9).

This trend continued into the 21st century. According to Ogunwusi (2012), there were 1325 sawmills in the country in 2010 out of which 300 were in the swamp forest zone of Lagos and Rivers states; 945 were located in the moist rainforests of Edo, Delta, Ondo, Ogun, Osun, Oyo and Cross River states; and 80 were located in the savanna zone in Taraba, Adamawa, Benue, Kwara, Kogi, Kaduna and Niger states (Table 2.10a & 2.10b). The 1325 sawmills had a total installed capacity of 11,734,000m^3 per year (Table 2.10c) and a utilization capacity of 3,800,000m^3 per year. What these statistics reveal is that, a great deal of timber that previously would have been exported is now processed and consumed locally. This is a very healthy development which has encouraged large capital investments in the industry over time. Factories for the manufacture of furniture and other wooden artefacts for the home market were established early at Sapele, Epe, Calabar and Ondo.

In fact, a number of wood treatment and preservation factories were established in support of the internal timber trade, but almost all of them suffered from low capacity utilization for many reasons, including the practice among wood suppliers to rush products to the market without bothering about the treatment and preservation of their stock of wood. This has compounded the problems of the timber industry in Nigeria as infestation and damage of timber trees by pests is more prevalent nowadays than in the past. This is because, on the one hand, more and more secondary timber species are being exploited, while on the other hand, the felling cycle of most timber trees, whether primary or secondary species, has been shortened to 50 years and less making them less mature and lower in durability when they are felled for the market. The Federal Department of Forestry initially set the felling cycle at 100 years and minimum girth 60-90dbh.

However, Shasanya *et al.* (2015) noted that the suitability of timber does not depend only on its working stresses but also on its natural durability, that is, its ability to resist attack by insects, fungi and other forms of wood decaying agents. The common causes of deterioration in building and other constructional timbers in Nigeria are (1) decay fungi, (2) termites (white ants), and (3) wood-boring beetles. Unfortunately, while the need for the preservative treatment of wood has become more critical as a result of decrease in the ages and diameters of the remaining timber species in the country's forests, the wood treatment plants are shutting down by the day as a result of out-dated equipment. There is increasing utilization of hand brushed wood in the sector as opposed to more effective mechanical techniques such as spraying, dipping and the hot and cold tank process (Ogunwusi, 2012, Shasanya *et al.*, 2015), and various processes of wood drying and seasoning (Ohagwu and Ugwuishiwu, 2011).

Unfortunately, since the 1980s sawmills and large-scale wood industries have been feeling the adverse effects of the depletion of the favoured merchantable timber species. As shown in Table 2.10c, wood utilization in the country in 2010 was only 32% of total installed capacity. However, as mentioned earlier, one beneficial outcome of this decline in available merchantable timber species has been the increased use of secondary timber species, which had previously been ignored or indiscriminately destroyed. But, much of the tree felling and trade is poorly regulated because of corruption at the local government level.

The significance of the internal timber trade to the national economy can be judged from the booming building and construction industries. There are many wood-based industries in the country making wood panels, plywood and veneers, particle boards, standardized wooden doors and windows, house furniture, safety matches, and tooth picks. There are many artisans producing coffins, mortar and pestles, and wood carvings. This wood-based industry is important also because it engages many people in the lower income groups, many of whom, because they have only had basic education, have found it difficult to find employment in any other industry.

Unfortunately, in the opinion of some, the domestic wood-based industry is no longer contributing significantly to Nigeria's foreign exchange earnings. The reason for this is partly because the industry is dominated by small scale operators with low capital and technical inputs (Ogunwusi, 2012). Indeed, Nigeria has been importing a lot of wood and wood products to meet the high demand in the country, especially in the burgeoning urban centres. These foreign imports put the internal wood trade and industry at a competitive disadvantage: the foreign imports are cheaper and of higher quality than what the local people bring to the market. The Nigerian wood industry as a whole has declined in performance, efficiency, productivity and quality of products.

TABLE 2.8 Distribution of sawmills 1975 and 1980

State	Number of Mills 1975	1980
Ogun	67	192
Ondo	54	140
Oyo	65	195
Bendel	46	135
Lagos	88	104
Kwara	9	45
Cross River	6	8
Imo & Anambra	4	–
Rivers	2	10
Northern States	7	45
Total	348	874

Source: Extracted from Huntanen, H. J. I. (1975), *Report on FAO/UNDP High Forest Development Project,* Federal Department of Forestry and Alviar G. O. (1983), *Report on Forest Industries in Nigeria.* UNDP/FAO/NIR/NIGERIA Federal Development of Forestry.

TABLE 2.9 Number of plants, workers and annual production of forest industries in Nigeria, 1981

Industry sector	No. of Workers	No. of Plants	Annual Production
Sawmill	12,460	1,030	2,929,000 m^3
Plywood	1,800	6	89,000 m^3
Veneer	300	6	5,000 m^3
Particle Board	250	2	55,000 m^3
Match Factories	1,200	5	14,000 boxes
Pulp and Paper	1,000	1	121,000 tons

Source: *Preliminary Report on forest Industries in Nigeria* Federal Department of Forestry Ibadan (Document No. 12) (After: Nigeria Forestry Statistics (1983).

TABLE 2.10A Total Number of Sawmills in Nigeria, 1990 – 2010

Year	Total Number of Sawmills
1990	1617
1992	910
1996	1252
2002	1259
2010	1325

TABLE 2.10B Zonal Distribution of Sawmills in Nigeria, 2010

Ecological Zone	Number of Sawmills
Swamp forests of Lagos and Rivers States	300
Rainforests of Edo/Delta/Ondo/Ogun/Oyo/Osun/Cross River States	945
Savanna Zone in Taraba/Adamawa/Benue/Kwara/Kogi/Kaduna/Niger States	80
Total	1325

The use of forest lands

In the attitude of the local people to their forest surroundings lies the key to the misuse of forest resources in Nigeria as Ladipo (2010) had remarked. To many people forest lands represent 'vacant' land which, someday, will be brought into use for farming, or the establishment of housing or industrial estates. This is not to say that the people do not realize the economic potential of the forest lands, especially as sources of timber, firewood, fruits and bush meat. The point is that, mostly, forests are not seen as permanent features of the landscape or as likely to be more economically important than other forms of land use in the future. Faniran and Areola (1976) pointed out that this type of thinking was a reflection of the prevailing poor socio-economic status of the rural areas in Nigeria. The country dwellers, no longer satisfied with the drudgery of their rural life, had tended to underrate the resources of their immediate environment and to aspire to have the same modern amenities and economic benefits as the people in the big urban centres.

TABLE 2.10C Installed capacity and utilization (round log equivalent) in the sawmill industry between 1988 and 2010

Year	No. of Sawmills	Total Installed Capacity m^3/year	Utilization Capacity m^3/year	Capacity Utilization %
1988	N/A	8,831,750	6,994,000	79
1992	910	15,793188	6,031,922	38
1996	1252	10,900,000	4,200,000	39
2002	1259	14,684,000	5,177,700	35
2010	1325	11,734,000	3,800,000	32

Source: Source: Ogunwusi A.A., 2012 The forest industry in Nigeria. *African Research Review: An international multidisciplinary Journal, Ethiopia*, Volume 6 (4), Serial No. 27, October 2012, pages 191–205.

While in the past forest lands suffered from traditional agricultural practices, in modern times they are in danger of being lost through four kinds of human activity: (1) food crop cultivation and the establishment, by both individuals and government agencies, of tree crop plantations (cocoa, citrus, rubber, oil palm); (2) the establishment of commercial tree plantations; (3) urbanization, road construction and red earth quarrying; and (4) uncontrolled logging bedevilled by corruption especially at the local level.

As a result of the government agricultural promotion campaign of the mid-1970s tagged "Operation Feed the Nation" many people took to farming either as a pastime when they were still fully employed or as a survival strategy when they retired from their jobs in the public or private sector. Local farmers, workers' unions and modern-day entrepreneurs formed or joined co-operatives in order to engage in large scale production of certain crops including cassava, maize, rice and various vegetables. To accommodate these newcomers in the rural areas much of the formerly undisturbed or secondary forest regrowth has been and is being cleared. The same easily acquired lands are also preferred for government projects, because they can be exploited without displacing the local farmers. In Benin, Ondo-Okitipupa and Ijebu areas many hectares of former high forests are now occupied by plantations of rubber, cocoa and oil palm owned by government or by private citizens. For instance, in the Delta State, rubber dominates the vegetal cover of large areas of the Urhobo plains and adjacent lands. Commercial tree plantations have been restricted largely to the erstwhile forest reserves but local farmers also are being encouraged to plant exotic trees along with their tree crops. Furthermore, in some states, such as Lagos, for example, large areas of secondary forest regrowth are being reserved for eventual establishment of commercial tree plantations.

Urbanization, road and building construction and the attendant high demand for stones, sand and red earth from the quarrying industry, which is very wasteful of land and forest, have been responsible for much forest clearance everywhere, but, especially in Lagos, Ogun, Ondo and Edo States. According to the Food and Agriculture Organization (FAO) of the United Nations, as of 2005, Nigeria had the highest rate of deforestation in the world with an average annual rate of loss of 3.5%. Between 1990 and 2005, Nigeria lost 35.7% or 6,145,000 ha. of its remaining forest cover. Today, therefore, the area of surviving unreserved forests is only about five million hectares and is dwindling rapidly. The secondary forest regrowth is also suffering from various forms of human activity. Since the 1980s, there has been a phenomenal increase in the burning of wood to make charcoal for sale in the cities. At Ibadan, for example, there is a booming trade in charcoal obtained from nearby secondary forests. The trade flourishes particularly in times of shortage of cooking gas and kerosene.

A very important factor militating against the maintenance of a semblance of forest vegetation as part of the rural landscape in many parts of Nigeria is the annual bush fire. Apart from the pastoralists who set the bush on fire to promote new growth of grass for their livestock, villagers looking for game also cause many bush fires. Fallow lands are set on fire even though they are potential areas for the regeneration of forest tree species if the vegetation is left undisturbed for long enough. But the fallow period is not as long as it used to be, and tree seedlings that manage to survive the bush fires do not have much time to grow before the land is cultivated again. Areas which have been subjected to burning for a long time, especially in the drier margins of the forest region, may not revert to forest even if they are protected against fire for a great length of time. At best, a dense community of savanna woodland may develop in which there are very few, if any, timber tree species. This is the situation in parts of the derived savanna zones in south-eastern Nigeria, and in the northern parts of Oyo, Ekiti, Ondo and Kogi States. Forest survivals in these areas are restricted largely to rocky ground, deep gullies and river valleys.

Forests, however, have made positive gains in some districts. Some water corporations have adopted the practice of reforesting the catchment areas of their water reservoirs in order to protect them against erosion and excessive sedimentation. Such catchments are normally acquired along with land inundated by the reservoir and are declared forest reserves. Furthermore, with the encouragement of the Federal Department of Forestry, several towns in the northern states embarked on tree planting programmes in the 1960s in order to ameliorate local climatic conditions. The pace increased after the Sahelian drought of 1968-73. Maiduguri, the capital of Borno State, and many of the major towns in the same zone benefited from the tree planting programmes. Also, for a long time, it has been the practice in the savanna regions, to plant fruit bearing trees around the settlements and on farmlands. The common fruit trees include *Parkia clappertoniana* (West African locust bean), *Ceiba pentandra* (cotton tree), *Mangifera indica* (mango), and *Tamarindus indica* (tamarind). Furthermore, as part of the aftermath of the recurrent droughts in the zone, a shelterbelt tree (Green Wall) planting programme was vigorously pursued in the Sudano-Sahelian zone to check desertification and the attendant wind erosion and southward advance of sand dunes. These sand dunes had been stabilized for a long time until changes in climatic conditions set them in motion again during the second half of the 20th Century.

Logging and its ecological consequences

Sawmilling is the oldest and best-established wood-based industry in Nigeria and the one which has made the greatest impact on the country's unreserved forests. Logging in the forests is done by logging firms, sawmilling establishments and middlemen who obtain licences to fell the trees. The activity has always been dominated by (a) small sized units with a production capacity less than 283 cubic metres per year; and (b) medium sized units with yearly production in the range 566-2830 cubic metres (Enabor, 1973).

Wood conversion methods

Pit-sawing used to be the commonest mode of log conversion. The FAO report (1966, p. 240) commented as follows:

> *"pit-sawing, a wasteful and out-dated method of log conversion is probably more widespread in Nigeria than in any other major country in the world. Apart from the heavy labour requirements, waste occurs because of inaccuracy in sawing and the scattered nature of working, leading to delay and time-consuming collection of production. Most of the material so sawn is completely unsuitable for export even though cut from choice logs of the most valuable species."*

The report also noted that although several sawmills were highly efficient and produced excellent sawn wood, none had been completely modernized.

Enabor (1973), also noted that the average wood recovery rate coefficient in Nigeria was only 50 per cent and that the timber industry had even adopted this low figure in negotiating concessions with governments. He further noted that the use of obsolete machinery, out-dated methods and poorly trained operatives was primarily responsible for the inefficiency of Nigerian sawmills. Apart from the large-scale mills at Sapele and Epe which used modern equipment to produce graded lumber of export quality, most other sawmills produced rough green sawn wood for the domestic market.

Nowadays, however, pit-sawing is no longer practised and sawmills have been modernized considerably both in terms of equipment and management. Still, according to Ogunwusi (2015), the sawmilling industry has not improved but deteriorated over time noting that "*the saw mill industry is characterized by small scale operatives (who) constitute more than 90% of the entrepreneurs in the sector. A major characteristic of the subsector is increasing number of operatives and decreasing performance. The capacity utilization in the industry is averaged 37% and the lumber recovery rate 40 – 60% as a result of old equipment.*" More than 70% of the workforce is made up of manual labourers, while less than 10% of sawmills use advanced technologies. Thus, this resource process has been highly inefficient in its techniques and wasteful of the product itself. In the past the wastage arose indirectly from the practice of selective logging. There are many tree species in the forest which yield fairly good if not first-class timber but which were regarded as 'secondary species' by the industry and were therefore, largely unexploited. This practice led to the over-exploitation of the favoured species to the extent that certain forests were almost totally denuded of them. The impression created in the minds of the local people that the secondary species were not useful encouraged farmers to kill such trees on their arable farmlands and tree crop plantations by firing at the base. Also, in an effort to persuade government to site development projects in their areas, local communities sometimes put forward the argument that there was plenty of vacant forest land available, which did not contain any valuable trees! Another indirect consequence of the practice of selective logging was that as the favoured species of suitable size diminished loggers were tempted to break regulations regarding the age and size of trees that may be felled. This, in the long run, has further inhibited the survival of those merchantable species.

The scarcity of the primary timber species, which has resulted in increased use of secondary species and the decrease in log diameter consequent upon the reduction in the felling cycle contributed to the low wood conversion efficiency of the sawmills because the mills were structured to utilize large diameter logs. Indeed, according to Larinde (2010), round wood processing has reached the limits of available forest resources and the future lies in further processing of sawn wood rather than in expansion in sawmill capacities and exploitation of wood resources. The future lies in integrating all wood industries in Nigeria so that wood waste and wood materials which are not suitable for use in any sector of the wood industry may be channelled to other sectors.

Impact on the forest ecosystem

The impact of all this on the forest ecosystem itself has been considerable. Selective logging reduces the floristic diversity and structural complexity of a forest. Poor production methods in the past did not mean that the chips and wastes were returned to the ecosystem; in fact, more often than not, they were burnt or allowed to rot at the mill sites; this is still the case today. Thus, altogether the forest ecosystem has been losing a great deal of energy and nutrient-rich material in the wood that is removed or wasted by the logging firms. Such losses greatly impair the ability of the forest to regenerate naturally. Furthermore, when a single tree is felled a sizeable patch of the surrounding forest is laid waste and young trees are destroyed. Gaps created by the felling of trees do not have the same beneficial effect on the growth of young emergents as those caused naturally when dead trees fall. Keay (1952) suggested long ago that regeneration of emergent species was far less certain in gaps made by felling because the break was sudden and was not preceded by several years of battering by falling branches.

With the removal of most of the emergents and the opening up of the forest canopy, the ecosystem suffers, to a greater or lesser extent. Microclimatic conditions are altered; there is an increase in temperature and a lowering of the relative humidity of the air within the forest. The organic content of the soil, being more exposed to insolation, undergoes rapid decomposition and oxidation, and the soil water regime is altered.

The negative feedbacks of forest degradation on the environment contribute to the slow pace of natural regeneration of emergent forest species. The faunal population is also somewhat affected as their habitat is destroyed or greatly modified, and many wild animals are wounded, killed or captured during the logging operations. The opening up of the forest canopy and subsequent growth of impenetrable thickets which hinder the free movement of some animals force them to migrate. The overall effect of the exploitation of the forests for timber is therefore a general simplification of the ecosystem and a degradation of its ecological status and resource value. Only by combining logging operations with careful resource management and a planned programme of replanting depleted forestlands, can the adverse effects of timber extraction be reduced.

The future of forests and biodiversity conservation in Nigeria

Ladipo (2010) noted that while in the past many rural dwellers treated our forest resources as inexhaustible, nowadays they realise that the forest is "finished". Yet, in spite of this realization poverty forces the people to continue to exploit even the remaining forest relics. The relentless pressure of human activities has led to high rates of forest depletion and environmental degradation in the forest ecoregions of Nigeria. Ladipo (2010 *op. cit.*) pointed to another serious consequence of forest exploitation and depletion, which is loss of biodiversity manifest in the depletion of plant and animal genetic resources which would be critical for plant and animal breeding in future. Forests and forest resources are vital ingredients of rural development and they need to be conserved. "*Biodiversity is required for the recycling of essential elements. It is also responsible for mitigating pollution, protecting watersheds, and combating soil erosion. Controlling deforestation will ensure that biodiversity exists and can help reduce the impacts of climate change and thus act as a buffer against excessive variations in weather and climate. It can then protect us from catastrophic events*" (Ladipo, 2010 *op. cit.*) Much would depend on how seriously and consistently the federal and state forest agencies implement the new National Forestry Policy and the National Document on Biodiversity Conservation Action Plan in such a way that they do not suffer from party politics, executive fiat in decision making, and changes in government from time to time.

It is clear from the preceding discussions that the sustenance of forest resources in Nigeria will depend on several factors including: (1) enlightened management and exploitation of the government-owned forest reserves and intensification of silvicultural activities in them to promote the growth of both exotic and indigenous species (see Chapter 4); (2) promotion of awareness and interest among the local communities about the economic potential and ecological desirability of forestry as a component of rural land use; (3) encouraging farmers in the derived savanna zone to protect the seedlings, not only of the so-called merchantable economic trees, but also, of the erstwhile derided 'secondary species' which are now the mainstay of the internal wood trade; (4) the protection of existing forests from annual bush fires and destructive agricultural practices; and (5) encouragement of logging firms to invest in re-afforestation programmes, perhaps, with the participation of local communities and local governments.

Most government rescue efforts had been directed at the forest reserves as well as those parts of the country suffering from certain environmental hazards, such as the drought-affected parts of the far north and the badly eroded lands of south-eastern Nigeria. Operations in the forests outside the reserves are still largely uncontrolled. It is true that a licence is required before any economic tree can be felled, but this measure has turned out to be more of a way of generating revenue for the local government than as a means of actually regulating the use and management of the resource. There is a lot of corruption and lack of accountability in the issuance of licences for tree felling by the local governments.

The system of land ownership in the country makes it more difficult for governments to control forests outside the reserves. Thus, there appears to be a general feeling in official circles that in order to give them authority over the use of a forest it has to be designated a forest reserve. It is difficult to see how this policy can be sustained in the face of opposition from the rural populace. As mentioned in Chapter 1, development initiatives in the rural areas must be tailored to meet the goal of poverty alleviation at the grassroots level. The solution that has long been canvassed by forestry and wildlife experts is to adopt the participatory community-based forest resources management approach. According to Borokini *et al.* (2012), the community-based approach is one that regards local communities whose livelihoods depend very much on available local resources, as stakeholders in conservation efforts and projects rather than as a threat to them. This approach requires that conservationists work with local farmers, hunters, lumbermen and such like to promote forest resource conservation from which they also derive appreciable benefits. Local communities are active participants in conservation projects and they are the first and primary beneficiaries. The practice whereby local farmers themselves are encouraged to engage in the establishment of tree plantations for their own gain seems to have yielded very good outcomes.

In the case study of the Ngel Nyaki Forest Reserve on the Mambila Plateau, Borokini *et al.* (2012), listed three key areas where the local communities benefited including (1) establishment of tree plantations, whereby several local communities were trained in planting, seedling and field management of fast growing trees (e.g. *eucalyptus*) which they harvested in 6-10 years for timber, pole wood and fuelwood; (2) employment of local inhabitants in the Ngel Nyaki Forest Reserve and by the Nigerian Montane Forest Project a biodiversity conservation and forest ecology research project of the University of Canterbury, Christchurch, New Zealand; and (3) community development projects funded by an oil company in partnership with the biodiversity conservation and forest ecology project. The benefit to researchers and forest reserve managers has been protection of the wildlife and forests against poaching and illegal tree felling. In some other parts of the country, proceeds of the participatory community-based conservation projects are shared between government and the local communities with the latter having the larger share.

Moreover, it is important to promote greater multiple use of forests in order to make forestry and forest management more profitable, more competitive and more socially acceptable. In the long term this would help to change the traditional view that a forestland is just so much vacant land. Forest uses that can be combined with logging by the local people include wildlife conservation, controlled harvesting (hunting) of game, water conservation, grazing of wild and domesticated animals, fishing in forest lakes and rivers; recreation, and the collection of firewood and other non-timber products. The concept of multiple land use had long been accepted in government circles but it has not been promoted among the local people. Rather, as early as the late 1960s government-controlled forests were being managed for multiple use including production or extraction of fodder and forage, gum, wild rubber, palm nuts, wildlife and other products. Agroforestry was also embraced as a policy and as a forest management system to cultivate quick-maturing trees of economic value as sources of pulp, fruits, pole wood, fuelwood and other products.

But as was shown earlier in this chapter, the strategy that would bring the most positive outcomes would be to focus on *non-timber forest products* and create economically viable and socially acceptable and beneficial NTFP value chains with species of plants and animals that that can be cultivated, commoditized and mainstreamed into the market economy or even developed to have value chains across international borders. With a responsive governance framework, these mainstreamed NTFPs and global niche NTFPs may be able to satisfy the twin purposes of poverty alleviation at the grassroots level and conservation of the environment and natural resources in the various ecological zones in Nigeria.

Moreover, Nigeria will do well to reduce the rate of consumption of fuel wood by people at the grassroots level. Nigeria can follow the example of Kenya which has successfully implemented a project on the adoption of sustainable biomass energy practices and technologies by institutions (schools and hospitals) and medium-small scale businesses (restaurants, hotels) in the rural and urban areas of Kenya (Areola & Wanyamai, 2008). The pillars of the project are (i) the adoption of a highly energy efficient locally fabricated wood burning stove (*jiko*) by schools, hospitals, hotels and local people; (ii) a programme that encourages institutions and individuals to establish woodlots of eucalyptus trees in their premises or on land acquired for the purpose; (iii) government forestry department raising millions of seedlings of fast growing eucalyptus trees for distribution to woodlot owners; and (iv) the establishment of a revolving fund credit scheme to disseminate energy-saving stoves to institutions in Kenya on a commercial basis.

The wood-burning stoves require 70% less wood for the same cooking task; and reduce emissions of products of incomplete combustion which have higher global warming potentials than CO_2. There are local institutions to ensure the sustainability of the project; they include (i) the Renewable Energy Technology Assistance Programme (RETAP) Green Capital, that runs the revolving loan scheme mentioned earlier; (ii) the Kenya Improved Stove Association (KEISA), a cooperative that produces and sells the improved stoves; (iii) the Kenya Tree Propagators and Tree Growers Associations and Networks of School Clubs and Teachers; and (iv) Kenya Forest Service that raises the seedlings of the fast growing eucalyptus trees.

The project which is being funded by the Global Environment Fund (GEF) under the auspices of the United Nations Development Program (UNDP) country office, Kenya is based on the premise that Kenya is among the countries of the world classified as being "forest poor." Nigeria, with its huge population making increasing demands on forest and woodlands for firewood should begin to see itself as a "forest poor" country and take necessary steps to reduce fuel wood consumption and carbon emissions. At the beginning of the century, the Forestry Research Institute of Nigeria also experimented with locally made energy efficient stoves and commissioned some consultancy projects to introduce and demonstrate the use of the clay stoves to the rural communities. But the project does not appear to have been sustained and carried through to full implementation.

In concluding this section on the future of forest resources and associated resource processes in Nigeria, it is becoming more and more clear that climate change has to be factored into future policy formulations and development planning in this sector. Forest Machine Magazine of May 2020 has a piece on sustainable logging and noted that future tree cutting and replanting schedules will have to take into account climate change, by taking into account how growing trees absorb carbon from the atmosphere:

- Combating climate change – younger trees absorb carbon quickly whilst growing but as a tree ages a steady state is reached where this slows down. A well-managed forestry system harvests the trees at this time in their growth cycle;

- Replanting or natural regeneration can ensure the forest remains a "net sink" of carbon, which means that the trees absorb the maximum amount of carbon possible.

Also, as noted by Ahmed (2019) and Ingram (2014), the future of the non-timber forest products (NTFPs) is threatened by global warming, drought and desertification. Climate change is exacerbating the impact of the thinning or outright disappearance of both forest and savanna vegetation thereby increasing exposure of the land to desiccation and accelerated soil erosion. NTFPs are already reduced in number and diversity in many areas and the livelihoods of the rural peoples who depend on them are under serious threat.

3

Savanna and livestock grazing

Grazing is a major resource process in Nigeria and one which has had a profound effect on the country's savanna ecosystems in particular. The data presented in Table 1.1 on page 19 in Chapter 1 show clearly that the changes that had taken place in the vegetation cover had greatly favoured the spread of grazing throughout the country. For example, between 1975 and 2005, the area of "grassland" increased by about 581% from 1196.74km^2 to 8146.74km^2; the area of "discontinuous grassland dominated by grasses and bare surfaces" increased by about 64% from 7614.72km^2 to 12,517.23km^2; and "montane grassland" increased in coverage by approximately 56% from 2,502.27km^2 to 3,88.15km^2. Thus, the prospects of grazing in rural land use seemingly have increased progressively over the years as the forest gave way to savanna and as the Fulani herdsmen expanded their areas of operation from the far north to the coastal areas and parts of the Niger delta. But, as noted in Chapter 1, arable crop cultivation has also greatly expanded its coverage throughout the country much more than grazing. This simultaneous expansion of both grazing and crop cultivation has been accompanied by conflict as the competition for land between farmers and pastoralists reached breaking point in many parts of the country resulting in violence far destructive of agricultural produce, livestock and human life and settlements beyond anything the country had experienced in the past. The conflict rose to a new level in 2019 when Fulani militia brought in from surrounding countries took over the fight, among other things, to force access to grazing land by their kinsmen herders in a manner that has been likened to terrorism. These bloody conflicts have threatened the political stability and the very survival of Nigeria as a nation.

The chief product derived from the grazing resource process is meat; milk, blood, hides and skins are of secondary importance. As in other parts of the tropics grazing in Nigeria is characterized by a lack of range management and improvement measures. More is taken out of the environment than is put into it; artificial chemical fertilizer input is nil, the only nutrient inputs being from animal manure. The livestock depend entirely on native vegetation and there is virtually no cultivation of forage crops to supplement this. In the literature, a distinction is sometimes made between nomadic and commercial grazing (ranching). This distinction is rather tenuous in the case of Nigeria; while the livestock trade is certainly a lucrative enterprise organized on a commercial basis, the rangeland and animal management techniques are still largely nomadic and archaic in nature.

Grazing of domesticated animals is done in a wide variety of ecological/climatic zones with marked differences in the total annual rainfall; the duration of the wet season and the availability of dependable sources of water for the animals. The one element that is common to all the zones is the availability and prominence of grass in the vegetation community.

The savanna of Nigeria

Savanna vegetation covers nearly 75 per cent of the country (Fig. 3.1). The 1143mm isohyet which marks the northern limit of the former dry forest zone may be regarded as the southern limit of natural savanna vegetation in Nigeria. But south of this line is a broad irregular belt of derived savanna vegetation which, over the years, has been brought into the grazing economy and is important as an all-season grazing area for far-ranging Fulani cattle and the local *muturu* and *n'dama* species. The derived savanna belt comprises a mosaic of forest, dense thickets and savanna woodlands whose luxuriant growth is enhanced by the prevailing humid climate and good moisture-retentive soils. The grasses which are tussocky and broad-leaved, reach up to heights of 3-4 metres. They include *Andropogon, Pennisetum* and *Hyparrhenia* species, and *Loudetia arundinacea* and *Ctenium newtonii.* The weed grass *Imperata cylindrica* (West African spear grass, *lalang*) is often present on fallow lands.

The other savannas are usually subdivided into the *Guinea*, the *Sudan* and the *Sahel* savannas. This is a climatically determined ecological series of vegetation communities in which the woody component becomes progressively less abundant while the herbaceous and grass component gets more and more prominent to the north, until the situation is reversed on the approach to the desert margins.

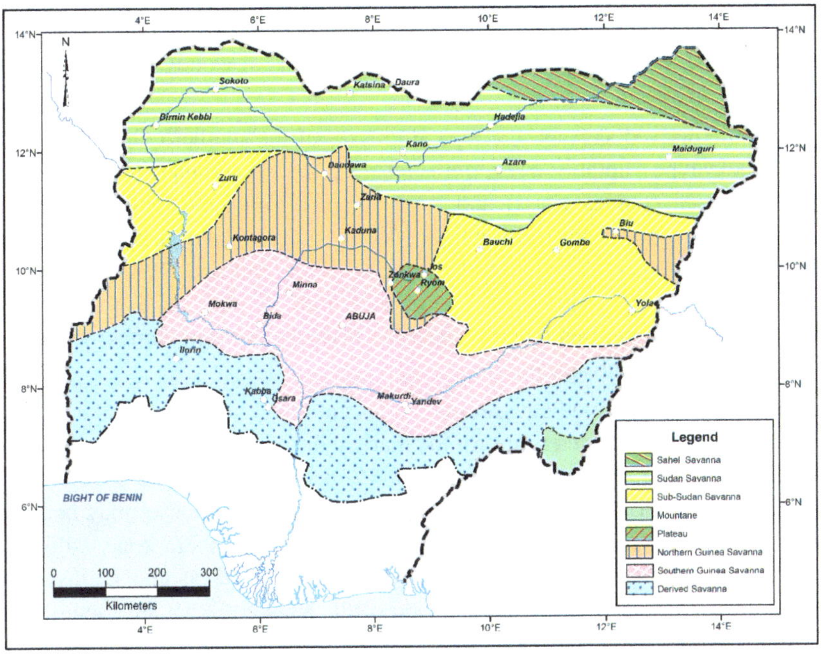

FIGURE 3.1 The savanna of Nigeria

The **Guinea savanna** is easily the most extensive vegetation zone in Nigeria spanning the middle of the country and covering the Niger –Benue troughs and the foot slopes and sections of the high plains and plateaus to the north and south (Fig. 3.1). Except for the alluvial and sedimentary rock basins, the region is underlain largely by highly ferruginous tropical soils with iron concretions (*Lixisols, Nitisols, and Luvisols*). Its northern limit is marked by the 1000mm isohyet. The wet season is at least six months long and may be longer in the south. The Guinea savanna is usually subdivided into the northern and southern zones chiefly on the basis of their floristic differences.

The **southern Guinea savanna** is a transition zone between the forest and the true savanna. As we saw in Chapter 2, many of the tree species have a close affinity with forest species among them *Lophira lanceolata* and *Terminalia glaucescens*. Other tree species typical of this zone are *Daniellia oliveri (West African copal), Hymenocardia acida, Afzelia africana* (mahogany bean tree), *Vitex doniana* and *Detarium microcarpum* (tallow tree). Most of the trees have thick, fire-scarred and darkened barks which are able to withstand the annual bush fires. They possess remarkable powers of recovery, sending out fresh suckers after the fires have passed. Gallery forests occur along the water courses while the flood plains of the Niger, the Benue and their big tributaries have thick growths of woody vegetation or tall grass and reed swamps.

Like the derived savanna the typical grasses of the southern Guinea zone are still tussocky types, notably, *Beckeropsis uniseta, Monocymbium ceresiiforme, Panicum maximum* (guinea grass) and *Andropogon, Pennisetum* and *Hyparrhenia* species. These grasses have the unique ability to sprout and grow quickly after the dead aerial parts have been burnt off during the dry season. Although they grow luxuriantly (3-4m. in height) and have broad leaves, they do not form a complete cover. The ground is often bare between the tussocks but, sometimes, sedges and microflora occupy such spaces.

The **northern Guinea savanna** is, under natural conditions, similar in many ways to the southern zone except for the marked difference in flora and the shorter vegetation. Trees range in height from 10-15 metres while in the southern zone most trees are up to 16 m tall and a few grow up to 30 m. Grasses may be absent under dense tree and bush growth, but again, as in the southern zone, cultivation and bush burning have opened up the vegetation thereby increasing the grass cover. The same grass species now characteristic of the southern zone are found also in the northern zone.

In some places, especially on slopes, overgrazing and frequent bush burning have led to a reduction of the grass cover to the advantage of the woody plants, under which the land surface is bare in some places. These shrubby species have prospered as the intensity of the annual bush fires decreased.

Gallery forests occur along the water courses but unlike the southern zone the tree species have savanna rather than forest affinities. The most abundant and characteristic trees are those of the miombo savanna woodlands *Isoberlinia* and *Brachystegia* which occur throughout eastern, central and southern Africa. They *include Isoberlinia doka* (doka), *I. tomentosa* (faradoka), *Monotes kerstingii* and *Uapaca togolensis*. On fallow lands and near settlements valuable fruit trees are usually preserved and sometimes planted, including, *Parkia clappertoniana* (West

African locust bean tree), *Ceiba pentandra* (cotton tree), *Vitex doniana, Mangifera indica* (mango) and *Tamarindus indica* (tamarind).

The **Sudan savanna** zone is marked by increased seasonality of rainfall. The dry season is about 6-7 months long and rather severe with the relative humidity of the air falling below 30 per cent. The drier parts of the Sudan zone have a mean annual rainfall of only about 500 mm. The whole region is underlain by well-drained, light soils such as the loams, sandy-loams and sandy-clay loams (cambisols) derived from drift deposits. Highly ferruginised soils with clay pans and lateritic crusts frequently occurring in the profiles are also common.

The vegetation is markedly xeric: the grasses are short, less than 3m high and feathery in contrast to the stout Guinea grasses. The trees are smaller and have shorter boles and small leaves or leaflets. The larger trees attain heights of only 8-16 metres; the smaller ones range between 5 and 7 metres in height. The most abundant of the larger trees include *Anogeissus leiocarpus* (anogeissus), *Sclerocarya birrea, Balanites aegyptiaca* (soap berry tree, desert date), *Lannea microcarpa* and *Rosopis africana* (ironwood). The smaller trees include *Combretum glutinosum* and *Acacia spp.* There are also numerous woody climbers. As in the Guinea savanna zone certain tree species are preserved or planted near settlements, for example, *Parkia clappertoniana* (West African locust bean tree), *Butyrospermum parkii* (shea butter tree), *Acacia albida* (acacia), *Adansonia digitata (baobab), Balanites aegyptiaca* (soap berry tree, desert date) and *Tamarindus indica* (tamarind).

The **Sahel savanna** is a 'thorn scrub' with little grass understorey, the ground being bare and sandy under the shrubs. The boundary between the Sudan and Sahel savannas roughly coincides with the 500 mm isohyet. The acacias are the most abundant tree species especially *A. raddiana* and *A. senegal.* Since grass is not an important component of the vegetation, it is the shrubs that support the grazing economy of this zone. They are browsed chiefly by goats but also by cattle and sheep.

The **upland grasslands** of the Jos Plateau and the Adamawa and Mandara highlands in the eastern borderlands are worth mentioning because they provide excellent feed and a healthy environment for grazing animals. The seasonal movement of Fulani cattle to these uplands is as important as the movement to the Guinea and derived savanna zones. The uplands are free of the tsetse fly so that cattle can graze and breed without fear of trypanosome infection.

The vegetation of the Jos Plateau is predominantly open grassland. Tall hedges of cactus are found around villages while many streams are lined by gallery woodlands and bamboo thickets. The commonest species of grass are *Hyparrhenia spp., Andropogon gayanus* (gamba grass) and *Pennisetum americanum* (bulrush millet). *Imperata sp.* (spear grass, *lalang*) may be predominant on young fallows. The grasses appear towards the end of the dry season sprouting remarkably soon after the annual bush burning.

In the eastern borderlands, the vegetation on the plains and the valleys is savanna woodland but on many plateau surfaces and granitic hills it is predominantly open grassland as it is on the Jos Plateau.

Quality and nutritive value of Nigeria's grasses and browse

The ecology of the savannas is very complex particularly because of the wide variation in climatic and edaphic conditions and in the composition of the vegetation. The vegetation pattern is further complicated by burning, grazing and cultivation. Apart from general impressions based on scanty evidence, very little is known yet about energy flow and nutrient cycling in these grasslands. However, studies of the nutritive value and carrying capacity of the grasslands for grazing purposes have been carried out at the Shika Agricultural Research Station of the Ahmadu Bello University near Zaria and on some government-owned cattle ranches.

The grass and leguminous components of the savanna communities described above are important in the grazing economy of those areas. But the nutrient status and, therefore, the grazing value of the natural pasturage vary very much depending on local conditions of rainfall, soil and air temperature. The same plant species may show marked differences in quality and digestibility from place to place in response to changing environmental conditions.

Grass

As the growth of grass is regulated by season and climate it is only available to grazing animals for a limited period of the year. The Guinea and derived savanna zones provide much more adequate grazing during the dry season than other zones, hence the seasonal migration into these zones by the Fulani herdsmen. In the middle belt the most important dry season grazing area for long distance cattle is eastern Niger State including the part now constituting the Federal Capital Territory, and the adjoining areas of Kaduna, Nassarawa and Plateau States.

However, most of the grasses are of the stout, broad-leaved tussocky type, which in terms of nutritive value, are at their best at the beginning of the rainy season when the leaves are still young and succulent. As the season progresses the grasses grow coarser and less digestible, while their great height also makes them less accessible. In addition, the luxuriant growth of the grasses suppresses the leguminous plants in the community thereby further lowering the nutritive value of the pasturage. Fortunately, however, when these grasses are burnt off during the dry season, they respond quickly by sprouting new fresh growth which provides succulent and digestible feed for the animals. This is unlike the Sudan savanna grasses which do not sprout immediately on being burnt but delay until the coming of the first **rains.**

Many of the grass species present in the Guinea zone are also found in the Sudan savanna but the more abundant ones are Sorghum and other turf species *including Pennisetum americanum* (bulrush *millet), Axonopus compressus* (axonopus grass) and species of *Brachiaria, Chloris, Aristida, Digitaria* and *Eragrostis.* These grasses support the largest concentrations of domesticated animals in the north, north-west and north-east of Nigeria. Like the Guinea grasses their nutritive value decreases as they mature and get coarser. In the dry season they turn into 'standing hay' which can still be grazed although the nutritive value of the dry grasses is rather low. In particular, the dry grass contains little protein while the abundance of fibrous materials reduces its digestibility.

The seasonal nature and the relatively low quality of the natural pasture, commonly known as 'bush grazing', cause serious problems for the herds and flocks. Many animals suffer from weight loss and sometimes from stark starvation. This makes them highly susceptible to intestinal parasites and debilitating diseases. The situation is compounded by the fact that the animals expend a lot of energy in trekking long distances in search of food and water.

In Nigeria, as in other parts of Africa, there is a search for suitable leguminous species which will form a stable association with the grasses and thereby increase the nutritive value of the pasture. The problem is that the legumes are easily choked off by the more aggressive grasses and also tend to be more sensitive to overgrazing and trampling. Up till today the pastoralists do not practice cultivation of grass and legumes, which has been restricted mostly to the government-owned cattle ranches and research institutes.

Browse

The significance of browse, that is, the leaves, twigs and young shoots of woody plants in traditional pastoralism is now better known and recognised. The weeds, shrubs, and smaller trees of the grassland communities provide feed not only for goats, the natural browsers, but also for the traditional grazers, cattle and sheep. The quantity of browse consumed varies, of course, with the availability or otherwise of grass in the pasture. The browse woody plants remain green and nutritive during the dry season when the annual grasses and herbaceous species have dried out. In addition, leguminous browse species are useful because they fix atmospheric nitrogen, which contributes to their increased crude protein (CP) content (Aregawi and Melaku, 2008). However, the presence of tannins and other phenolic compounds in a large number of these feed resources limits their utilization as animal feed (Tolera, Khazaal and Ørskov, 1997; Ogunbooye, 2015).

TABLE 3.1 Frequency distribution (%) of crude protein content (%DM) in grasses and browse, Zaria region

	Grasses			Browse		
CP content	Sahel dry	Guinea dry season	Sudan wet season	Sahel	Guinea year round	E. Africa
0.0–2.5	27	25	15	–	–	–
2.6–5.0	50	71	41	2	–	–
5.1–7.5	36	3	21	4	10	6
7.6–10.0	6	1	21	8	24	9
10.1–12.5	1	–	2	22	27	26
12.6–15.0	–	–	–	15	17	18
15.1–17.5	–	–	–	16	15	16
17.5–20.0	–	–	–	11	2	9
Over 20.0	–	–	–	21	5	16
Number of samples	176	273	53	107	41	97

Note: % DM = Per cent of dry matter.
Source: Brinckman W.L. & de Leeuw P.N., 1975.

In earlier studies, Brinckman and de Leeuw (1975) analysed both grass and browse and showed that the crude protein content of a browse averaged 15 per cent as compared to 2 per cent in dry grasses from December until the coming of the rains in May (Table 3.1). The browse also contained higher levels of calcium, magnesium, potassium and phosphorus during this period (Table 3.2).

TABLE 3.2 Mineral content of grass and browse in West Africa

Source	Ca	P	Mg	K	Ca/P	Sample size
Grasses						
Kapu, 1975*	0.34	0.22	0.23	2.56	1.55	9
Kapu ***	0.22	0.06	0.11	0.37	7.33	5
Boudet & Ellenberger, 1971	0.44	0.14	0.32	1.35	3.12	34
Bartha, 1970**	0.31	0.22	0.22	2.12	1.41	36
Rippstein & de Peyre, 1972	0.43	0.11	0.24	1.43	3.91	66
de Peyre, 1967	0.46	0.15	0.18	1.43	2.75	20
Audru, 1966	-	-	0.19	0.94	2.48	32
Browse						
Kapu, 1975***	0.70	0.21	0.28	1.66	3.33	10
IAB, 1947	1.62	0.26	-	1.63	6.23	47
Bartha, 1970	1.28	0.15	0.40	1.65	8.04	20
Rippstein & de Peyre, 1972	2.20	0.22	0.57	2.32	10.22	35
Audru, 1966	1.53	0.15	0.60	1.45	10.45	27
Average grasses	0.35	0.14	0.21	1.46	3.22	202
Average browse	1.47	0.20	0.46	1.79	7.65	139
Estimated needs (ARC, 1965)	0.28	0.19	0.07	0.25	1.47	-

Source: Brinckman W.L. & de Leeuw P.N., 1975. Note: * Wet season; ** Late wet season; *** Late dry season; All other data year-round.

TABLE 3.3A Chemical composition (%DM) of legume and non-legume browse plant foliage

Browse species	CP	CF	EE	Ash	NFC	NDF	ADF	ADL	HC
Parkia biglobosa	19.90cd	29.36a	3.98bc	8.25abc	6.59e	61.32a	49.58a	9.14a	11.73
Gliricidia sepium	20.61cd	16.20de	3.25c	9.14a	22.15a	44.84cd	31.25d	6.96d	13.59
Albizia odoratissima	24.43b	13.87e	3.51bc	9.48a	20.18a	42.41d	29.51d	5.71e	12.90
Prosopis africana	16.20e	17.54d	4.02bc	7.15c	20.14a	52.07abc	41.71bc	7.68bcd	10.36
Non-legume									
Daniella oliveri	15.34e	21.64c	6.30a	7.23c	13.41bc	57.72a	45.26ab	8.80ab	12.48
Eclipta alba	15.80e	22.05c	6.23a	7.36c	10.44d	59.87a	48.81ab	9.27a	11.06
Ficus polita	20.74d	24.23bc	3.86bc	7.84bc	11.74cd	54.82ab	43.84ab	8.24abc	10.98
Ficus cogenesis	18.22de	25.56b	3.68bc	6.77c	12.39bcd	58.94a	47.38ab	8.96a	11.64
Polyalthia longifolia	17.33e	26.74ab	4.05bc	6.68c	14.04b	57.90a	45.95ab	8.77ab	11.95
Moringa oleifera	29.60a	15.87c	4.62b	7.91bc	20.06a	46.81cde	35.83cd	7.51cd	10.94
SEM	0.82	0.97	0.22	0.21	0.93	2.88	2.14	0.36	0.21

Note: abcde= means on the same column with different letters are significantly different ($P < 0.05$), NDF- Neutral detergent fibre, ADF- Acid detergent fibre, ADL- Acid detergent lignin, NFC= Non fibrous carbohydrate, HC= Hemicellulose

More recent studies confirm the significance of browse, especially the leguminous plant species, to livestock nutrition. Ogunbosoye *et al.* (2015) noting that many potentially useful leguminous species were not being utilized, emphasized the need to expand the variety of leguminous browse because they have great promise as a protein supplement to low quality feed such as grasses and crop residues due to their high levels of crude protein of all ten selected browse plant species analysed in the southern Guinea savanna zone. Tables 3.3a and 3.3b provide data respectively on the chemical composition (%DM) of legume and non-legume browse plant foliage; and macro and micro mineral composition (DM ba sis) of browse plants leaves collected in Nigeria during the dry season. Table 3.4 on page 72 shows some of the elements that limit the utilization of these plant species as browse.

It is important to know that the browse does not, by any means, make up for

TABLE 3.3B Macro and micro mineral composition (DM basis) of browse plants leaves collected in Nigeria during the dry season

Browse species	Ca, %	Mg, %	P, %	Na, %	K, %	Zn, mg/kg	Fe, mg/kg	Mn, mg/kg	Cu, mg/kg	S, %
Parkia biglobosa	1.05abc	0.54d	0.35b	0.45abc	2.78abcd	34.10a	295.8a	61.70ab	6.10bcd	1.79
Gliricidia sepium	0.73def	0.41abc	0.34b	0.25de	1.93f	29.10bc	273.6a	47.10c	5.40cde	1.54
Albizia odoratissima	0.83f	0.34bc	0.25c	0.30bcde	2.45bcdef	29.10bc	271.3a	52.40bc	6.20bcd	1.69
Prosopis africana	0.88cde	0.25cd	0.30bc	0.47ab	2.24ef	27.50c	270.4a	46.80c	5.05bcd	1.69
Non-legume										
Daniella oliveri	1.22a	0.63a	0.45a	0.53a	2.96ab	35.10a	301.5a	64.40a	7.10ab	1.89
Eclipta alba	1.25a	0.63a	0.49a	0.56a	3.02a	35.50a	304.5a	65.55a	7.45a	1.92
Ficus polita	0.93bcd	0.47abc	0.27c	0.40bcde	2.55abcde	31.45ab	287.5a	56.70ab	5.45cde	1.57
Ficus cogenesis	1.14ab	0.58ab	0.38b	0.49ab	2.85abc	34.80a	196.4b	62.40ab	6.40abc	1.69
Polyalthia longifolia	0.78de	0.31ab	0.18d	0.27c	2.36cdef	27.70bc	269.6a	67.50a	4.80e	1.52
Moringa oleifera	0.68ef	0.35bc	0.15d	0.19e	2.26def	28.40bc	271.4a	46.40c	5.20de	1.55
SEM	0.05	0.04	0.05	0.03	0.08	0.74	6.12	2.22	0.17	0.05

abcdef= means on the same column with different letters are significantly different ($P < 0.05$)

Source: Ogunbosoye Dupe O., G. O. Tona and F. K. Otukoya 2015 Evaluation of the Nutritive Value of Selected Browse Plant Species in the Southern Guinea Savannah of Nigeria for Feeding to Ruminant Animals. *British Journal of Applied Science & Technology 7(4): 386-395, 2015, Article no. BJAST.2015.156* ISSN: 2231-0843 SCIENCEDOMAIN international www.sciencedomain.org

the dry season deficiencies of protein and mineral, especially phosphorus, in the diet, but it does help to ameliorate the worst effects of undernourishment during the dry season. Year-round supplementation of animal diet with protein and mineral mixtures rich in phosphorus is routine practice on the ranches. The Fulani herdsmen are fully aware of the seasonal and spatial variations in the distribution of fodder sources and are therefore, well able to adjust their movements to this ever-changing supply situation.

Apart from the supply of feed, two other issues are relevant to an understanding of the ecology of grazing in Nigeria. These are water supply and animal health. The latter issue is discussed in Chapter 7.

TABLE 3.4 Anti-nutritional factor (%DM basis) of browse plants leaves collected in Nigeria during the dry season

Browse specie	Tannin	Saponin	Phytate	Oxalate
Parkia biglobosa	0.49a	0.57a	0.30a	0.19a
Gliricidia sepium	0.28b	0.33c	0.20b	0.10b
Albizia odoratissima	0.29b	0.31c	0.21b	0.11b
Prosopis africana	0.28b	0.33c	0.20b	0.10b
Non-legume				
Daniella oliveri	0.49a	0.58a	0.31a	0.20a
Eclipta alba	0.50a	0.59a	0.31a	0.20a
Ficus polita	0.48a	0.57a	0.29a	0.21a
Ficus cogenesis	0.33b	0.45b	0.28a	0.21a
Polyalthia longifolia	0.33b	0.45b	0.28a	0.21a
Moringa oleifera	0.27b	0.31c	0.19a	0.09b
SEM	0.04	0.033	0.01	0.01

abc= means on the same column with different letters are significantly different ($P < 0.05$) *Source*: Ogunbosoye Dupe O., G. O. Tona and F. K. Otukoya 2015 Evaluation of the Nutritive Value of Selected Browse Plant Species in the Southern Guinea Savannah of Nigeria for Feeding to Ruminant Animals. *British Journal of Applied Science & Technology 7(4): 386-395, 2015, Article no. BJAST.2015.156* ISSN: 2231-0843 SCIENCEDOMAIN international www.sciencedomain.org

Water supply in grazing

The concentration of the bulk of domesticated animals in the sub-humid region of the country creates grave water supply problems. During the rainy season the supply is very fair; the streams are in full flow and, in addition, there are numerous incipient drainage channels, inundated pans and depressions where animals may drink and wallow; and the pasture itself is fleshy and contains a lot of water. During the dry season, however, water is available only in the beds of a few depressions and streams. It may also be obtained from shallow wells, especially in Borno where the surface sands are underlain by less permeable clay beds which act as traps for water percolating from the surface. Such sources of water are not widespread and vast areas of grazing land are rendered unusable by lack of water. The movement of cattle southwards during the dry season, is due as much to the scarcity of water as it is to the poverty of the pastures.

In order to combat the problem of water shortage, dams, ponds and wells have been constructed in many states including Borno, Kano, Katsina and Sokoto. In Borno State wells used to be dug with crude implements and were therefore tapping only the small perched aquifers, close to the surface. Over the years mechanical drilling equipment have been introduced to dig deeper, more solidly walled wells whose improved yields have so far been maintained.

The most spectacular water supply schemes in the 1970s were developed in Kano State where many dams were constructed to provide water for use in irrigation agriculture and for watering livestock. The dams are sited on the Kano, Challawa and Hadejia rivers. The largest, the Tiga dam on Kano River, was completed in 1973. Unfortunately, climate change and the attendant uncertain and irregular rainfall give much cause for concern in present times. Flash floods are more frequent than in the past in spite of the dams that have been built on quite a number of rivers in the Sudano-Sahelian zone. Dam bursts are an ever-present danger and livestock, crops, settlements and human life have been lost to both flash floods and dam bursts.

Grazing as land use in Nigeria

In Nigeria as in other tropical areas pastoralism has come to be associated with a particular ethnic group to whom nomadism is a way of life and who have little or no concept of land tenure. All land is looked upon as open grazing and its limits are defined only in terms of how far a nomadic group can travel. Although there may be an understanding among the nomadic communities that specific areas are more or less reserved for particular sections of a group, the same understanding does not exist with non-pastoralists.

Thus, in Nigeria, there have always been conflicts between cultivators and the Fulani herdsmen. In the past, cultivators complained particularly about damage done to crops by cattle and the indiscriminate burning of bush by the Fulani. The independence of the herdsmen on the one hand, and the unwillingness of cultivators to accommodate grazers, on the other, has delayed needed improvements in both farming and pastoral practices. This applies particularly to grazing as a resource process since no serious range improvement can be embarked upon until it is accepted as an integral part of rural land use. Furthermore, plans to supplement animal diets by commercial feeds cannot be implemented when the animals are continuously roaming about. These problems can only be solved when specific areas are set aside for grazing and such demarcations are respected by cultivators.

Grazing reserves

The same conditions that had favoured the large increases in livestock populations and the expansion of grazing throughout Nigeria favoured arable farming even more. As noted in Chapter 1, of all land use types, agriculture has made the greatest gains in land as it has encroached not only on forests but also on grazing areas in the savanna zones. Apart from traditional agriculture, agricultural development projects, first, by state governments and, later, by the river basin development authorities, took more land out of the reach of the pastoralists. The phenomenal growth of urban centres, which also fuelled developments in industry, roads and other infrastructure, increased the pressure of agriculture on grazing and grazing lands. As Iro (2015) explained it, as development spilled from the urban core to the rural hinterland, farmers escalated their incursions on the dry-season grazing land (*Hurumi*), the rainy-season grazing land (*Mashekari*), and the cattle-routes (*Burtali*). Scarcity of pasture is also the reason why the Fulani take their herds to the uncontested forage in the outskirts of towns and large villages. Unfortunately, cows, sheep, and goats scavenging around school playgrounds, golf courses, government residential areas, street shoulders, and railway sidings in the cities obstruct traffic flow; endanger street users and cause annoyance and resentment among city dwellers and travellers.

It can be seen, therefore, why farmer/herdsmen conflicts that began during the colonial era had intensified over time to such a dangerous level as to threaten the political stability and survival of the country today. The primary concern of both federal and regional or state governments from the colonial period has been to avoid conflict between herdsmen and arable farmers. Grazing reserves were conceived as a way to put an end to farmers' encroachment on grazing land which always resulted in conflicts. The Northern Nigeria regional government passed the Grazing Reserve Law of the Northern Nigeria in 1965, which carved out some grounds for the exclusive use of pastoral Fulani. The Law merely formalized a process that had begun since the 1950s during the colonial era.

A *grazing reserve* is a piece of land that the federal, regional or state government acquires, gazettes, develops and releases to the pastoral Fulani. In order to make things easy for the Fulani pastoralists and encourage them to stay within the grazing reserves, the relevant government tasks itself to provide water, water pumps, roads, schools, markets, hospitals, motor parks, veterinary clinics, and other facilities to service the game reserves and Fulani settlements (camps) within them. In order to accommodate the grazing reserves, government could not take land out of agriculture or other forms of land use hence it took the line of least resistance which was to convert many existing forest reserves to grazing reserves. Deregulation of forest reserves has been a contentious issue among conservationists. It is not surprising then that government has not been able to deliver on the number of grazing reserves which it identified for development totalling about 28 million hectares of land. Apart from the well-known official ineptitude in project implementation and management, the grazing reserve project has raised serious political issues, which have stultified progress. These issues came to the fore in 2019 when the Federal Government moved to force every state in the country to set aside land for Fulani herdsmen, so-called RUGA settlements. Most states, especially in the south and middle belt were not disposed towards the idea of ceding their traditional lands to any 'stranger' elements. The social and political atmosphere was already much polluted by months of brigandage and terror unleashed by herdsmen militia groups on local inhabitants in many parts of the country.

It is clear that the problem of the pastoralists cannot be resolved until the archaic pastoral nomadic system of animal husbandry is transformed: there is not that much room for nomads to roam around with their livestock in a densely settled territory like Nigeria. The present open grazing system inhibits range and stock management whereas the various established grazing reserves have provided opportunities for range improvement and better management of livestock. A sore complaint of the herdsmen is that the grazing reserves have inferior fodder, consisting of low-protein *Andropogon*, *Brachiaria*, and *Loudetia because of their poor soil and terrain characteristics*. The problem is one that can only be solved through range management.

Types of grazing animals in Nigeria

Table 3.5 column 2 gives the population estimates for different types of livestock in Nigeria by the FAO in 1983. The Resource Inventory and Management (RIM) Limited in conjunction with the Federal Department of Livestock and Pest Control carried out a more thorough estimation of the population of livestock using low level aerial surveys combined with ground level estimates at selected villages, urban centres and commercial livestock holdings (Bourn *et al.*, 2018). The figures shown in column 3 of Table 3.5 are the combined results of a 1990 traditional sector survey and the 1992 survey of commercial livestock enterprises. Column 4 of the Table shows the results of a National Livestock Sample Survey carried out in 2011 by the Federal authorities.

The three main types of domesticated animals involved are cattle, sheep and goats, but poultry outnumbers all other types of livestock and is the most widely produced by commercial enterprises especially in and around the big urban centres. The history of animal breeding in tropical Africa had been widely discussed (Williamson & Payne, 1965; Owen, 1973). It appears that early animal breeders in Africa placed great emphasis on the selection of kinds that were resistant to environmental constraints especially disease and drought. Considerations of their food value were rather secondary. In the case of cattle, this was at least partly due to their importance in social and religious functions. Goats and sheep were allowed to breed freely and become locally adapted to the climatic and biotic conditions of the environments in which they lived. As a result of this lack of breeding policy most breeds of domesticated animals today are of rather low quality and productivity compared to those in temperate regions.

Oyenuga (1967) had identified three important factors responsible for the low productivity of Nigerian and West African livestock, namely, low level of nutrition, lack of genetic ability and incidence of diseases, pests and parasites, particularly of trypanosomosis. Domestic animals in Nigeria and West Africa, according to Oyenuga, had suffered from chronic hunger for centuries. Up till today sheep and goats roam about the streets picking up domestic waste and foraging in the surrounding bush in search of food, and in the process expend a great deal of energy. The cattle, too, still feed on herbage of relatively low quality.

None of the indigenous animal breeds have the hereditary characteristics needed to make efficient use of the environment. Oyenuga noted that while the Fulani are excellent herdsmen who seek to improve their herds in fineness of body, appearance and colour, they pay less attention to such economic factors as their fitness as draught beasts or for meat and milk yield. Thus, animal breeding among the Fulani

TABLE 3.5 Nigeria Livestock Population Estimates, 1983–2011

Species	1983 (FAO)	Number, 1990/92 (AGR, Nigeria)	2011(National Livestock Sample Survey)
Poultry	150,000,000		
Chicken		82,400,000	145,000,000
Goats	26,300,000	34,500,000	72,500,000
Sheep	12,850,000	22,100,000	41,300,000
Cattle	12,300,000	13,900,000	19,500,000
Donkeys	700,000	900,000	974,499
Horses	250,000	200,000	
Camels	18,000	90,000	28,000
Other Poultry (incl. pigeons, ducks, guinea fowl, turkeys	31,900,000	11,600,000 ducks	1,200,000 turkeys
Pigs	1,300,000	3,600,000	7,100,000
Dogs		4,500,000	
Cats		3,300,000	
Rabbits		1,700,000	
Guinea pigs		500,000	
Giant rats		60,000	

Sources: 1. FAO Production Yearbook, Volume 37, 1984; 2. Bourn, D., W. Wint, R. Blench and E., Woolley, 2018. *Nigerian Livestock Resources Survey* Nigeria: Animal Genetic Resources www.angr.org.ng ; 3. Minister of Agriculture, Federal Department of Livestock, 2011

has failed to develop their most productive capabilities.

Cattle

There are over nineteen million heads of cattle in Nigeria (Table 3.5) of which the most important is the *zebu*, a hardy breed well adapted to the savanna environment. Four breeds of zebu cattle are widespread in the north: (1) the Sokoto *gudali*, docile and relatively good milk producers, found mainly in the north-west; (2) the Adamawa *gudali*, found in the east; (3) the *bunaji* or white Fulani cattle found throughout the north except in the extreme north and north-east where (4) the *rahaji* or red Fulani cattle, which yield less milk, are adapted to the harsher conditions. In addition, the Shuwa keep the red-brown *wadara* cattle, which although smaller than other zebu breeds, are nevertheless relatively good milk producers.

In the humid tsetse-infested south of the country, the *muturu* or dwarf, short-horned cattle and the larger humpless *n'dama* make up the bulk of the cattle population. They are naturally immune to trypanosome infection. Unfortunately, their productivity and the quality of their meat and milk are much lower than those of the zebu. A muturu-zebu cross, the *keteku*, is kept by the Fulani in the derived savanna of Oyo and Kwara States and southwestern Nigeria in general both for themselves and for the local farmers.

In order to raise good quality animals, cattle ranches were established in different parts of the country. The best-known are those at Obudu Plateau in Cross River State; Gombole, near Maiduguri in Borno State; Shika near Zaria in Kaduna State; Ilorin in Kwara State; Mokwa in Niger State; Fasola near Oyo in Oyo State; and Ado-Ekiti in Ekiti State. The breeding policy in each of these ranches was designed to achieve three objectives: (a) improvement through selective breeding of the more promising indigenous breeds for beef, milk and work and a combination of all three characteristics in one animal; (b) improvement of local indigenous zebu breeds through crossbreeding with imported zebus from other countries such as the Sudan, Egypt and India; and (c) upgrading of certain local animals by crossbreeding with European cattle, for example Friesian cows from Britain, to improve milk yield and early-maturing properties (Oyenuga, 1967, p. 235). Even now, at the beginning of the third decade of the 21st century, Nigeria is grossly deficient in milk production to the extent that it imports milk and milk products valued at US$1.3 billion annually to make up the shortfall. While a cow in Saudi Arabia or Brazil produces 30 – 40 litres of milk per day, a Nigerian cow produces only 1 litre per day. The dismal performance of the country in upgrading animal husbandry practices, particularly, the failure to develop ranching in most parts of the country devoid of religious and ethnic undertones has been the bane of her livestock development programme.

Sheep and goats

These domestic animals are important primarily for the supply of meat, especially during Muslim festivals but they are also milked for human consumption, especially in the north. The sheep numbering over forty-one million (Table 3.5) are used for wool even though the wool does not compare with what obtains in the temperate latitudes. Three main types of sheep are found in Nigeria: the *y'ankasa;* the long-legged *udo; balami* and the West African *dwarf sheep.*

The goats of which there are over seventy-two million are browsers and like the sheep are most undiscriminating in diet. The three main types of goats are the

Kano brown, the *Sokoto red* and the hardy short-legged *dwarf goat,* usually black and white, commonly found in the middle belt and southern humid forest regions.

The dwarf sheep and goats which are more characteristic of the humid south and are tolerant of trypanosomosis. In general, the pastoral sheep and goats are more productive than the free-range village stock.

Poultry

Chickens are by far the most common not only in traditional holdings but also among commercial livestock enterprises. There are now about 145million chickens in the country (Table 3.5). Bourn *et al.* (2018) estimated that commercial enterprises, which experienced a dramatic boom in the 1970s and 1980s following the Green Revolution campaign by the military administration, accounted for about 11 per cent of the estimated total population of 82.4 million chickens in 1990-92. The commercial enterprises provide meat and eggs to the teeming populations of the urban centres. Turkey (population: 1.2 million in 2011), raised by commercial enterprises, ducks (11.6 million) and guinea fowl raised by traditional poultry farmers in the northern parts of the country are the other major poultry resources of Nigeria. The traditional farmers raise not only live birds but also the eggs of guinea fowl for sale.

The vast majority of poultry in Nigeria are raised under traditional, low input, free-range management systems. The major constraint on traditional chicken production is the Newcastle disease to which local breeds are highly susceptible. The fortunes of the commercial poultry enterprises rise and fall with the vagaries in the supply of feedstuff and drugs. In the 1970s and 1980s, the enterprises benefitted from government subsidies on feeds and drugs but these subsidies have been removed since late 1980s.

Pigs

FAO (2018) *Africa Sustainable Livestock 2050* projections noted that population growth, urbanization and gains in per capita income would result in phenomenal increases in demand for livestock products in Nigeria between 2012 and 2050. Aggregate demand for livestock products would more than triple and, demand for pork would be second only to poultry in overall percentage growth and annual growth rate (Table 3.6a). Table 3.6b shows production projections for different animal source foods and it shows clearly that pork production will not outstrip demand until year 2050.

Consequently, pig farming has great potential in Nigeria. Piggeries are common near the major urban centres in southern Nigeria. But the local traditional free range black hairy pigs are not favoured by modern piggeries and are rapidly being replaced by the exotic Large White, Landrace, Hampshire and Duroc varieties from overseas.

TABLE 3.6A Nigeria: Current and projected consumption of livestock products, 2012-2030-2050

	Estimated Consumption ('000 tons)			Growth 2012–2050	
	2012	2030	2050	Percentage	Annual Rate %
Milk	1384	2804	4226	262	3.2
Beef	392	800	1393	255	3.7
Mutton & Goat	467	878	1473	216	3.3
Poultry	296	602	602	266	3.8
Eggs	591	1251	2069	250	3.6
Pork	248	494	899	262	3.7

According to Bourn *et al.* (2018), "intensive pig rearing is economically viable on the periphery of large cities because of the availability of industrial by-products, particularly brewers' grain. Units of between 50 and 200 pigs kept in concrete pens are common, especially in the densely populated regions of the south. Commercially managed piggeries with more than five breeding sows account for about 3 percent of the total estimated pig population of 3.5 million."

Traditional animal husbandry practices and their ecological impact

Grazing in Nigeria is a rough blend of nomadism and transhumance. The Fulani are nomadic people with a natural disposition to roam about with their cattle. This way of life has been imposed by the semi-arid conditions and the scarcity of pasture. Nomadism in its true form can be said to be found in the drier northern, north-western and north-eastern extremities of the country, in Sokoto, Katsina, Kano and Borno areas. These are the main centres of operation of the Fulani during the rainy

TABLE 3.6B Nigeria: Current and projected production of animal source foods, 2012-2030-2050

	Estimated Consumption ('000 tons)			Growth 2012–2050	
	2012	2030	2050	Percentage	Annual Rate %
Milk	579	1372.2	2241.6	287	3.9
Beef	392	846.4	1306.1	233	3.5
Mutton & Goat	468	850.8	1214.7	159	2.8
Poultry	297	671.0	671.0	532	5.4
Eggs	642	1630.0	4857.2	657	6.0
Pork	249	444.4	928.5	273	3.8

Source: FAO, 2018 *Africa Sustainable Livestock 2050: Transforming Livestock Sector: Nigeria What do long term projections say?* Brief prepared by Wisser D., Odunze E., Boussini H., Mickecz O. and U. Pica-Ciamara October 2018

season. Their movement is determined by the availability of pasture and by the distribution of watering points.

Transhumance is a dry season phenomenon when there is a mass movement of cattle to the south and the upland areas of the Jos Plateau and the eastern borderlands. Up to 70% of Nigeria's traditional (Fulani) pastoralists used to engage in transhumance activities, moving along well-defined cattle routes or 'cattle trails (ILCA, 1979). Today, in spite of disruptions by arable cultivation and urbanization the cattle still move along well marked routes, which pass through grazing reserves, open grass fallows and river valleys where water may be obtained. Some of the cattle move as far south as the derived savanna areas of Oyo, Ekiti, Ondo, Edo and Anambra States where some Fulani have established camps and have settled more or less permanently. The bulk of the cattle, however, are found in the Guinea savanna and on the upland grazing lands. In addition, herds and flocks are brought in from neighbouring Chad, Cameroon, Niger and Benin Republics in the dry season, and taken back in the rainy season; covering between 200 and 700 kilometers either way (Babalobi *et al.*, 2000).

One of the ecological effects of grazing in Nigeria is associated with this dry season movement of cattle from the Sudan zone to the south. Studies have revealed that cattle are responsible, among other factors, for the spread of savanna tree and grass species to the humid and forested south. George Jackson (former lecturer in Botany at the University of Ibadan) traced the spread of certain plant species along the cattle trails from the north to the south. The cattle eat the seeds and nuts of savanna plants in the north and deposit these with their faeces as they move along the trails. Some of the seeds then germinate and grow up in their new environment. It is reasonable to assume that species from the south are transferred northwards in the same way. Cattle movements thus blur the distinctions in the floristic composition of the various vegetation communities.

A major concern about these dry season cattle movements is that the extensive transhumance across regional, national and international boundaries by Nigeria's traditional pastoralists constitutes a problem which has great implications for the design and implementation of effective programmes of disease prevention and control. Accordingly, unless there is a national and internationally coordinated programme of disease prevention and control, any local programme would be futile (Babalobi *et al.*, 2000). Table 3.7 indicates major diseases and their seasonality at a grazing reserve surveyed in the northern Guinea zone of Niger State.

Overgrazing

The greatest impact of grazing on the environment can be directly attributed to overgrazing and bush burning. Overgrazing can be traced to a number of root causes such as, the semi-arid environment and the scarcity of pasture which have been discussed already. Another important factor is that, because of the scarcity of water and the need to avoid areas liable to tsetse infestation, there is a tendency to concentrate livestock in a few suitable places, with the result that animal populations in these favoured places are too large.

Among the Fulani herdsmen, a man's wealth and social status are usually judged by the size of his herd. This encourages them to accumulate stock without regard to the carrying capacity of the available grazing areas. But carrying capacity is, in any case, not a concept of which the Fulani are aware. The accumulation of stock has also been encouraged by the high mortality rate in the herds and flocks. Accumulation of stock is therefore an insurance against reduction in total numbers that sometimes occur as a result of starvation, disease and drought.

Overgrazing occurs in areas which have fairly adequate water supplies while many good grazing lands are left untouched. One such overgrazed area, pointed out by de Leeuw (1976), extends from Nguru district to Geidam in northern Yobe State. Apart from the Hadejia River, water is available from artesian boreholes. Other areas which show signs of overgrazing may be found in the Sokoto-Rima basin and on the Jos and Mambila Plateaux.

TABLE 3.7 Main diseases and their seasonality, Bobi Grazing Reserve, Nigeria

Disease	Season
Malnutrition	Dry Season
Contagious Bovine Pleuropneumonia	Dry Season
Trypanosomosis	Both Seasons
Helminthosis	Both Seasons
Leech Infestation	Both Seasons
Foot-and-Mouth Disease	Mainly Rainy Season
Dermatophylosis	Rainy Season
Black-Quarter	Rainy Season

Source: Field Survey 1991 – 1996. From Babalobi O.O., Ogundipe G.A.T. and G.O. Esuruoso, 2000 **Livestock Population Dynamics and Problems in a Grazing Reserve in Northern Nigeria** *Proceedings of the 9th International Symposium on Veterinary Epidemiology and Economics, 2000* www.sciquest.org.nz

Overgrazing in some areas of Borno State and Sokoto, Kebbi and Zamfara states in the Sokoto-Rima basin has led to the development of 'patched' land surfaces due to the exposure of the land to intense insolation and increased rates of evaporation. The result is that the hardened soil surface can no longer be colonized by plants. With the removal of the vegetation and the hardening of many surfaces, infiltration of water into the soil is greatly reduced. The increased runoff thus generated has been responsible for erosion of slopes in the grazing areas. In the Sokoto-Rima basin, for example, sheet and gully erosion take place on the slopes overlooking the broad river valleys. Gully erosion also occurs on the overgrazed lands of the Jos Plateau.

Moreover, reduced infiltration of water into the ground has resulted in a general lowering of the regional water table causing streams and springs to dry up, or to become more markedly seasonal, as a result of the increased dependence of their flow regime on surface runoff from local rainfall. Even where patched surfaces have not emerged, the soils have developed hard clay pans in their profiles as a result of increased desiccation.

On the dry-season grazing lands of eastern Niger State evidence of overgrazing can be found. First, the tussocky grasses are widely scattered with a great deal of bare ground in-between as, for example, in the Izom and Kau plains in the Federal Capital Territory. Secondly, while shrubs and small trees are injured by constant gnawing by the livestock.

Bush burning

Bush fires caused by the Fulani herdsmen in the dry season have been a major cause of conflict between them and the cultivators. Crops have often been inadvertently destroyed. More seriously, many fallow lands are now at rest only in the sense that they are not being cultivated. Burning destroys the restorative fallow vegetation annually and this probably reduces the rate at which soil fertility is restored during the fallow period (Nye & Greenland, 1965; Areola 1977). The Fulani are not solely responsible for this; setting fire to dry vegetation is a pastime of the cultivators too. The point of conflict appears to be not so much the burning itself but who does it.

In the dry Sudan and Sahel zones burning in some places has led to the disappearance of the grass vegetation leaving only fire-resistant thorny shrubs which are unpalatable to grazing animals. The same is true of sections of the Guinea savanna zone which now only carry a woody vegetation with little grass understorey. In addition, the weed grass *Imperata cylindrica* (spear grass, *lalang*) dominates large areas of once good grazing land.

The merits and demerits of the savanna bush fires have long been debated (e.g. Darling, 1960). To the Fulani herdsmen the fires serve a very useful purpose because they destroy dead vegetation and promote the growth of new grass. Furthermore, insect vectors of diseases are destroyed by the fires. Apart from these obvious advantages other positive roles of fire have been suggested, for example Owen (1973) argued that on the whole fire is less destructive than cultivation, because the plants that reappear after the fire are in general the same species that were there before. By contrast, after cultivation, complex successional patterns are generated often involving colonization by what may be regarded as weed species. The compromise solution lies in controlled burning in order to maximize the advantage to the resource process and minimize the adverse effects on the natural ecosystem.

One area where the problem of bush burning may not be amenable to a simple solution is in relation to wildlife conservation, where fire appears to be a double-edged sword. On the one hand, it promotes the growth of fresh green foliage which attracts game but, at the same time, it kills many wild animals. Burning becomes a major concern in those areas declared as grazing reserves which are also wildlife conservation zones where hunting is forbidden. The Fulani frequently set these grazing reserves on fire. Fortunately, riverine vegetation in such reserves is hardly affected and the valleys provide havens of wildlife refuge in the grazing reserves. The controlled use of fire by trained conservationists appears to be the answer since it would eliminate much of the destruction associated with uncontrolled and irrational bush burning by the herdsmen.

Selective grazing

Selective grazing by domestic animals is also said to be responsible for marked ecological shifts in grazing areas. But there is some doubt as to the truth of this claim in Nigeria; this is because domestic grazing animals feed on a wide variety of forage containing many species of grass, herbs and woody plants. They appear also to make as efficient use of savanna grazing resources as wild herbivores do (de Leeuw, 1975). Thus, degradation of the vegetation of grazing areas is caused not so much by selective grazing as by mismanagement on the part of the herdsmen leading to overgrazing. It is true that selective feeding tends to be encouraged by open grazing but when confined to a fenced area, or when there is not much choice of forage, cattle and sheep tend to consume the majority of plant species present. Low selective grazing especially of the grasses may also be due to the uniform growth of the dominant grasses so that one is not less mature and therefore more succulent than another at any given time.

Climate change and grazing

The defunct Natural Resources Conservation Council Action Plan (1991) had listed rangeland management as one of the targets of the action plan on combating drought and desertification. Before the current strident cry for climate change mitigation, conditions in the Sudano-Sahelian parts of the country had given cause for concern about the impact of the recurrent droughts in the region since about 1968. The drought episodes posed the risk of exacerbating the environmental impacts of existing problems of overgrazing, incessant bush burning, scarcity of water for livestock watering, and irregular rainfall for dryland farming, which was exposing the soils to desiccation. Soil desiccation had resulted in the, hardening and compaction of the lateritic clayey soils and the disruption of the cohesion and stability of the sandy soils making them more susceptible to wind erosion. Long stabilised sand dunes had been set in motion again and were encroaching on and burying oases, farmlands, grazing areas and human settlements. In other words, in the Sudano-Sahelian zone, climate change impacts had set in before the world took notice. The shrinking of Lake Chad to near total disappearance is a stark evidence of how far climate change has advanced in the region. The increasing incidence and ferociousness and cruelty of farmer-herdsmen conflicts and the rising incidence of conflicts between grazing and wildlife conservation in game reserves are living evidences of the desperation of the people whose livelihoods have been touched to overcome the vagaries of climate change. In Chapter 1 of this book, we alluded to the expansion of agriculture to new areas because of the fall in soil fertility and productivity in the old farmlands. This is one of the major factors responsible for the increased conflict between farmers and herdsmen in the savanna lands (cf. Odoh and Chidozie, 2012). The faulty steps that governments at both state and federal level have taken to deal with farmer-herdsmen issues are driven very much by the pressure from the herdsmen and their kinsmen in the face of dwindling grazing resources. Idowu (2018) explained the situation as follows:

> The Fulani herdsmen are nomadic and habitually migratory. They annually move from north to south with their cattle in search of grazing fields. The movement is seasonal. Now with climate change, the movement pattern has been markedly altered. Due to expansive desertification, drought and unchecked deforestation in northern Nigeria, the herdsmen naturally seek greener pasture further south. As the resultant migration has intensified, so too has violent clashes over grazing lands between local farmers and pastoral herdsmen, whom the former accuses of wanton destruction of their crops and forceful appropriation of their lands. The emerging conflict is further compounded by the shrinking of Lake Chad from 45,000km2 to 3000km2 in less than three decades. The consequence according to the United Nations, is the displacement of about 10.5 million people. It's a combination of these factors that has pushed herders from north-eastern Nigeria, the region closest to Lake Chad, to the southern parts of the country... The government's response has ignored climate change as the source of conflict exacerbating the herdsmen's grazing crisis. Historically, since the existence of Nigeria, the Fulani herdsmen have grazed their herds in the north and intermittently in other areas. But incremental drought with resultant desert encroachment forced them to regularly look southwards for greener grazing areas.
>
> Idowu, 2018 `https://www.climatescorecard.org/who-we-are/`

Nothing positive can be achieved by government fiat. The Water Bill being rushed through parliament to ensure right of way and access to water for Fulani cattle herders cannot bring lasting peace. It is nothing more than a short-term palliative that would not solve the problem in the long term. Long term solutions would require interventions that would promote guided adjustments and changes in age old systems of both crop and animal husbandry. Steps to promote ranching and mixed farming would be more productive and effective in stemming farmer/herder conflicts in the long run. Although, some experts had expressed reservations about making nomadic herdsmen to take to ranching, there is no better alternative that would bring peace both to people and the environment and resources in the long run. Nigeria also needs to join international organizations with common challenges and interests in climate change impacts. According to Idowu (2018), "Nigeria also needs to change its policy on climate change from a vision into action. It is distressing that Nigeria is not yet a member of the Climate Vulnerable Forum (CVF)—a 43-nation group of most vulnerable countries that negotiate as a bloc at the United Nations Framework Convention on Climate Change (UNFCCC)." Membership of the CVF would give Nigeria the opportunity for knowledge-sharing with countries facing similar challenges.

The grazing reserves already established could be the centres of innovation, training and extension services in the progressive move towards ranching. Grazing reserves can be managed the same way that game reserves are being managed and provided with watering points and mineral salt licks (Iro, 2015). The rangelands need to be managed to improve pastures and avoid land degradation. Farmers in the same ecological zones can be encouraged to adopt mixed farming; this would be an extension of the age-old collaboration between herdsmen and farmers in the Kano closely settled zone whereby the herders are allowed to camp on harvested farm plots so their cattle can feed on crop stubble and at the same time dump their waste as farmyard manure on the crop field (Mortimore, 1967; see Chapter 6).

The future

The future of open grazing or pastoralism has become a very contentious issue in Nigeria in the 21st Century. The unresolved issue of Fulani herdsmen and their accommodation in rural land use in different parts of the country has become an albatross on Nigeria's neck. The issue has assumed dangerous political and religious overtones that have the potential of tearing the country apart if not properly handled. The simple truth is that Nigerian governments over the years have failed to develop policies and statutes to guide and regulate the activities of herders and herder/farmer relationships. As Ogboru and Adejonwo-Osho (2018) noted, Nigeria lacks the required laws or statutes for the regulation of grazing land use and the protection of the environment against pastoralism, which has intensified in the last two decades or so. The herdsmen have become more desperate and violent in forcing their access to grazing lands on local peoples all over the country but especially in the Middle Belt.

The perception of many Nigerians is that the Fulani herdsmen have created an armed militia to enforce their access to grazing areas. The last few years have witnessed vicious attacks by herdsmen militia in many parts of the country characterized by the killing of villagers, burning down of settlements, destruction of farmers' crops, abduction of hostages for ransom and such like acts of violence. Highway armed robbery and kidnapping of people for ransom have become the order of the day and are thought to be the strategy adopted by the herdsmen to raise money to pay for the sophisticated guns and ammunition being used against the people. Every attempt by local inhabitants to retaliate invariably leads to more vicious attacks by the herdsmen. The herdsmen accuse state governors and officials of going back on agreements made with the herdsmen while campaigning for office to keep open the traditional cattle trails, which had been approved and gazetted since colonial times. Changes in land use, including the expansion of settlements and infrastructure have blocked a number of the cattle trails and former open grazing lands. The vicious circle of violence in the herder/farmer interactions has been compounded by the terrorist activities of Boko Haram and similar groups that have risen up in recent years.

At some stage, some form of sedentarization of the pastoralists will have to take place. It would be in the best interests of the whole country to pursue this goal through a gradual, open and peaceful process. A revolutionary or dictatorial approach imposed by government fiat would inexorably lead to conflict far bigger and more destructive than the erstwhile herdsmen/farmer clashes.

The Federal Government has adopted a strategy of appeasement of the herdsmen through their umbrella organization, Miyetti Allah Cattle Breeders Association of Nigeria, whose patrons are the Sultan of Sokoto and the Emir of Kano. The strategy has been to pay the organization huge sums (billions of naira) of money to cover their supposed losses. This is clearly an unsustainable solution to the problem of accommodating herders, who have refused to modernize their pastoral practices and way of life, in rural land use in this day and age. The present Buhari administration thought it could force state governments to appropriate land to create ranches to accommodate the Fulani herdsmen and their livestock. But the move was vigorously resisted by the general public, who questioned why government would single out the Fulani herdsmen for special treatment. After all, grazing is private enterprise just like farming, fishing, lumbering, and other activities engaged in by other Nigerian citizens. The rationale for the huge sums of money being paid by government to pacify the herdsmen is one that has been difficult for most citizens to understand.

As stated earlier in this chapter, there are a few modern cattle ranches in different parts of the country established to produce high quality animal products. The cattle are fed not only on native grasses but also on fodder crops, cultivated pastures and concentrates with chemical additives to supply mineral deficiencies in the forage.

Unfortunately, these cattle ranches have made little or no impact on grazing generally. Management efforts have been geared more towards the provision of high-quality meat and milk production than to the dissemination of modern, restorative animal grazing and range management techniques. Thus, traditional animal husbandry practices have not improved much over the years.

What is sorely needed now is a concerted effort to carry modern range management and improved techniques beyond the confines of the ranches to the open grazing lands. The grazing reserves are the obvious points at which such innovations could be introduced and perfected. An end to the open grazing system must follow. Although the Fulani herdsmen, like all pastoralists, are difficult to administer and control and although their recalcitrance has so far frustrated attempts to settle them, experience in the derived savanna zone indicates that they can be persuaded gradually to settle at least semi-permanently. There are Fulani herdsmen camps in the derived savanna zone where some of them have lived more or less permanently for nearly a century tending cattle for themselves and for the local inhabitants. These camps are visited seasonally by the long-distance herdsmen and their flocks.

In the Sudan zone attempts are being made to induce some of the Fulani to settle and adopt a mixed farming system especially those among them who have expressed the desire to do so, if suitable areas can be found for them. Such an area must have good pasture and water supply and be large enough to accommodate the number of cattle in the herd. This was the principle upon which The Agangara Fulani Settlement Scheme in Sokoto State was established in 1951. The extensive land requirement of cattle ranching is likely to be one of its more serious obstacles because it limits the land available for other land users.

It is pertinent to point out that, as emphasized by Harris (2000), it has been notoriously difficult to formulate and implement practicable management policies for grazing land, particularly with a progression from humid to arid environments. Management systems must take grassland dynamics into account and factor seasonal and spatial and temporal variability of forage supply into the equation. Harris (2000) also observed that "the worldwide attempt to replace land extensive pastoralism by ranching based on North American, Australian or South African models has generally failed." He says the underlying assumption that ranching is sustainable and extensive pastoralism is not, is false. He thinks the movement of livestock in consonance with annual and seasonal variations in availability of water and grazing resources protect the environment better than ranches, as there are many overgrazed ranches all over the world. Harris suggests that what is needed is to provide legal pastoral tenure for areas such as: water points, arable lands, transhumance routes, trees, and various types of grazing land. It is, however, difficult to see how long such a system can be sustained especially given the rapid population growth, high rates of urbanization and the expansion of agriculture throughout the country. The type of legal framework proposed by Harris (2000 *op. cit.*) can only be provided in an atmosphere of mutual trust between herders and farmers and a government with the political will to enact such laws and the ability to convince all parties of their impartiality in arbitrating between herders and farmers devoid of ethnic or religious hegemonic undertones.

Further settlement of the Fulani must contend with and find solutions to three other problems which have been mentioned already. First, there is the need to carry out more detailed studies on the ecological relations and establish the carrying capacity of the grasslands (de Leeuw, 1976). According to de Leeuw, as of the 1970s, the tsetse-free pastures were being overstocked and overgrazed while the tsetse-infested pastures were grossly under populated (Table 3.8). Therefore, the herdsmen can still be helped by creating more tsetse-free areas through a well organised tsetse eradication programme. It is important to pay more attention to the relative proportions of the different grazers and browsers making use of a grazing area in order to forestall overgrazing.

Secondly, it is important to realize that one of the reasons the animals are moved about is the uneven distribution of water in the main grazing areas of the Sudan zone. Stock watering facilities need to be distributed more widely throughout the region. This in itself will make for a more equal spread of stock and avoid over concentration of cattle in a few areas.

The third problem is that of disease. Aliu (1975) had argued that dry season Fulani cattle were partly responsible for the persistence of trypanosomosis. According to him Fulani herdsmen mobility in all its various forms, could be a menace to public health and jeopardize the vital economy of the livestock industry, both by assisting in the maintenance and spread of trypanosomosis (Fig. 3.2), and by thwarting measures to eradicate it. However, Bourn (1983), believed that deforestation and the removal of wildlife had greatly reduced the natural habitats and wild hosts of the tsetse fly over much of Nigeria. Today, the concentration of over half of the country's cattle herd in the middle belt is an obvious indication that tsetse fly infestation is no longer as serious a constraint as in the colonial and early independence

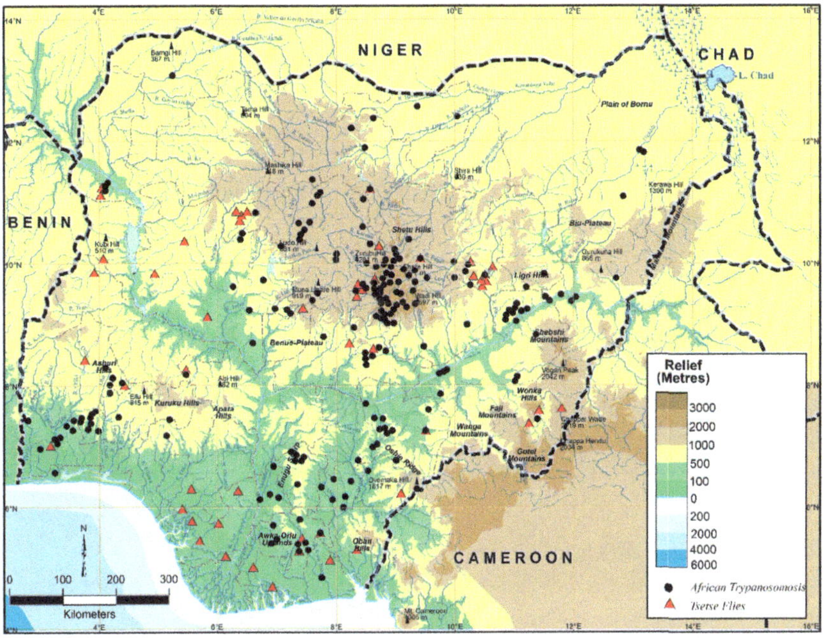

FIGURE 3.2 Distribution of tsetse flies and African animal trypanosomosis in Nigeria

periods.

In his study of fodder resources and livestock development in north-east Nigeria, de Leeuw (1976) had shown that the eradication of tsetse fly could considerably increase the grazing potential of Nigeria's grasslands, and suggested that the total available resources were adequate to feed the existing herd, provided a vigorous attempt was made to redistribute the livestock population, taking advantage of the areas cleared of the tsetse fly. Table 3.8 shows that the estimated stock population of the tsetse-infested part of the study area was far below the potential carrying capacity of the land, while the tsetse-free zone was overstocked and overgrazed. Bourn *et al.*, (2018) believe that there are indications of a progressive move towards development of mixed farming in Nigeria, which would greatly improve livestock productivity. The signs include: (a) a marked reduction in pastoral nomadism; (b) increasing sedentariness of pastoralists and their adoption of crop cultivation in addition to keeping livestock (van Raay, 1975); (c) the taking up of animal husbandry and livestock fattening by arable farmers; (d) the utilization of crop residues by livestock in exchange for dairy products and/or manure; and (e) the spread of the

TABLE 3.8 Estimated and potential wet-season cattle population in North East Nigeria (in thousands)

	Grassland Association	Estimated	Potential	% Difference
	Tsetse-free zone			
1	Andropogon & Aristida	81	73	+11
2	Aristida & Cenchrus	555	295	+88
3	Brachiaria	162	171	−5
4	Schizachyrium & Schoenefeldia	308	243	+59
5	Schoenefeldia	112	153	−27
6	Schizachyrium & Loudetia	405	279	+45
7	Sorghum	94	–	–
8	Alluvial complex (north)	191	–	–
	TOTAL	1,987	1,214	+63
	Tsetse zone			
9	Loudetia	94	405	−77
10	Pennisetum & Schizachyrium	215	500	−57
11	Hyperrhenia	44	735	−94
12	Andropogon	46	180	−74
13	Loudetia & Aristida	63	227	−72
14	Hyperrhenia (Sotuda)	70	305	−77
15	Alluvial Complex (north)	52	–	–
	TOTAL	584	2,352	−75
	Total estimated cattle population	2,751,000		
	Total potential cattle population	3,566.000		

Source: de Leeuw P.N., 1976

practice of animal traction for ploughing and carting.

4

Protected ecosystems and biodiversity conservation

Nigeria has had a long history of human occupation; many empires and kingdoms had come and gone; and some of the largest ethnic groups in Africa had settled there for centuries. Thus, a marked feature of the Nigerian landscape is the degree and extent of human interference with the natural environment. The evidence is found not only in favourable terrains and long-settled, thickly populated regions but even in such difficult and sparsely populated areas as the coastal creeks and swamps, the Niger delta mudflats, the waterlogged and thickly wooded river floodplains and the rugged uplands of the Jos Plateau and the eastern borderlands. Thus, it can be said that there is no true wilderness as such in the country in the sense of natural, undisturbed wild land. The vegetation, even in the middle of the high forest belt, and in the so-called "undisturbed forests" shows signs of having been disturbed in the distant past and is at best a re-established climatic climax. In more recent times certain areas have been designated 'reserves' in order to protect what is left of the natural landscape and to manage it to achieve given ends. Thus, there is virtually no nature reserve in the country that is not subject to some form of human management or interference. Hence, they all fall, more or less, in the category of *protected ecosystems*, that is, nature reserves that are managed to achieve given ends.

The protected ecosystems or nature reserves include the few areas of natural forest vegetation still to be found as well as other areas of secondary forest and savanna vegetation (Fig. 4.1). Most of these occupy 'buffer zones' between former tribal principalities and kingdoms which were devastated and left virtually empty

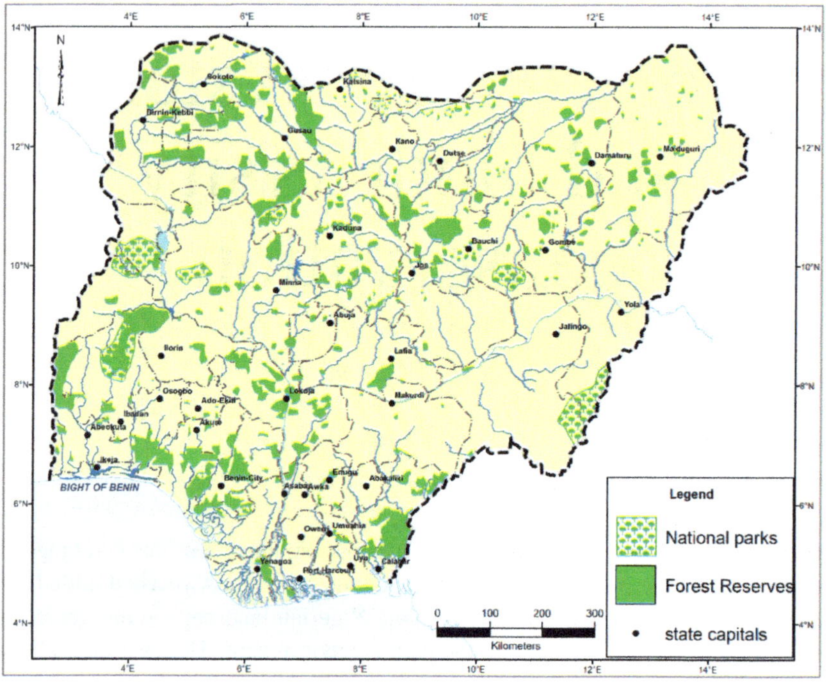

FIGURE 4.1 Forest and game reserves and other conservation areas in Nigeria

by the civil wars of past centuries. The buffer zones were designated nature reserves not so much because the natural environment was still intact but rather as a means of protecting the land and vegetation from wasteful exploitation by what the colonial masters regarded as the primitive agricultural practices of the local inhabitants. It is true, however, that as a result of protection the vegetation succession in some reserves advanced to the extent of approaching the stage of maturity.

Legacy of traditional nature protection

It is important to note that many of the conservation areas in the country today were bequeathed by the indigenous peoples in various parts of the country through their traditions, cultural and religious practices. It is true that many existing nature reserves were depopulated and left empty of human occupation by the internal wars that raged in the country just before colonial rule was imposed on the country. Even so, these conserved areas are in reality legacies preserved for long by the indigenous ethnic groups in the country. Almost everywhere in the country forests, woodlands, lakes, rivers, rocks, mountains and picturesque landscapes were protected as sacred ground with limited access and protection against exploitation of the available natural resources. Many of these landscapes and features were objects of worship and forbidden ground to the common people. Local taboos protected even individual burial grounds, shrines, as well as, individual species of trees, plants, birds, animals, lakes, rivers and springs against human interference. Indeed, as Kala (2017) argued, sacred landscapes, sacred groves and sacred species by various cultural groups are the live manifestations of historical, cultural and emotional attachment of people to nature and natural resources.

FAO (2020) recognizes the value of indigenous territories and lifestyles to biodiversity conservation and the fight against climate change because indigenous peoples throughout the world have adapted their lifestyles to fit into and respect their environments (e.g. New Vision, 2014; Rim-Rukeh *et al.*, 2013). This has been true of Nigeria's indigenous peoples from time immemorial. As we saw in the discussion on forest resources, the non-timber forest products, which are now highly regarded as providing leverage for promotion of viable livelihoods in Nigeria, were preserved over the ages by the indigenous peoples. In spite of complaints about the destructive tendencies of traditional crop and animal husbandry practices biodiversity conservation in the tropics is a legacy bequeathed by the indigenous peoples. Long before Europeans came to promote the protection of watersheds, indigenous peoples in Nigeria religiously protected the catchment areas of the springs from which they fetched drinking water. The author remembers an incident in his own local community in his younger days when the community enquired of the oracle why a certain spring had dwindled to a trickle. The answer came promptly that the (god of the) spring complained that people had insulted it by building on its head! How about that for a lesson on what happens when a watershed is deforested and built up!

As Rim-Rukeh *et al.* (2013) noted, the environmental wisdom and ethics expressed through these religious beliefs are very useful tools in natural resource management. Based on the classification of the IUCN Red List, and the plant and animal species found in the Delta State communities studied Rim-Rukeh *et al.* (2013) adjudged the sacred groves as havens of abundant plant and animal species that are not in any way endangered. The practices identified in a study in Tanzania as promoting conservation of wild plants in sacred forests hold true for most communities in Africa. They are (1) domestication; (2) beliefs in sacredness of trees; (3) beliefs in sacred forests; (4) respect of cultural forests; (5) protection of plants at the burial sites; (6) selective harvesting; (7) secrecy; (8) collection of dead wood for firewood. The only practice listed that is not common to all local communities in Africa is the use of energy-saving traditional stoves.

Types of protected ecosystems

The protected ecosystems may be grouped into three geographical zones: (1) the high forest reserves in the humid south including the Benin, Ondo and south-eastern Nigeria forests; (2) the dry forest reserves, most of which are probably secondary in nature although some are approaching maturity, found in the drier hinterland margins of the tropical rain forest zone, which were transformed into the derived savanna zone; and (3) the savanna forest reserves which are mostly woodlands but which also include forest outliers in the southern Guinea savanna zone. The forest reserves are subject to different types and intensities of use and management and hence varying degrees of human interference with the natural environment.

The Forest Resource Situation Assessment of Nigeria project carried out by Olaleye and Ameh (1999) under the EC-FAO Partnership Programme (1998-2002) to aid sustainable forest management, recognized three categories of forestland in Nigeria, namely, forest reserves, free areas and conservation areas (see Table 4.9, page 113). But, in the following discussion, four recognizable types of protected ecosystems are identified each with its associated resource processes. Each type can be correlated with the three categories of forestland recognised by Olaleye and Ameh (1999 *op. cit.*)

First there are the *natural* or *semi-natural, self-maintaining forest reserves* being exploited for timber and firewood and enjoying little or no management other than legislative protection against unauthorized encroachment. This category of forest reserves is not much different from the category of "free areas" recognised by Olaleye and Ameh (1999), and described as forested areas that are not under strict management by the State Forestry Departments but where permission to exploit trees in them still has to be obtained from the forestry departments. Secondly, there are the government-owned forest reserves being used for silviculture (tree plantations) under intensive care and management. They correspond to the class of *forest reserves* recognised by Olaleye and Ameh (1999), which are managed by the State Forestry Departments (SFD_S) that have the required professional and technical staff, including uniformed guards, for performing their various responsibilities. These first two types of protected ecosystems (free areas and forest reserves) provide the bulk of the forestry products and services obtained from the management of nature reserves in the country including sawn wood, veneer, poles and fuelwood. Thirdly, there are the *game reserves*, which include "the National Parks, Game and Wildlife sanctuaries, proposed Game Reserves/Wildlife Sanctuaries and Strict Nature Reserves", all of which are grouped together and designated *conservation areas* by Olaleye and Ameh (1999). Fourthly, there are the savanna 'forest' reserves designated 'grazing reserves' for the use of long-distance pastoral Fulani cattle. Before describing in detail each of these protected ecosystems' resource processes, we will look briefly at the history of forest reserves in Nigeria and the legislative measures by which they were created.

Background to the creation of the protected ecosystems

The creation of forest and other nature reserves in Nigeria was achieved through a series of proclamations and bills by the British colonial administration. The creation of these protected ecosystems reached its zenith in the immediate post-Independence period in the 1960s, when the nature reserves attained their greatest spatial extent of about 10% of the total land area of Nigeria, that is, about 10million hectares (Table 4.1). Thereafter, the efforts of both federal and state governments to expand the area of nature reserves did not yield positive results. Indeed, quite a number of nature reserves were de-regulated under pressure to make way for grazing and arable farming. A comparison of Tables 4.1 and 4.2 would show that the total area of forest reserves declined approximately by 17% from 92,740km^2 in 1968 to 77,063.23km^2 by 1981. As seen in Table 4.9 (page 113) by the mid-1990s, total area of all conservation areas in the country was only 3,989,871ha (or 39898.71km^2), which translated to a 48.22 per cent decline from the 1981 hectarage. What the various governments were adept at doing was transforming many erstwhile forest reserves into game reserves, national parks, grazing reserves, wildlife sanctuaries, and the like. Some of the initiatives were at the instance of such international organizations as the IUCN, the World Bank, the FAO and the European Commission, which also provided much needed technical and financial support to the forest departments and forest research institutes. The objective in each case was to provide vital technical data for the planning and management of the various types of protected ecosystems.

TABLE 4.1 Forest Composition in Nigeria, 1968

Territorial Unit (States)	Area of Forest Reserves						
	Mangrove (km^2)	High Forest (km^2)	Savanna (km^2)	Total State Forest Reserves (km^2)	High forest as % of State Forest Reserve	High forest as % of total national forest reserve	State Forest Reserve as % total national forest reserve
Benue/Plateau/Nassarawa	–	139.9	4,006.7	4,146.6	3.4	0.7	4.4
Imo/Anambra/Enugu/Abia/Ebonyi	–	367.8	36.3	404.0	9.1	1.9	0.4
Kano/Jigawa	–	–	2,178.2	2,178.2	–	–	2.3
Kwara/Kogi/FCT	–	945.4	10,041.4	10,986.8	8.6	5.0	11.8
Lagos	5.2	15.5	–	20.7	7.5	0.08	0.02
Edo/Delta	20.7	5,029.8	676.0	5,726.5	87.8	26.7	6.2
Kaduna/Katsina	–	114.0	9,484.6	9,598.5	1.2	0.6	10.2
Borno/Yobe/Bauchi/Gombe	–	80.3	16,063.2	16,143.5	0.5	0.4	17.4
Adamawa/Taraba							
Sokoto/Zamfara/Kebbi/Niger	–	64.7	24,053.3	24,118.0	0.3	0.3	26.0
Akwa Ibom/Cross River/Bayelsa/Rivers	116.5	6,156.4	139.9	6,412.8	96.0	32.7	6.9
Ogun/Oyo/Ondo/Ekiti	–	5,946.6	7,057.7	13,004.4	45.7	31.6	14.0
TOTAL	**142.4**	**18,864.0**	**73,737.3**	**92,740.0**	**–**	**100.0**	**100.0**

Source: Compiled from *Annual Reports on the Forest Administration of Easter, Midwestern, Northern and Western Regions* (Lagos: Government Printers). Note: After Adeyoju S.K. (1975)

The history of Nigeria's protected ecosystems in Nigeria began in earnest in 1897 when a Forest Department was established by the colonial administration in the Colony and Protectorate of Lagos. The mandate of the department was to protect the existing forests and to formulate policies and plans for the re-afforestation of denuded areas. Right from inception, the Forest Department had to contend with the issue of land tenure, which greatly slowed down the process of creating forest reserves in the country. Early legislation, which began about 1902 and included the Forestry Ordinance, 1916 laid down the corner-stone of Nigerian forest policy as the creation of forest reserves. At first, areas selected for reservation were those where this did not clash with the immediate interests of any landowner or coincide with areas controlled by a chief or local council sufficiently powerful to effect an adjustment (Adeyoju, 1975, p. 20).

TABLE 4.2 Forest Reserve Area by State, 1981

States	Area of Forest Reserve (km^2)	% Total
Anambra/Enugu/Ebonyi	345.94	0.45
Bauchi/Gombe	6,710.20	8.70
Edo/Delta	6,513.93	8.45
Benue/Plateau/Nassarawa	1,003.93	1.30
Borno/Yobe	7,845.76	10.18
Cross River/Akwa Ibom	6,419.86	8.33
Adamawa/Taraba	2,400.10	3.11
Imo/Abia	104.89	0.14
Kwara/Kogi/parts of Niger	11857.85	15.39
Oyo/Osun	6,512.16	8.45
Ondo/Ekiti	3,550.36	4.61
Ogun	2,809.83	3.65
Rivers/Bayelsa	863.19	1.12
Sokoto/Zamfara/Kebbi/Niger	20,125.19	26.12
Total	77,063.23	100.00

Source: Federal Department of Forestry: Nigerian Forestry Statistics, 1983 Forest Projects Monitoring and Evaluation Unit (World Bank Project)

In the 1920s a new reservation policy was put in place to secure 25 per cent of the available forest land in each province, for reserves. At the same time the restrictions of the forestry law were removed from the remaining lands in each province. This policy was modified in 1934 by making the area of land put under reservation proportional to the size of the available forest land and compatible with political conditions in each province. The ownership and administration of the reserves was transferred to the local government authorities, which collected taxes on trees felled in the forests and shared the revenue with the regional governments. In 1941 the policy of encouraging the provinces to set up 'communal forest areas' was adopted. Many provinces willingly accepted this and, unfortunately, the establishment of communal forests took precedence over forest reservation in many areas.

The success with which the forest policies were implemented varied from region to region depending on the needs of the local people for arable farmland and their willingness to cede land for reservation. For example, forest reservation was most difficult to accomplish in eastern Nigeria partly because of the high pressure of population on the land and, partly, because the people considered the idea an abomination claiming that their ancestors never placed any restriction on family land. By contrast, forest reserves were quite easily created in the northern region because there were many sparsely populated pockets of land, which had previously been buffer zones between city states. Furthermore, the colonial administration often won over the support of the emirs who had sole authority over affairs pertaining to their emirates.

The basis of present-day forest reserve management in Nigeria was laid down by the 1946 Forest Administration Plan, which was subsequently modified on a number of occasions. It laid down the principle that forest reserves were to be placed under planned management and regarded as inviolable. This legislation thus gave legal backing to the creation of the forest reserves.

The area of forest reserves increased rapidly over the years until after the Second World War when the pace slowed down. After independence the states took over the control and management of the forest reserves from the federal government. But, the Federal Department of Forestry, primarily through the Forestry Research Institute of Nigeria, has remained very active in the field of research in collaboration with international organizations and university Departments of Forest Resources and Wildlife Management. It continues to promote forestry throughout the country especially by encouraging tree planting programmes in the states and in individual towns; and by training personnel to manage the forest reserves and the numerous game reserves, nature reserves and wildlife sanctuaries established in different parts of the country.

Unfortunately, the number and area of forest reserves have declined since the late 1960s especially in the savanna lands. Many forest reserves were de-regulated and converted to 'grazing reserves' to meet the needs of the nomadic Fulani. There are at least 109 of such grazing reserves now in the savanna areas. De-reservation of land has been done also to make way for man-made lakes, irrigation and other modern agricultural development projects. A survey carried out by the Federal Department of Forestry (1984) showed that de-reservation was spreading in the northern states. The decline in the area of reserved lands was up to 3-5 per cent in Kano State; 8-14per cent in Kaduna (incl. Katsina) State; 11-15 per cent in Bauchi State; and 16-21 per cent in Sokoto State. As indicated earlier there was a 17% reduction in the total area of forest reserves between 1968 and 1981. Forest reserves had become virtually non-existent in some states by 1981.

Natural and semi-natural forest reserves

These come closest to satisfying the concept of wilderness in the sense that they are not deliberately managed. But unlike wilderness areas elsewhere, the forest reserves in Nigeria are heavily exploited for wood. The area of natural forest reserves in the humid south is dwindling year by year. The reserves are under constant pressure from lumbering and from the establishment of commercial tree, tree crop (notably, oil palm, rubber) and fruit crop (e.g. plantain) plantations, all of which are officially approved and encouraged.

The exploitation of forest reserves for timber had gathered momentum since the 1970s. Outside the reserves, as we saw in Chapter 2, the exploitation of the forests is not adequately monitored and controlled. Thus, many forests have been heavily exploited and the standing volume of valuable species of suitable size has dwindled to an appallingly low level. This has led to great pressure on the government-controlled forest reserves for the supply of timber. These timber-yielding forest reserves are a mere 2 per cent of the country's land area.

Timber-yielding forest reserves

Redhead (1971) grouped these timber-yielding forest reserves into four types. (1) the dry forest reserves of Akure, Ala, Ara, Eggua, Gambari, Ifon, Ijaiye, Ikerre, Ogotun, Ilaro, Ise, Little Osse, Ogbesse and Owo, which receive less than 1524mm rainfall annually; (2) the forests in the rainfall zone 1524-2032mm per annum including AkureOfosu, Idanre, Iiebu-Ife, Onishere, Oluwa, Otu and Omo; (3) the Benin forest reserves including Ebue, Obaretin, Sapoba, Urhonigbe, Okomu and Ologbo receiving 1778-2286mm annual rainfall; and (4) the high rain forests of

Cross River State including Afi river, Cross river and the Oban group.

Since the 1970s, many state governments had granted large concessions to timber companies to exploit the reserves, some of which were denuded of their merchantable species in the process, a typical example being the Aramoko Ekiti forest reserve.

The earliest concessions were granted to the United Africa Company in western Nigeria to exploit the forests south of Omifunfun in Ife division and near Okeluse in Owo division. The dense Ondo and Benin forests were also being exploited under such government concessions. The African Timber and Plywood Company factory at Sapele built during the colonial era and the wood factory at Epe built in the 1970s were established to make use of the timber resources of the adjacent forestlands. Later, the Nigerian-Romanian Wood Industry (NIROWI) was also established at Ondo to exploit the timber resources of the nearby forests.

Impact of lumbering on the forest reserves

The ecological impact of such lumbering activities is not as devastating as in the unprotected forests outside the reserves; logging is done usually by experienced crews using modern equipment for cutting and hauling the trees. Also, they are required to follow rules about the type and size of trees to fell. One aspect of lumbering, which was not restricted to the reserves, although one wondered why it was permitted under government supervision, was the practice of selective logging. Wood industries and the export market favoured certain types of trees over others which were consequently regarded either as useless or as 'secondary species' which might become useful in certain situations. Consequently, in time there were large tracts of forest reserves which had been denuded of the so-called merchantable timber species.

As natural regeneration of these merchantable species is slow and uncertain, forestry departments all too often cleared and burnt the entire remaining forest, after the merchantable species had been exhausted, in preparation for the establishment of tree plantations. This entailed a great deal of wastage usually of B- and C-storey tree species. When, in the 1970s, the country began to suffer from timber shortage and was unable to meet domestic demands, the industry belatedly turned to these erstwhile disregarded secondary species. Forestry divisions tried in the southern Guinea zone to remedy the situation by establishing commercial tree plantations of both exotic and indigenous species.

Lumbering in the savanna 'forest' reserves

The forest reserves in the savanna zones serve two main purposes. First, they help in ameliorating local microclimatic conditions and protecting the land against natural hazards such as soil erosion and desiccation; and, secondly, they provide a ready source of firewood which is a relatively scarce commodity in great demand especially in the northern savanna regions. In addition, the savanna forest reserves supply some quantities of poles for building purposes, from the less twisted, fire-scarred trees. *Borassus aethiopum* (fan palm) provides a useful quantity of durable split wood for construction (roofing) purposes.

However, as we have already seen, in the Guinea savanna zone some of the reserves are forest outliers or gallery forests with some valuable timber. Thus, there is a thriving sawmilling industry in the Nigerian middle belt. Sawmills are located in Suleja, Jos, Zaria and Makurdi among other places. The other reserves in the savanna regions are woodlands and, strictly speaking, most of them carry a crop fit only for firewood. These are cut into pieces of suitable length and piled up beside the roads for collection and sale in the urban centres.

Many of these savanna woodland reserves, especially those which lie along the numerous cattle trails, suffer from annual bush fires. However, apart from this burning, the use of the reserves for firewood does not appear to greatly disrupt their ecological balance in the short run. New shoots spring up from the stumps and roots; the soil is left undisturbed and the plant roots, leaves and twigs of the felled trees are left *in situ* to augment the organic content of the soil. Because the vegetation is more open and the trees smaller, the felling of a tree does not cause as much destruction of the surrounding vegetation as one finds in the high forest zone.

However, in the long term the accumulated effects of firewood harvesting; grazing by livestock and annual bush burning has significantly reduced the vegetation cover in the savanna zones. Hence, the savanna zones are suffering more from the effects of global warming and climate change. It cannot be denied either that reserves lose a great deal of energy-rich material as a result of the removal of the trees for fuel. Although there is no available data on the quantity of wood removed, the huge piles of firewood on many roadsides in the savanna areas suggest that it is considerable. The woody component of the vegetation is very important in nutrient cycling in savanna grasslands because the trees store nutrients much more efficiently than the grasses. The greater the proportion of woody plants the greater the nutrient capital of the vegetation. This is advantageous because in the soil the nutrients are more susceptible to leaching and erosion losses.

However, as stated above, over-exploitation of the woody plants can lead to vegetation degradation and loss of a valuable source of domestic fuel in the savanna zones. Reforestation may also be more difficult and expensive to achieve in the degraded vegetation communities which have lost much of their nutrient reserves. In establishing commercial tree plantations, for example, forestry departments use the nature of the existing vegetation communities as one of the criteria for locating suitable plots of land. More seriously, outside the forest reserves large-scale firewood collection goes on without any control. Fuel wood is taken from the fallow communities around the farms and villages. This activity is certainly detrimental to the long-term fertility of the soil and reduces the restorative value of the fallows. The present open grassland vegetation of the Jos Plateau arose from just such heavy exploitation for firewood in the original savanna woodland vegetation in the past (Udo, 1970).

Forest reserves used for silviculture

The 1960s and early 1970s witnessed the widespread establishment of tree plantations in many forest reserves. It says much for the impact which these commercial tree plantations made on the Nigerian landscape and in the minds of the people that up till today, the term forest reserve is commonly applied almost exclusively to the exotic tree plantations under government management. Perhaps, this is because the plantations are everywhere evident along the roads. Travellers apparently do not realize that many of the plantations do not extend more than a few hundred metres from the verge of the road while beyond them the rest is still natural or semi-natural vegetation. Of course, there are a few more extensive reserves such as at Onigambari forest reserve on the Idi Ayunre-Ijebu-Ode road.

Table 4.3 provides data on the area of plantations in different states of southern Nigeria up to 1976 at the height of their ascendancy in the forest landscape. The species composition of the plantations varies from place to place according to the different local climatic and soil conditions. Table 4.4 lists the species planted in south-western Nigeria up to 1976. The area covered by tree plantations in each region has fluctuated over time but the table does give an idea of the variety of exotic and indigenous species planted. In the northern states the common plantation species include *Pinus caribaea* (Caribbean pine), *Azadirachta indica (neem), Eucalyptus tereticornis; E. camaldulensis, E. saligna, E. citriodora* and *E. robusta* (eucalyptus). *Gmelina arborea* (gumbar, gmelina) and *Tectona grandis* (teak) which flourish in the south have not been successfully cultivated because of the drier environmental conditions.

TABLE 4.3 Total area of plantations in the southern states (in ha[*])

State	Up to 1960[a]	1961-65	1966-70	1971-5	1976[b]	TOTAL
Ogun	997	497	3,718	7,957	1,664	14,833
Ondo/Ekiti	–	1,316	3,262	10,474	2,276	17,328
Oyo/Osun	123	485	2,530	2,918	1,661	7,717
Kwara/Kogi	320	285	426	1,229	924	3,184
Edo/Delta	7,177	4,732	11,438	18,364	4,205	45,916
Anambra and Imo	1,192	2,868	1,410	2,726	2,278	10,474
Cross River/Akwa Ibom	1,300[c]	3,400	-	8,880	2,310	15,890

Notes:

[*] Figures likely to be too high by a factor of perhaps 20%
[a] Figures before 1960 are less reliable than from 1960 onwards. They have therefore been grouped together.
[b] Figures for 1977 are not from surveys of stocked area. They have therefore been omitted.
[c] It has been assumed that 1300ha of the hard woods in Cross River State were planted before 1960 and that nothing was planted between 1966 and 1970

Source: Ball, J.B. and Daniyan, C.O. (August 1977), *Summary of Plantation Areas in the Southern States of Nigeria.* (Ibadan: Federal Department of Forestry)

The drive to establish tree plantations has been due, primarily, to three factors. The first is the paucity of the merchantable indigenous species in the existing forests and the second is that the merchantable species of suitable size do not grow naturally in pure stands like those in the temperate coniferous forests. A great deal of time must therefore be allowed for searching them out, and this greatly increases labour and production costs. Thirdly, the felling cycle for natural forest trees is too long being of the order of 100-150 years. This has brought about a shift to fast growing species with a shorter felling cycle of about 40-50years. Indigenous timber species are at a disadvantage in this regard. Thus, while efforts are still being made to grow indigenous species on the plantations, more emphasis is placed on hardy and/or fast-growing exotic species, chiefly, teak and gmelina. The deciduous nature of some of the exotic species makes them quite suited to the existing environmental

TABLE 4.4 Total area of plantations in the southern states showing species and planting year (ha)

Species	Up to 1960	1961–5	1966–70	1971–75	1976	1981[c]	TOTAL
Gme. arb.	1,404	4,377	5,345	20,930	7,361	4,006	43,423
Tec. gra.	358	1,634	5,411	12,705	4,164	1,278	25,550
Tec. gme.	190	1,349	2,095	1,838	–		5,472
Gme mixtures	204	657	80	85	–		1,026
Tec. mixtures[a]	1,531	2,062	1,646	5,323	607		11,169
Nau did.	585	337	1,451	5,952	839		9,164
Nau. mixtures[b]	5,813	1,883	4,867	2,311	643	3,285	15,517
Term. spp.	86	635	1,384	1,761	865		4,731
Pin. car.	–	–	–	49	198		247
Other Hardwoods	839	341	225	1,323	566		3,294
Other Mixtures	99	308	280	271	75		1,033
TOTAL	11,109	13,583	22,784	52,548	15,318		120,305

Notes:

[a] This includes 1300ha of hardwoods in Cross River State. They are mainly pure teak, but since it is impossible to estimate the proportion they have been classified as teak mixtures

[b] Nauclea is mainly with Meliaceae

Gme. arb.: Gmelina arborea (gumbar, gmelina)

Tec. gra.: Tectona grandis (teak)

Nau. did.: Nauclea diderichii (West Africa boxwood)

Term. spp.: Terminalia spp. (afara)

Pin. car.: Pinus caribaea (Caribbean pine)

[c] Source: *Nigerian Forestry Statistics* (1984): Federal Dept. of Forestry Projects Monitoring & Evaluation Unit (World Bank Project)

Source: Ball and Daniyan (1977)

conditions in the dry forest and savanna areas. Furthermore, the boles of *Tectona grandis* (teak), for example, are resistant to insect (particularly termite) attack. The standing crop can also withstand light surface fires which are used in clearing the litter and undergrowth.

Plantation establishment techniques

In the high forest reserves in the south the *taungya* system was used in preparing the ground for planting and to provide some protection for seedlings in the early years. Under this system local farmers were given licences which permitted them to cultivate and grow food crops on land earmarked for commercial tree planting. The farmer was sometimes allowed to go on cultivating the land even during the first few years of the planting of the tree seedlings. This arrangement helped the forestry department in clearing the land initially and in keeping weeds down during the early life of the plantation. As Aladejana (1971), pointed out, the tree crops benefitted for two years if the farmer took care of them so that the forester needed not to start tending the plantation until the third year. Over time the need for this kind of arrangement grew with the spread of the Siam weed *Chromolaena (eupatorium) odorata.* The people learnt to plant cassava to combat this weed, and experience showed that this did not adversely affect the trees if the cassava seedlings were not planted too close.

A possible risk of *taungya* to the tree seedlings was that arable cropping could impoverish the soil. However, experience soon showed that the two-year period of cropping had little effect, particularly, since land freshly cleared of its original vegetation usually had high natural fertility and structural stability. Aladejana (1971) also showed that tree growth in an average *taungya* plantation compared favourably with that of direct plantation (without *taungya*) and was, sometimes, better.

By the late 1960s land clearing and preparation in the savanna areas was increasingly being done by mechanical means, although manual labour was still prevalent (Iyamabo and Ojo, 1971). Mechanical clearing resulted in a more complete removal of the vegetation, although the felled trees were not burnt off as they were in the high forest, because they were required for firewood. Ploughing and cultivation helped in breaking up the soil, which in most cases contained lateritic crusts or hard clay pans. They also helped to suppress weeds and reduce competition from grasses and coppice shoots of perennial root stocks. The soil was ploughed to a depth of 15-22 cm thus improving its infiltration capacity and reducing evaporation losses.

The tree seedlings were raised in nursery beds usually with the aid of fertilizers and irrigation which was one of the reasons why most nurseries were located in valley bottoms close to streams. Once transplanted, fertilizers were not much used and there was no irrigation of the growing trees at all. For this reason, water retentive soils such as clay-loams and sandy-clays were highly favoured. Indeed, the soil factor is very crucial to the successful establishment of a commercial tree plantation. In the savanna, most of the lands in the reserves were adjudged to be unsuitable for commercial tree planting largely because of unfavourable soil conditions (Iyamabo and Ojo, 1971).

Plantation management

Management techniques on the plantations included thinning, pruning of superfluous branches, occasional use of insecticides and burning to clear the litter and undergrowth. The Plantations in the savanna areas are often fenced to protect them against grazing and browsing animals. Goats which may browse on the young trees are particularly troublesome. Protection against fire is also provided through a system of fire breaks and patrols during the dry season. The removal of litter and clean weeding of the plantations also reduce fire risk.

The deciduous nature of the trees and the practice of burning the litter and undergrowth also tend to expose the soil to torrential rains in the early rainy season. A well-established teak plantation, for instance, may have little or no undergrowth and the litter layer may be so thin that there is little or no protection for the ground surface in the early rainy season. Thus, sheet erosion and, sometimes, gullying had been known to occur in some forest plantations especially those located on hill slopes (Faniran and Areola, 1974). Plantations where soils have been weakened by previous cultivation under the *taungya* system could be particularly susceptible to sheet erosion.

Furthermore, the ecological impact of the tree crops, for instance, on the underlying soil, has not been fully studied. A study by Adejuwon and Ekanade (1988) on soil changes consequent upon the replacement of the tropical rain forest by plantations of teak, gmelina and Terminalia superba, did not find any significant differences between the soils under the tree plantations and those under the natural forest (Table 4.5). It would be interesting to analyze nutrient cycling in these plantations, which are largely monocultures, and compare results with cycling in the natural forest. It is known that tree plantations sometimes encourage the spread of insect pests as monocultures are apt to do. It has been claimed by the Federal Forestry Division that in the savanna regions tree plantations within or near human settlements ameliorate local microclimatic conditions. While there are a few records on the growth rates of tree seedlings on the plantations, there is very little data on their productivity. But, the Forest Resource Situation Assessment of Nige-

TABLE 4.5 Means and standard deviations of the topsoil elements under tropical rain forest and plantations, Ikere-Ekiti, Nigeria

Soil Elements	Rain Forest		G. arborea		T. grandis		T. superba	
	$\bar{x}$	σ	$\bar{x}$	σ	$\bar{x}$	σ	$\bar{x}$	σ
Sand %	77.2	2.43	74.3	3.28	76.8	2.81	77.0	3.28
Silt %	8.1	1.01	9.5	1.40	9.2	1.63	6.7	0.73
Clay %	14.7	1.82	16.2	2.45	14.0	2.05	16.3	3.02
Moisture content %	11.0	1.14	11.9	1.34	11.0	1.45	10.6	0.91
pH	5.8	0.72	6.0	0.27	6.1	0.62	6.3	0.29
Organic Matter %	5.5	1.34	4.6	0.91	4.7	1.06	4.7	0.91
Available Phosphorus (ppm)	34.8	18.7	23.6	8.32	31.7	6.31	25.2	5.09
Nitrate-Nitrogen (ppm)	3.1	0.70	2.7	0.63	2.7	0.46	3.6	0.81
Calcium (ppm)	1052.0	136.61	1358.0	143.51	1112.0	107.48	1148.0	164.24
Sodium (ppm)	66.4	14.32	49.1	4.83	58.0	8.48	61.6	6.72
Potassium (ppm)	248.3	38.13	145.3	24.30	174.60	29.91	180.3	17.08

$\bar{x}$ = mean; σ = standard deviation
Source: Adejuwon & Ekanade, 1988.

ria project carried out by Olaleye and Ameh (1999) provided preliminary data on the productivity of forest and tree plantations in the southern parts of the country (Tables4.6–4.8).

The state and federal forestry departments urge local inhabitants to plant exotic trees. The response has been encouraging particularly in the northern states where the trees are grown to provide shade against intense insolation. In the south a few farmers have planted these species along the borders of their tree crop plantations. Many of them have realized that the exotic trees are valuable cash crops. While they are not likely to make much impact in the primary and secondary wood-based industries, perhaps with the exception of teak (*Gmelina arborea*), the exotic plantation trees provide the needed raw materials for pulp and paper manufacturing in the country.

TABLE 4.6 High forest gross timber volumes (excluding bark) by forest designation and forest type

Forest Land Designation	Forest Type	Area (ha)	Gross Volume (m^3)
Forest Reserve	Freshwater swamp	224,369	23,353,102
	Lowland rain	757,740	98,599,957
	Riparian	2,547	169,101
Free Area	Freshwater swamp	1,424,739	150,814,914
	Lowland rain	912,094	109,544,813
	Mangrove	5,314	443,860
	Riparian	80,368	4,526,678
Game Reserve	Lowland rain	12,365	1,633,706
	Riparian	5,492	386,513
National Park	Lowland rain	369,412	46,878,597
Total		**3,974,440**	**436,351,241**

Source: Beak Consultants 1998. *Forest Resources Study*, Vol. II page 9-3.
Source: Olufemi A. Olaleye and Ameh, C. E: *The EC-FAO Partnership Programme (1998-2002)*. Project GCP/INT/679/EC Data Collection and Analysis for Sustainable Forest Management in ACP Countries - Linking National and International Efforts. `http://www.fao.org/3/ab578e/ab578e00.htm`

Game reserves and related conservation areas

African savannas are generally very rich in wild life, but compared to the grasslands of Eastern and Southern Africa, the stock of wildlife in Nigerian savannas is very low indeed as a result of many centuries of settlement and domination of the landscape by ancient empires and kingdoms. However, compared to other vegetation zones in the country the savanna lands are better stocked. It is in these savanna areas that efforts were first made to create game reserves, which, in the long run, could be used for the rearing and cropping of wild grazing animals.

A number of factors were touted as making this type of resource process a viable proposition. In the first place it was argued that wild grazers make better use of the roughage because they are not as selective as the domesticated grazers in the choice of plants they feed on. Thus, the ecological shift in the floristic composition of the grasslands, which often accompanies grazing by domesticated animals,

TABLE 4.7 Stand volumes table for Gmelina arborea

Age (years)	Volume (m^3)	Age (years)	Volume (m^3)
2	115.7	40	647.8
4	156.4	42	648.5
6	205.8	44	649.0
8	262.5	46	649.3
10	323.4	48	649.5
12	384.5	50	649.7
14	441.5	52	649.8
16	491.5	54	649.9
18	532.4	56	649.9
20	564.7	58	650.0
22	589.2	60	650.0
24	607.1	62	650.0
26	620.1	64	650.0
28	629.2	66	650.0
30	635.6	68	650.0
32	640.1	70	650.0
34	643.2	72	650.0
36	645.3	74	650.0
38	646.8	76	650.0

Source: Beak Consultants 1998. *Forest Resources Study*. Appendices (Draft).
Source: Olufemi A. Olaleye and Ameh, C. E: *The EC-FAO Partnership Programme (1998-2002)*. Project GCP/INT/679/EC Data Collection and Analysis for Sustainable Forest Management in ACP Countries - Linking National and International Efforts. `http://www.fao.org/3/ab578e/ab578e00.htm`

TABLE 4.8 Yield (m^3/ha) projection table for Tectona grandis, site index 28, site class1

Age (year)	Basal Area (m^2/ha)																				
	4	6	8	10	12	14	16	18	20	22	24	26	28	30	32	34	36	38	40	42	44
4	91.6	108.3	125.0	141.7	158.4	175.1	191.9	208.6	225.3	242.0	258.7	275.4	292.1	308.8	325.6	342.3	359.0	375.7	392.4	409.1	425.8
6	97.8	114.5	131.3	148.0	164.7	181.4	198.1	214.8	231.5	248.2	265.0	281.7	298.4	315.1	331.8	348.5	365.2	382.0	398.7	415.4	432.1
8	104.0	120.7	137.4	154.1	170.8	187.6	204.3	221.0	237.7	254.4	271.1	287.8	304.5	321.3	338.0	354.7	371.4	388.1	404.3	421.5	438.3
10	110.1	126.8	143.5	160.2	176.9	193.6	210.3	227.1	243.8	260.5	277.2	293.9	310.6	327.3	344.0	360.8	377.5	394.2	410.9	427.6	444.3
12	116.1	132.8	149.5	166.2	182.9	199.6	216.3	233.0	249.8	266.5	283.2	299.9	316.6	333.3	350.0	366.7	383.5	400.2	416.9	433.6	450.3
14	122.0	138.7	155.4	172.1	188.8	205.5	222.2	238.9	255.7	272.4	289.1	305.8	322.5	339.2	355.9	372.6	389.4	406.1	422.8	439.5	456.2
16	127.8	144.5	161.2	177.9	194.6	211.3	228.0	244.8	261.5	278.2	294.9	311.6	328.3	345.0	361.8	378.5	395.2	411.9	428.6	445.3	462.0
18	133.5	150.2	166.9	183.6	200.4	217.1	233.8	250.5	267.2	283.9	300.6	317.3	334.1	350.8	367.5	384.2	400.9	417.6	434.3	451.1	467.0
20	139.2	155.9	172.6	189.3	206.0	222.7	239.4	256.1	272.9	289.6	306.3	323.0	339.7	356.4	373.1	389.9	406.6	423.3	440.0	456.7	473.4
22	144.7	161.4	178.2	194.9	211.6	228.3	245.0	261.7	278.4	295.1	311.9	328.6	345.3	362.0	378.7	395.4	412.1	428.8	445.6	462.3	479.0
24	150.2	166.9	183.6	200.4	217.1	233.8	250.5	267.2	283.9	300.6	317.3	334.1	350.8	367.5	384.2	400.9	417.6	434.3	451.1	467.8	484.5
26	155.6	172.3	189.1	205.8	222.5	239.2	255.9	272.6	289.3	306.0	322.8	339.5	356.2	372.9	389.6	406.3	423.0	439.8	456.5	473.2	489.9
28	161.0	177.7	194.4	211.1	227.8	244.5	261.2	278.0	294.7	311.4	328.1	344.8	361.5	378.2	394.9	411.7	428.4	445.1	461.8	478.5	495.2
30	166.2	182.9	199.6	216.4	233.1	249.8	266.5	283.2	299.9	316.6	333.4	350.1	366.8	383.5	400.2	416.9	433.6	450.3	467.1	483.8	500.5
32	171.4	188.1	204.8	221.5	238.3	255.0	271.7	288.4	305.1	321.8	338.5	355.3	372.0	388.7	405.4	422.1	438.8	455.5	472.2	489.0	505.7
34	176.5	193.2	209.9	226.7	243.4	260.1	276.8	293.5	310.2	326.9	343.6	360.4	377.1	393.8	410.5	427.2	443.9	460.6	477.4	494.1	510.8
36	181.6	198.3	215.0	231.7	248.4	265.1	281.8	298.5	315.3	332.0	348.7	365.4	382.1	398.8	415.5	432.3	449.0	465.7	482.4	499.1	515.8
38	186.5	203.2	219.9	236.7	253.4	270.1	286.8	303.5	320.2	336.9	353.6	370.4	387.1	403.8	420.5	437.2	453.9	470.6	487.4	504.1	520.8
40	191.4	208.1	224.8	241.6	258.3	275.0	291.7	308.4	325.1	341.8	358.5	375.3	392.0	408.7	425.4	442.1	458.8	475.5	492.2	509.0	525.7

Source: Beak Consultants 1998, Forest Resources Study of Nigeria, Vol. IV (Appendices).

Source: Extracted from Olufemi A. Olaleye and Ameh, C. E: *The EC-FAO Partnership Programme (1998-2002) - Project GCP/INT/679/EC Data Collection and Analysis for Sustainable Forest Management in ACP Countries - Linking National and International Efforts.*

would be reduced. Secondly, because they consume more roughage, wild grazers are better energy converters producing more protein and less fat than domesticated grazing animals. However, it is relevant to point out that some studies later suggested that well managed domesticated livestock could be as efficient in the use of roughage as wild grazers (de Leeuw, 1975; Brinkman and de Leeuw, 1975).

At the initial stage, effort was concentrated on developing the Yankari Game Reserve in Bauchi State and the Borgu Game Reserve in Kwara State, which was later combined with a number of smaller reserves to constitute the Kainji National Park. The Yankari and Borgu game reserves were, for long, the dominant centres of safari-type tourism in the country. Since then both the federal and state governments have worked with international organizations, research institutes and environment advocacy groups to develop other budding game reserves in different parts of the country. These were formerly forest reserves which were later converted to game reserves, wildlife sanctuaries, nature or biosphere reserves (Table 4.9). In the present stage of their development these game/nature reserves are not much more than visual amenities for tourists.

TABLE 4.9 Updated List of Conservation Areas in Nigeria

	Conservation Area	Size km^2
	National Parks	
1	Kainji Lake	534,082
2	Old Oyo	251,200
3	Okomu	11,200
4	Kamuku	112,700
5	Cross River	280,000
6	Gashaka-Gumti	640,248
7	Yankari	224,410
8	Chad Basin	225,800
	Sub-Total	2,279,640
	Games Reserves/Wildlife Sanctuaries	
1	Alawa	29,620
2	Dagida	29,422
3	Falgore	92,300
4	Gilli-Gilli	36,200
5	Kambari	41,400
6	Kwale	340
7	Kwiambana	261,400
8	Lake Chad*	38,000
9	Lame-Burra	205,967

Continued on next page

Table 4.9 – continued from previous page

	Conservation Area	Size km^2
10	Ologbo	19,440
11	Opara	110,000
12	Orile	5,440
13	Pai River	70,000
14	Pandam	22,400
15	Sambisa	51,800
16	Udi-Nsukka	5,600
17	Wase Rock	92
	Subtotal	1,019,421
	Current proposed Games Reserves/Wildlife Sanctuaries	
1	Ifon	28,270
2	Akpaka	19,400
3	Ebba-Kampe	121,730
4	Moko	96,610
6	Taylor Creek	30,000
7	Ohosu	47,100
8	Okeluse	11,440
9	Opanda	10,520
10	Stubbs Creek	21,000
11	Afi River	10,000
12	Kashimbila	139,600
13	River Benue	154,000
	Subtotal	689,670
	Strict Nature Reserves	
1	Akure	32
2	Bam Ngel zarma	142
3	Bonu	145
4	Lekki	78
5	Milliken Hill	49
6	Omo (Biosphere Reserve)	460
7	Ribako	170
8	Urhonigbe	64
	Subtotal	1,140
	Grand Total	**3,989,871**

Table 4.9 – concluded

Source: Adapted from World Bank, June 1992. Federal Republic of Nigeria: Forestry Sector Review. Confidential Report No. 10744-UNI and Nigeria National Parks, Annual Calendar (1998) 13pp.; and (2) Wikipedia: Natural Areas in Nigeria. `https://en.wikipedia.org/wiki/Natural_areas_in_Nigeria`

While it is yet too early to assess the ecology of the protected ecosystems and associated natural resource processes, a few studies have been carried out which deal with aspects of habitat management for wildlife conservation and their likely impact both on the environment and on the utilization of the reserves for game and recreation. As Table 4.9 shows, there are now 8 national parks, 17 game reserves/wildlife sanctuaries with 13 more under consideration, and 8 strict nature reserves under development. Data on animal populations, biodiversity and ecological relations within each reserve are still very scanty. There is so much disparity between those who have given data on animal populations in the game reserves. For example, figures given for the elephant population vary from 100-150 to 550. Table 4.10 is based on an aerial census of elephants and other large mammals in the Yankari Game Reserve between April 29 and May 2, 2011 (Bergl *et al.,* 2011), and it is perhaps, the most accurate animal population census available at the moment. Bergl *et al.* (2011 *op. cit.*) noted that all of the known large mammal species in Yankari were observed during the survey with the exception of hippopotamus and lion. They also excluded some species from total count calculations (e.g., warthogs, baboons), since total counts were unlikely to have been accurate and they are not species of conservation concern. As has always been the case the greatest concentration of sightings was along the Gaji valley.

TABLE 4.10 Results of aerial survey (total counts) of elephants and other large mammals in Yankari Game Reserve, 2011

BLOCK	Buffalo	Bushbuck	Duiker	Elephant	Elephant carcass	Hartebeest	Roan antelope	Waterbuck
Block 1	174	6	10	326	4	41	86	29
Block 2							16	
Block 3		1	1			1	6	
Block 4			3			5		
Block 5			3	35		22	63	
Total	174	7	17	361	4	69	171	29

Source: R.A. Bergl, A. Dunn, S. Haruna, J. Mshelbwala, and G. Nyanganji 2011 *Aerial Survey of Elephants and Other Large Mammals at Yankari Game Reserve, Bauchi State, Nigeria.* Wildlife Conservation Society, North Carolina Zoo, MIKE

Table 4.11 shows the results of road counts and aerial surveys at Borgu Game Reserve in the early 1970s. The animal species populations were much larger than

TABLE 4.11 Borgu Game Reserve estimate of animal population

Species	1971–2	1973–4
Elephant	400	650
Warthog	2,000	2,100
Buffalo	600	400
Roan Antelope	1,850	1,600
Hartebeest	3,650	2,350
Waterbuck	350	350
Kob	2,500	1,800
Reedbuck	200	150
Bushbuck	1,000	500
Oribi	2,500	1,400
Grey duiker	2,000	900
Red flanked duiker	500	500
Baboon	11,750	10,900
Green monkey	2,600	2,300
Red patas	1,550	1,100

Source: Afolayan (1975).

at Yankari Game Reserve even though species populations were low as compared to the safari of East and Southern Africa. There are no comparable current data on species populations in Borgu Game Reserve. However, Osunsina *et al.* (2018) recorded a total of 70 bird species from 31 families in a survey carried out in December 2014 and June 2015 being the early dry and early rainy seasons respectively. More recent animal counts around mineral licks give some idea of the variety of wild mammalian species in the Borgu Game Reserve (Lameed and Adetola, 2012; Table 4.12). Finally, some of the national conservation areas also contain lakes, fadama and other wetlands, and rock basins filled with pools where birds, reptiles and amphibians breed in large numbers.

Government initiatives in wildlife conservation

The conversion of some forest reserves to game reserves and wildlife sanctuaries, and the creation of so many national parks bear eloquent witness to the efforts

TABLE 4.12 Diversity of Wild animals sampled within the various transects leading to a salt lick, Borgu Game Reserve

Animal	A	B	C	D
Kobus kob	65	51	20	20
Hippopotamus	39	21	13	33
Roan antelope	1	10	4	-
Red flank duiker	5	13	1	-
Hare	1	2	-	-
Baboons	10	13	9	4
Silvet cat	2	3	5	2
Bush buck	9	2	-	-
Monkey	4	3	2	1
Lion	1	5	-	1
Flankolin	-	-	-	1
Crocodile	-	-	-	1
Total	137	123	54	63
Average	13.70	12.30	7.71	7.86

Source: Lameed A.G. and J. Adetola 2012 Species-Diversity Utilization of Salt Lick Sites at Borgu Sector of Kainji Lake National Park, Nigeria. In Gbolagade Akeem Lameed (2012) *Biodiversity Enrichment in a Diverse World, Chapter 2* IntechOpen, DOI: 10.5772/51089

of federal and state departments of forestry and wildlife management in the last four to five decades. The willingness of the government officials and institutions to team up with the universities and international organizations paid off. Several wide area studies have been carried out in many parts of the country which have provided valuable technical data and information for planning purposes. Wildlife conservationists have been trained both overseas and in local universities.

Many studies carried out or funded by international organizations have had Nigerian researchers participating fully not only in the data gathering and analysis but also in the policy formulation and development planning on forestry and wildlife management. But, the specific obligations of the federal government in many of the studies included provision of relevant existing documents to facilitate not only the study but also, the establishment of a study implementation unit for each study, secondment of appropriate counterpart staff and payment of study staff emoluments.

Game reserves and wildlife sanctuaries in the southern parts of the country are also receiving greater attention and efforts are being made to assess the animal populations in them. These assessments became necessary because of the fear that the survival of many species was under threat. Tables 4.12 and 4.13 show the results of a survey of large mammal species carried out in a part of Okomu National Park in 2008/2009 by Akinsorotan *et al.* (2011).

The survey clearly highlighted the low wildlife species populations common to most conservation areas in the country, particularly in the southern forest regions.

The upsurge of official interest in wildlife conservation since the 1960s and 70s was due, not only to the realization of the need to conserve nature but, also, to the immense social and economic benefits to be derived therefrom. Charter (1970) estimated that locally produced animal protein in southern Nigeria, included 19 per cent from wild game (bush meat) compared with 21 per cent from domestic animals. Fish accounted for the remaining 60 per cent. He put the value of bush meat consumed annually in southern Nigeria at about ₦20 million (₦1 = 2USD then). The FAO (1970) put the annual total value of bush meat in Nigeria then at ₦30 million and the total value of naturally produced protein food at ₦100 million. Thus, animal protein from wild animals already formed a significant element of the diet especially of the majority rural population in Nigeria. Bush meat contributed immensely to the eradication of protein deficiency in many rural parts of Nigeria. Demand for game or bush meat also grew phenomenally in the urban centres. Urban dwellers paid higher prices for the meat and this encouraged farmers and professional hunters alike to hunt or poach wild game for sale in the cities rather than to feed their own families

TABLE 4.13A Sightings of large mammals of Arakhuan Range in Okomu National Park, Southwest Nigeria (Field survey 2008/2009)

Animal species	Morning sightings	Evening sightings	Total no. of sightings	Sightings /km	Length of transect covered for 1 sighting /km
Mona monkey	23	14	37	0.22	4.5
White-throated monkey	12	7	18	0.11	9.3
Red-capped mangabey	8	8	16	0.10	10.4
Putty-nosed monkey	2	1	3	0.02	41.7
Maxwell duiker	14	10	25	0.15	6.67
Red-flanked duiker	17	13	28	0.17	5.96
Yellow-backed duiker	2	1	3	0.02	55.6
Kucusimance mongoose	7	8	15	0.09	55.6

Source: Akinsorotan O.A., Ogunjemite B.G. and T.A. Afolayan, 2011 Assessment of the large mammals of Arakhuan range, Okomu National Park, Nigeria. *Ethiopian Journal of Environmental Studies and Management* Volume 4 No. 3, pp. 25–37.

The problem which spurred officials to action in the area of wildlife conservation was that traditional methods of hunting were extremely crude and detrimental to the sustenance of the animal population in the forest. Many animals were wounded but only a few were actually captured; there was indiscriminate killing of young ones and pregnant females especially during the night hunts. Today, there is a marked scarcity of game in many once well-stocked forests. Forest degradation and burning destroyed the habitat of some species while many habitats in the savanna regions were destroyed by the annual dry season bush fires. A number of animal species such as the manatee and the Pygmy hippo were seriously threatened with extinction (Afolayan and Pelinck, 1975). The rural inhabitants themselves were complaining about the dwindling number of game animals in the bush. But, at the same time, they were relentless in their destructive hunting activities particularly

TABLE 4.13B Estimated populations of sighted large mammals of Arakhuan Range in Okomu National Park, Southwest Nigeria (Field survey 2008/2009)

Animal species	Density	Estimated population (troop)	Average troop size	Estimated population (individual)
Mona monkey	4.47±1.24	240.18±66.49	7	1441.05±398.96
White-throated monkey	2.66±1.52	142.93±81.55	6	657.2±489.45
Red-capped mangabey	2.15±0.88	115.7±47.36	3	347.3±142.29
Putty-nosed monkey	2.40±15.50	20.7±12.08	13	128.84±81.49
Maxwell duiker	20.54±5.44	–		1104±292.65
Red-flanked duiker	36.66±7.68	–		1970±412.93
Yellow-backed duiker	0.17±0.36	–		9.0±2.00
Kucusimance mongoose	8.16±3.96	–		439.43±212.64

Source: Akinsorotan O.A., Ogunjemite B.G. and T.A. Afolayan, 2011 Assessment of the large mammals of Arakhuan range, Okomu National Park, Nigeria. *Ethiopian Journal of Environmental Studies and Management* Volume 4 No. 3, pp. 25–37.

because demand from urban dwellers was on the increase.

Habitat management for wildlife preservation

The first assignment in the management of wildlife reserves is, of course, to carry out an accurate census of the animal populations and study their movements, nutritional habits, reproductive capacities and ecological relations, especially the prey-predator relationships. The most investigated reserves are Yankari, Borgu and later the Kainji Lake National Park. Census taking was hampered by the migratory nature of the animals, inaccessibility of certain parts of the reserves, and the general lack of suitable large-scale aerial photographs. Counts and scientific observations were most conveniently carried out along the roads or at the watering points and mineral salt-lick stands. Nowadays, low level, large-scale aerial surveys are carried out using low flying aircraft including unmanned airborne vehicles (UAVs).

Controlled bush burning, water supply and the provision of mineral licks are essential management aids in wildlife conservation. At Yankari some areas are burnt early in the dry season in order to eliminate the possibility of late hot fires consuming the whole reserve. Early burning also keeps down grasses so that they do not obstruct the view of tourists. The fire encourages the sprouting of new shoots of fresh grass which attract game. On the other hand, some areas are protected completely from fire including riverine vegetation which is the haven of many wild species. Protection of riverine vegetation is also essential to provide cover for game coming to the river to drink thus rendering them less exposed to the guns of poachers.

It is important to adopt a rational approach to burning because while some animals are grazers and are thus encouraged by fresh green growth of grasses, others are browsers who prefer woody vegetation. Early burning gives woody vegetation an advantage over the grasses. Late burning encourages the growth of grasses at the expense of woody vegetation whose dry nature at this period makes them more combustible; grazing animals are thus favoured. Burning should be planned to promote a just balance between the requirements of grazers and browsers alike. At Yankari, the adoption of a strategy of early burning seems to be working well: the woody species are not burnt out but coarse grasses are eliminated thereby allowing the sprouting of fresh shoots of both grasses and woody plants.

Salt licks are important in wildlife conservation. They are naturally occurring mineral outcrops in the soil or ground surface which are eaten or licked by herbivores (Ayeni, 1975). According to Lameed and Adetola (2012), the species of animals in Nigeria known to be making use of natural mineral licks include elephant (*Loxondata africana*, buffalo (*Syncerous caffer*), western hartebeest (*Alcelaphus buselaphus*), roan antelope (*Hippotragus equimus*) water buck (*Kobus defessa*), kob (*Kobus kob*) bushbuck (*Treqlaphus scriptus*), oribi (*Ourebi ourebi*), red flanked duiker (*Cephalophus rufilatus*), Grimm's duikers (*Sylvicarpra grimmia*), warthog (*Phacolchoerus aethiopicus*), and baboon (*Papio anubis*). All the known natural mineral lick users are herbivores. Table 4.12 shows the results of animal counts along 5 transects leading to mineral salt lick in the Borgu Game Reserve.

Animals regularly visit licks, which are composed primarily of common salt (sodium chloride). The licks provide sodium, calcium, iron, phosphorus etc. (Ayeni, 1972; Table 4.14). Salt is also known to be a carrier of trace minerals such as copper, magnesium and cobalt; since all herbivores have natural appetite for salt, the licks could serve as a source of trace minerals for them. Salt licks occur naturally in certain locations in the forest where mineral salts are found on the ground surface. Shortage of sodium in the plants which are eaten by wildlife could motivate the

TABLE 4.14 Chemical analysis of mineral contents in salt licks at Borgu Game Reserve

Salt licks	%Na	%Ca	%Mg	%K	Fe (mg/kg)	$\%PO_4$	$\%SO_4$
1	3.69	1.79	0.88	6.76	1149.00	0.73	0.35
2	4.18	3.54	1.23 4	4.46	992.81	1.93	0.55
3	3.79	1.92	0.81	6.81	1168.00	0.89	0.39
4	4.21	1.85	0.79	6.95	1173.00	0.87	0.43

Source: Lameed A.G. and J. Adetola 2012 Species-Diversity Utilization of Salt Lick Sites at Borgu Sector of Kainji Lake National Park, Nigeria. In Gbolagade Akeem Lameed (2012) *Biodiversity Enrichment in a Diverse World*, Chapter 2, IntechOpen, DOI: 10.5772/51089

game to eat a lot of soil at the lick. According to Lameed and Adetola (2012 *op. cit.*), the shortage is as a result of water-soluble sodium salts being leached out of the soil during heavy rains following a long period of desiccation. Some plants even substitute potassium ions for sodium ions uptake from soil without showing mineral deficiency symptom and many plants are naturally rich in sodium and potassium. It is when the supply in plants fails or is insufficient that wildlife resort to mineral salt licks. Also, based on studies of game reserves in the drier parts of northern Nigeria, Henshaw and Ayeni (1971) postulated that increased utilization of the licks was an indication of nutritional deficiencies caused by a degrading environment and overpopulation. This appeared to be the case at least at Yankari Game Reserve.

The presence of natural mineral licks in an area does not necessarily mean that wild animals will make use of them. There may be alternative sources of mineral salts as, for example, at Tsavo National Park in Kenya, where sodium can be obtained at salt pans, boreholes and reservoirs. The functions of licks also change during the rainy seasons to those of natural water holes from which wildlife drink and wallow. Some animals like the buffalo and waterbuck, prefer using the drinking holes but some like the warthog and hartebeest still eat the soil lick minerals as well as drink the lick at drinking holes (Lameed and Adetola 2012 *op. cit*). Utilization of mineral licks by most wildlife species increases during the rains. Salt licks provide sodium, calcium, iron, phosphorus and zinc required at the beginning of the rainy season for bones, muscles and accelerated growth.

A noteworthy advantage derived from wild game utilizing natural mineral licks is that the animals are concentrated in areas of good visibility for viewing and photographing by tourists and scientists. Artificial licks could well be sited at vantage points within a reserve and could prove equally useful in following the movements and distribution of the animals. At Yankari, hideouts are installed near prominent licks where the animals can be observed at close range.

However, there are many problems associated with the use of mineral licks. These include the spread of animal diseases; intensification of predation; vegetation destruction; and soil compaction and erosion. So much urination, defecation, wallowing and feeding takes place at the salt mineral licks, that the transmission of disease from one animal to another is easy (Ogunsanmi, 1997). Coupled with the problem of disease is the increased predation among the animals caused by their coming together in large numbers. At Yankari this is evident from the frequency of carcasses near the mineral licks (Lasan, 1975). Furthermore, poachers are naturally attracted to the lick sites although these are closely watched by anti-poaching patrols.

Degradation of the habitat through overgrazing, over-browsing, soil trampling and compaction is a very serious problem at lick sites. Some areas are now completely devoid of vegetation cover thereby exposing the soil to erosion and intense insolation. Huge volumes of soil are consumed by the animals and more is lost through wallowing and trampling which loosen the soil fabric and make it readily susceptible to erosion. Locating artificial licks in less environmentally sensitive areas and, in sufficient numbers to forestall the over-concentration of game would reduce the problem. It would, also, help in dispersing the tourists and minimizing the risk of environmental degradation.

Wild game obtain water from both natural and artificial sources. The natural sources include rivers, springs, pools and pans seasonally filled by rainwater. The artificial sources are boreholes and reservoirs. The availability and distribution of sources of water within a game reserve are crucial factors affecting the ecological impact of wildlife on the landscape. At Yankari, the major source of water is the Gaji River which runs through it. But, during the rainy season the animals can also obtain drinking water in shallow pools and rivulets. During the dry season the pools and small streams dry up so that the wildlife concentrates along the Gaji river valley. This seasonal congregation of animals causes a great deal of damage to the riverine vegetation through trampling, compaction and exposure of the soil to the heavy rains at the beginning of the season. According to Lasan (1975), elephants usually congregate in herds of 150-200 grazing, browsing, pushing trees down and trampling on the soil and smaller plants. As at the salt licks, mortality amongst the animals is high as a result of predation, fighting, disease and the activities of poachers. However, the concentration of the animals in the valley during the dry season facilitates viewing by tourists.

In contrast to the situation at Yankari, the less developed Lame Game Reserve in the same state has numerous rivers and streams that flow throughout the year therefore, the wildlife is more dispersed and evenly distributed. The disadvantage in terms of tourist requirement is that wildlife is rarely found congregating in large numbers and are more difficult to keep track of for observation purposes. In the Yankari Game Reserve such a total dispersal of the wildlife would not be desirable, but it has been suggested that the provision of boreholes would help to decongest the Gaji river valley. Although this would be ecologically sound there could be some amenity loss to tourists and conservationists. There will be less concentration of animals for viewing and more potential sites for poachers, requiring more extensive anti-poaching patrols.

The provision of waterholes in a game reserve needs careful planning to assess accurately the possible environmental impacts of such a measure. Ayeni (1975) suggested that research must precede the development and management of waterholes in order to ensure that the quality of the water supplied to game throughout the year would not adversely affect animal health and habits; that there are no prospects of encouraging pests and disease transfer through increased waterholes utilization; and that the creation of the waterholes will not adversely alter the spatial distribution of vegetation cover types, reduce soil permeability or increase erosion; and finally that the microflora and fauna in the reserves are not disturbed.

Protected ecosystem conversion to grazing reserves

As discussed extensively in Chapter 3, many savanna 'forest' reserves have been re-designated grazing reserves as a concession to the Fulani pastoralists to be able to graze their cattle as they move southwards from the far north. The activities of the herdsmen are not strictly controlled except that hunting is usually prohibited in such reserves. Since the same reserves are more often than not, used for firewood collection the combined effects of grazing, bush burning and firewood gathering can be considerable. However, destruction of vegetation and soil trampling by cattle do not usually occur throughout a grazing reserve as much as along the cattle trails running through it. Close to such trails the ground under the woody layer may be completely devoid of grass cover. The trees and shrubs are also heavily gnawed by the animals. These effects can be observed for instance in the Kudu Grazing Reserve along the Mokwa-Bida road in Niger State

Burning is a common feature of the grazing reserves. In the Kudu Grazing Reserve, which is fairly densely wooded, the combined effect of burning and grazing has reduced the ground cover to patches of tufted grasses. This is also the case on the Izom plains bordering the Gurara valley north-west of Abuja in the Federal Capital Territory.

Current issues in protected ecosystem management

The future has caught up with Nigeria's protected ecosystems with both positive and negative consequences. The country has been afflicted by much violence in the first two decades of the 21st century. All the conservation areas throughout the country stand the risk of environmental degradation and destruction of biodiversity of unimaginable proportions given the habit of terrorist groups, herdsmen militia, kidnappers and human traffickers to use them as hideouts and operational bases. Herdsmen militia have penetrated deep into the rain forests of southern Nigeria and wreaked havoc on the local people and travellers. So much of the advances that have been achieved in nature conservation in the country can be destroyed within a short period of time if care is not taken.

As predicted in the first edition of this book, the conservation areas in the country have witnessed much positive human activity in the form of planned, rational management of the natural resources so the country can derive maximum social and economic benefits from them while maintaining the intrinsic quality of the natural environment. But the area of natural or semi-natural forest reserves has declined almost everywhere in the country as more reserve areas have been given over to commercial tree plantations. Indigenous species have been more difficult to grow on the plantations. This fact, coupled with their long felling cycle, has led to a concentration on the exotic tree species. The danger lies in the fact that, apart from teak (*Tectona grandis*), the exotic species are not timber species and hence they cannot help reduce the increasing shortage of timber in the primary and secondary wood-based industries. However, non-timber species support pulp and paper, match stick and tooth pick manufacturing in the country.

Unfortunately, the exotic tree plantations that reached maturity during the long period of military rule (1966 – 1999) were massively exploited by the military administrations in the states. Some conservation areas, especially those close to the big cities or state capitals, were converted to other uses including housing, golf course, and establishment of educational and other institutions. The conservation areas were much targeted for the implementation of the Federal Government proposal to establish cattle ranches to accommodate the nomadic Fulani herdsmen in every state of the federation. The project had to be suspended as a result of loud protests from the states, particularly in the south. But, the spectre of a resurgence of this project in the future is still there.

The encroachment on the game reserves by the Fulani herdsmen appears to be an intractable problem. The recurrent drought in the northern Sudan and Sahel zones is forcing more and more Fulani herds to encroach on the game reserves particularly in the Guinea savanna zone. As ever, the Zugurma and Kwiambana game reserves in the Kainji National Park permanently accommodate large herds of cattle and flocks of sheep and goats. The game reserves cannot be effectively managed unless the problem of encroachment is fully resolved.

TABLE 4.15 Wildlife species involved in human – wildlife conflict identified by the host communities (n =69) in Yankari Game Reserve

Animal species identified English name	Scientific name	Local name	Frequency	%
Buffalo	Syncerus caffer	Bauna	39	57
Tantalus monkey	Chlorocebus tantalus	Kirka	49	71
Patas monkey	Erythrocebus patas	Jan biri	41	59
Roan antelope	Hippotragus equinus	Gwanki	18	26
Western Hartebeest	Alcelaphus buselaphus major	Kanki	5	7
Bush Buck	Tragelapus scriptus	Mazo	36	52
Waterbuck	Kobus ellipsiprymnus	Gwambaza	26	38
Elephant	Loxodonta Africana	Giwa	8	12

Source: Y. A. Magama, M. Babagana, A. U. Usman, A. A. Gujja, A. Adamu, Karachi A. E. 2018 Assessment of wildlife species mostly involved in human-wildlife conflict around Yankari Game Reserve, Bauchi State, Nigeria *International Journal of Contemporary Research and Review* ISSN 0976 – 4852 September, 2018 Volume 09 Issue 09, `https://doi.org/10.15520/ijcrr/2018/9/09/605`

But, the problem of the conservation areas is not limited to encroachment by grazing animals. There are serious conflicts with rural communities and arable crop farmers. Tables 4.15 & 4.16 give examples of the kinds of conflicts between game reserves and neighbouring farming communities.

The attraction of the conservation areas, especially the game reserves, as tourist centres continues to grow. There is a growing interest in recreation among the populace and there are non-governmental organizations, travel and tours enterprises, hotel groups and associations promoting tourism and rural outdoor recreational ac-

TABLE 4.16 Nature of human-wildlife conflicts around Yankari Game Reserve

Wildlife Species Involved	Nature of Conflict	Crop Raided	Property Destroyed	Season of Conflict
Elephant	Crop Raiding, Destruction of Properties, Attacks on Human and animals	All kinds of crops with long erect stems like maize, millet and corn	Housings, Granaries Farm fences	All year Round
Patas Monkey Crop raiding	All kinds of crops; standing like maize and underground such as peanut	Farm fences	Mostly wet season	
Tantalus Monkey Crop raiding	All kinds of crops; standing like maize and underground such as peanut	Farm fences	Mostly during wet season	
Roan Antelope Crop raiding, Attacks on Humans and animals	All kinds of crops with long erect stems like maize, millet and corn	Farm fences, huts	Mostly during dry season	
Buffalo Crop raiding, Attacks on humans and animals	All kinds of crops with long erect stems like maize, millet and corn	Granaries, Farm fences, huts	All year Round	
Western Hartebeest	Crop raiding	All kinds of crops with long erect stems like maize, millet and corn		Mostly during dry season
Bush Buck Crop raiding, Attacks on humans and animals	All kinds of crops with long erect stems like maize, millet and corn		Mostly during dry season	
Waterbuck Crop raiding, Attacks animals	All kinds of crops with long erect stems like maize, millet and corn		Mostly during dry season	

Source: Y. A. Magama, M. Babagana, A. U. Usman, A. A. Gujja, A. Adamu, Karachi A. E. 2018 Assessment of wildlife species mostly involved in human-wildlife conflict around Yankari Game Reserve, Bauchi State, Nigeria *International Journal of Contemporary Research and Review.* ISSN 0976–4852 September, 2018, Volume 09 Issue 09
https://doi.org/10.15520/ijcrr/2018/9/09/605

tivities within the country. At present there are not enough trained personnel to manage the game reserves and this will necessarily place a limit on the number of reserves that can be created, the enthusiasm of the Federal Department of Forestry notwithstanding. Hence, the large number of game reserves and wildlife sanctuaries that are still at the proposal stage as shown in Table 4.9 (page 113). The ecological studies currently going on will of necessity take some time to yield data and arrive at valid conclusions and recommendations that can be relied upon for planning purposes. However, the studies should provide useful guidelines on how best to manage and harness the wildlife resources of the protected ecosystems. The country will likely continue to benefit from close collaboration with and technical and financial assistance from such international organizations as the World Bank, the United Nations Food and Agriculture Organization (FAO), the International Union for the Conservation of Nature and Natural Resources (IUCN), the World Wildlife Fund (WWF), the United States Agency for International Development (USAID) and the European Commission (EC).

The WWF (2014), gave an insight into why the international organizations had been so concerned about creating nature reserves and wildlife sanctuaries in Nigeria. It noted that the Nigerian Lowland Forests ecoregion though relatively small was nevertheless

> "*one of the most densely populated areas in Africa, and already showed high levels of human activity before colonial times. As a result, the forest has been fragmented for a long period of time, a process that has accelerated considerably over the last four decades, leaving only a few remaining forest blocks*". Yet, this small and fragmented ecoregion contains "*five endemic animal species, including an endangered primate, the white-throated guenon (Cercopithecus erythrogaster), as well as chimpanzees (Pan troglodytes) and elephants (Loxodonta africana.)*"

Thus, the WWF (2014) determined that though a number of protected areas had been established they were too small and lacked adequate protection. Without increased protection efforts, it was unlikely that the viable sections of forest would survive long into the future.

Table 4.17 shows some of the biodiversity of Nigeria already declared "endangered" by the very first national biodiversity survey, 2001. The list of threatened species was even much longer.

TABLE 4.17 Threatened biodiversity species in Nigeria, 2001

Species		Main uses	Status
a. Plants			
Melicia excelsa	Iroko	Timber	Endangered
Diospyros elliotii		Carving	Endangered
Triplochiton scleroxylon	Obeche	Timber	Endangered
Mansonia altissima		Timber	Endangered
Masilania accuminata		Chewing stick	Endangered
Garcina manni		Chewing stick	Endangered
Oucunbaca aubrevillei		Trado-medical	Almost Extinct
Erythrina senegalensis		Medicine	Endangered
Cassia nigricans		Medicine	Endangered
Nigella sativa		Medicine	Endangered
Hymenocardia acida		General	Endangered
Kigelia africana		General	Endangered
a. Plants			
Crocodylus niloticus	Nile crocodile	Food/medicine/bags	Endangered
Osteolaemus tetraspis	African dwarf crocodile	Food/medicine	Endangered
Struthio camelus	Ostrich	Food/medicine Medicine/pet	Endangered
Psittacus erithacus	African grey parrot	Food	Endangered
Cercopithecus erythrogaster	White throated monkey	Food/Ivory	Endangered
Loxodonta africana	African bush elephant	Food	Endangered
Trichecus senegalensis	Manatee	Food/medicine	Endangered
Giraffa camelopedalus	Giraffe	Bags	Endangered
Python sabae	Rock python	Food	Endangered
Gazella Dorcas	Dorcas gazelle	Endangered	

Source: First National Biodiversity Report, July 2001
`https://www.cbd.int/doc/world/ng/ng-nr-01-en.pdf`

Biodiversity action strategy and plans

If Nigeria is to rescue the remaining conservations areas and biodiversity, it will have to work with the international organizations and the Nigeria Conservation Foundation to keep the tempo of the rehabilitation of our highly degraded forest and game reserves. In 2015 the Ministry of Environment revised and upgraded the National Biodiversity Strategy and Action Plan (NBSAP) to stem the worsening biodiversity conservation situation in the country. The revision was done in collaboration with the Convention on Biological Diversity (CBD), Global Environment Facility (GEF) and United Nations Environment Programme (UNEP). It was hoped that the revised NBSAP would guide the conservation and sustainable utilization of biodiversity, access to genetic resources and the fair and equitable sharing of the benefits arising from their utilization. It itemized 17 targets or *Global Strategic Goals*, that it hoped to attain between 2016 and 2020. It is still too early to know how far the implementing committee succeeded in meeting those targets. However, there have been significant biodiversity conservation projects in different places particularly those funded by the oil companies and international organizations in partnership with the Nigeria Conservation Foundation.

In Edo State, the Biodiversity Action Plan, begun in 2007, is being funded by the Shell Oil Company (NCF, 2020). In a community-based pilot participatory natural resources management project and livelihood support programme the Biodiversity Action Plan seeks to promote forest regeneration and rebuild the carbon stock of the forest reserves while at the same time helping the local communities develop livelihood enterprises that would not harm biodiversity regeneration. The participating communities are found around the degraded Urhonigbe and Gele Gele Forest Reserves which had suffered from heavy logging in the past. The local communities involved are Urhonigbe, Gele-Gele, Kolokolo, Ajumimogba and Abiala II. The livelihood enterprise development aspect of the project focuses on (i) poultry farming, and (ii) commercial plantations of the star apple tree, *Irvingia guinnensis*, combined with the cultivation of improved cassava and pine apple varieties, similar to the taungya system described earlier. The participants were provided with the tree seedlings and with improved cassava stems and pine apple seedlings provided by the International Institute for Tropical Agriculture (IITA), Ibadan. The Project established a micro-credit facility for the participants and also distributed cassava graters to help in processing their cassava harvests. The goal is to develop livelihoods that are compatible with the main goal of forest regeneration and biodiversity conservation. The Project established the Forest Management Committee (FMC) and the Grassroots Consultative Committee (GCC) in the thirty-six communities of the two Forest Reserves and instituted a governance structure designed to build the communities' local capacity to take over the management of the forest reserves. It is the kind of community-based participatory approach that had long been advocated by the moribund Natural Resources Conservation Council (1991). The rehabilitation of the forest reserves themselves is being pursued by raising nurseries of both indigenous and exotic tree species. The indigenous species being raised include *Khaya grandifolia* (Red Mahogany), *Khaya senegalensis* (White mahogany), *Terminalia ivorensis* (Black afara) and *Entandrophragma angolenses* (Lagos wood).

Similar forest rehabilitation initiatives are ongoing at Andoni Forest Reserve in Rivers State. The Andoni Island Forest Reserve in Rivers State, which is also involved in the African Elephant Database Project to monitor the dynamics of elephant population in different parts of Africa, is an area of the Niger river delta that has witnessed much deforestation. Eludoyin *et al.* (2019) highlighted the extent of deforestation in the Andoni Local Government Area between 1986 and 2015 based on land use/ land cover change analysis using Landsat images. The forest cover in the local government areas declined from approximately 40 per cent 1986 to about 23 per cent in 2015. However, the coverage of mangrove swamps increased over the period, which gives hope that the forest rehabilitation programme might yield positive results.

The 4000 km^2 Cross River National Park, established in 1991 (Wikipedia, 2020) is the target of rehabilitation efforts similar to those in Edo and Andoni forest reserves. The Cross River National Park, made up of the Oban and Okwangwo forest reserves is reputed to be the oldest moist tropical rainforest in Africa, "a naturally endowed biodiversity hotspot with diverse resources ranging from picturesque topography, rivers, fauna and flora" (Adekola and Adetoro, 2014). But, this treasure house of biodiversity is now under pressure from illegal and injurious activities including logging; fish and water poisoning by illegal fisherfolks using herbicides to kill and capture fish; pilfering of non-timber forest products (NTFPs); and encroachment by herdsmen who graze their cattle illegally in parts of the park. Moreover, the park is home to the threatened Cross River gorilla (Gorilla gorilla diehli), which has attracted the attention of the International Union for the Conservation of Nature and Natural Resources (IUCN). The Cross River gorilla is recognised by IUCN as critically endangered and as the most threatened taxon of ape in Africa. Its status is heightened by the fact that "it is the most westerly and northerly form of gorilla, and occurs in a limited area around the mountainous headwaters of the Cross River, straddling the border between Cameroon and Nigeria" (Oates *et al.*, 2007). The Cross River gorilla is now the focus of the "Regional Action Plan for the Conservation of the Cross River Gorilla (Gorilla gorilla diehli)."

Finally, it is worth mentioning that the Federal Government also signed an agreement with the United Nations Development Programme on the Niger Delta Biodiversity Project. This project's goal is to contribute to the conservation and sustainable use of globally significant biological diversity in the Niger Delta, which has experienced intense oil and gas exploration and exploitation activities since the 1950s. The project objective is "to mainstream biodiversity management priorities into the Niger Delta oil and gas (O&G) sector development policies and operations." What is most welcome about these new initiatives is that the development of sustainable livelihoods for local communities is given much needed priority. Otherwise, the biodiversity rehabilitation projects would have little prospect of achieving tangible results.

5

Water resources and water use

Water is the single most characteristic element of life. It is so basic to human existence on earth that any nation that fails to develop this vital resource is without question endangering the health and social and economic wellbeing of its people. The crucial importance of water in Nigeria's socio-economic development in the modern world was brought home to people by the recurrent droughts in Nigeria in the last three decades of the 20th century beginning with the Sahel drought of 1968-73. Moreover, in the first two decades of the 21st century, crop failure, degradation of grazing areas, and low water levels in rivers and reservoirs have become frequent features of the rural economy in the northern parts of the country due to either lack of rain or delayed onset of the rains at the beginning of the growing season. In addition, urban centres experience acute shortages of water caused by the growing demand for domestic and industrial purposes. Electricity supply is erratic during the dry season when the water level is low in the reservoirs supplying the hydro-electric power plants.

Yet, in spite of reduced rainfall and river discharge, the country has witnessed increasing flooding, especially, along its major rivers, the Niger and Benue. Because of the general reduction of vegetation cover throughout the country, more runoff is generated from each heavy rainfall during the rainy season. This swells up the river volume and results in destructive flash floods. Thus, altogether, the water resources in the country require more rational and comprehensive management to derive maximum benefits and minimize the negative outcomes of the water resource processes.

The sources of water

Nigeria's water resources include rainfall, surface and ground water. The recurring water shortage problems in Nigeria are not really the result of a lack of potential sources of supply; they are problems of development, distribution and quality control caused as much by human deficiencies as by natural forces. While, on the one hand, many parts of the country are experiencing shortages, on the other hand, there are areas that suffer annually from floods. Many rivers seasonally overflow their banks, even in those areas which are chronically short of water when and where it is needed. There are drought-hardened land surfaces almost bare of vegetation in the north juxtaposed with areas which are seasonally or permanently waterlogged. In some of those areas which experience the worst effects of the Sahel droughts, there are vast underground reserves of water that have remained largely untapped.

In spite of the numerous piped domestic water supply schemes serving large urban centres and their rural surroundings, the majority of the population still depends on water of doubtful quality from unprotected streams and shallow wells. Even those people living in areas with piped supply occasionally have to resort to unwholesome sources because the reservoirs and water treatment plants are grossly inadequate.

Rainfall

Fig. 5.1 is a schematic diagram of the hydrological cycle, which can be conceived as consisting of a series of water storage tanks interconnected by the transfer processes of evaporation, moisture transport, condensation, precipitation and runoff (Simmons, 1974). The starting point of the cycle, which is also the ultimate source of water found on the earth's surface, is precipitation, while the main cause of loss from the earth's surface is evaporation. These two processes, precipitation and evaporation, are thus crucial in the determination of Nigeria's water resources. An accurate assessment of these resources therefore requires an analysis not only of rainfall and its spatial and temporal distribution, but also of the nature and magni-

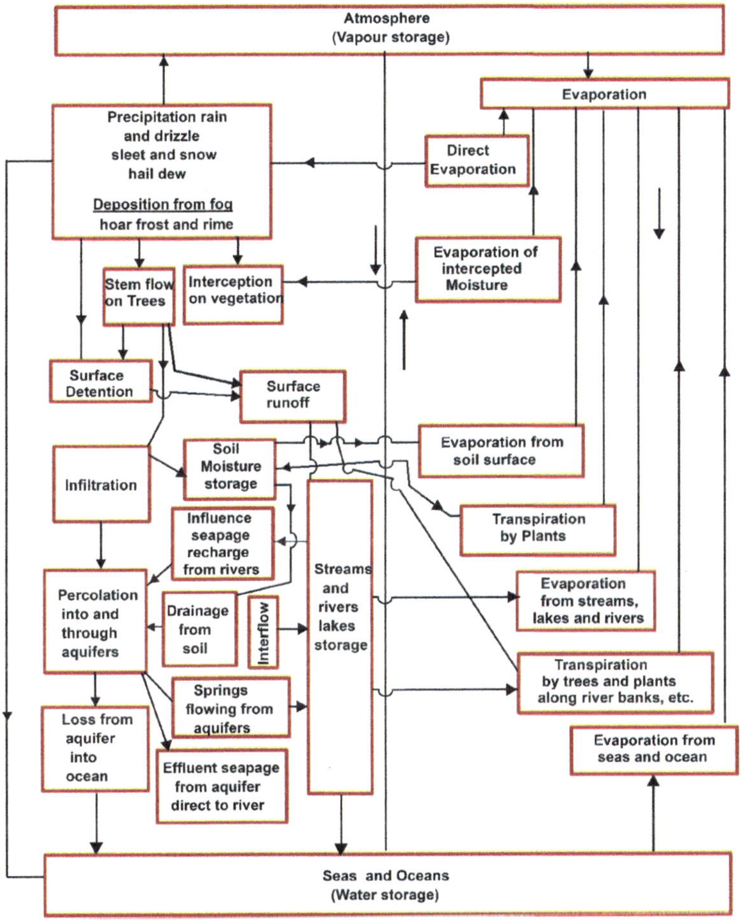

FIGURE 5.1 The hydrological cycle

tude of water loss through evaporation.

The distribution of rainfall in Nigeria is controlled by the northward and southward shifts of the Inter-Tropical Discontinuity (ITD), the boundary between the moist south-west monsoon air mass and the dry, north-east air mass. The position of the ITD varies not only between the seasons but also from one year to another and this accounts for much of the spatial variation in the type, amount and seasonal distribution of rainfall as well as the length of the wet season. In general, rainfall amounts decrease northwards from 3048mm in the Niger delta and the south-eastern coastal region to less than 508mm in the extreme north-east (Fig. 5.2). This trend is broken only in the Jos Plateau where the rainfall is relatively higher than for its latitude (1270-1524mm). In the south, rainfall amounts tend to decrease westwards due to differences in coastal alignment between the south-east and the south-west as shown in Fig. 5.2. This alignment of the coastline is thought to favour the routing of storms towards the south-east (Ayoade, 1974). Another reason for the lower rainfall in the south-west is that this area experiences 'the little dry season',

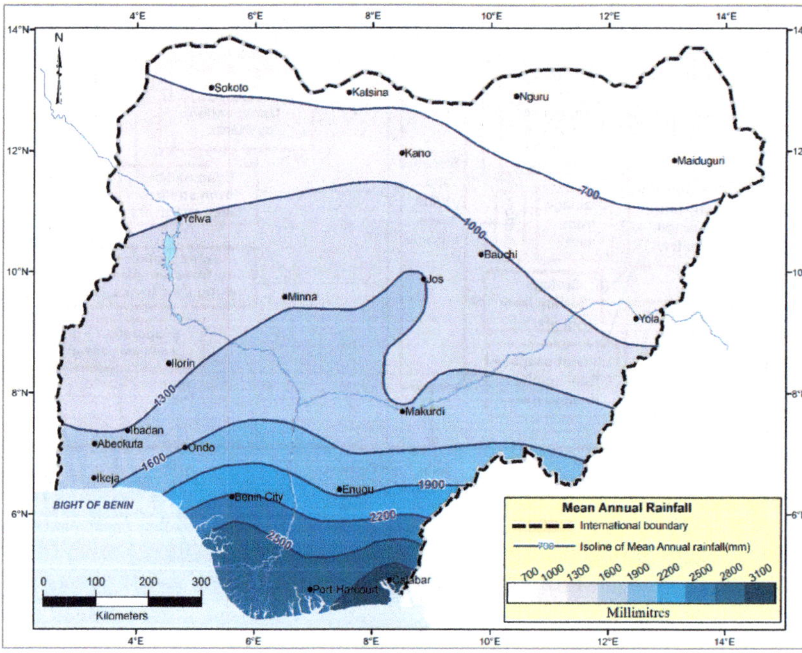

FIGURE 5.2 Mean annual rainfall (mm) in Nigeria

which is largely absent in the southeast.

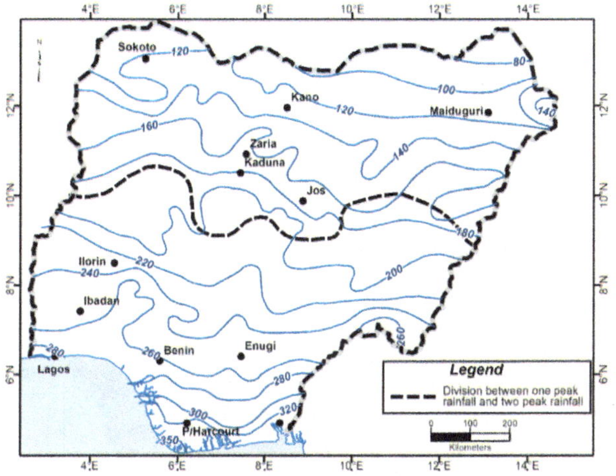

FIGURE 5.3 Duration of the wet season (no. of rain days) in Nigeria

Figures 5.3 and 5.4 show the number of rain days and mean annual potential evapotranspiration respectively. The greater aridity in the north is due to low rainfall, accentuated by the high rates of evapotranspiration (Fig. 5.4), which tend to increase northwards because of longer hours of bright sunshine, higher mean air

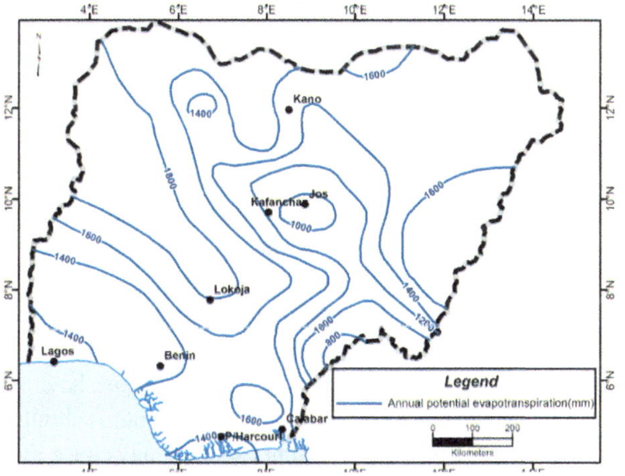

FIGURE 5.4 Mean annual potential evapotranspiration (mm) in Nigeria

temperatures (Fig. 5.5) and low relative humidity. During the wet season a water surplus is generally experienced in Nigeria but this is more so in the south than

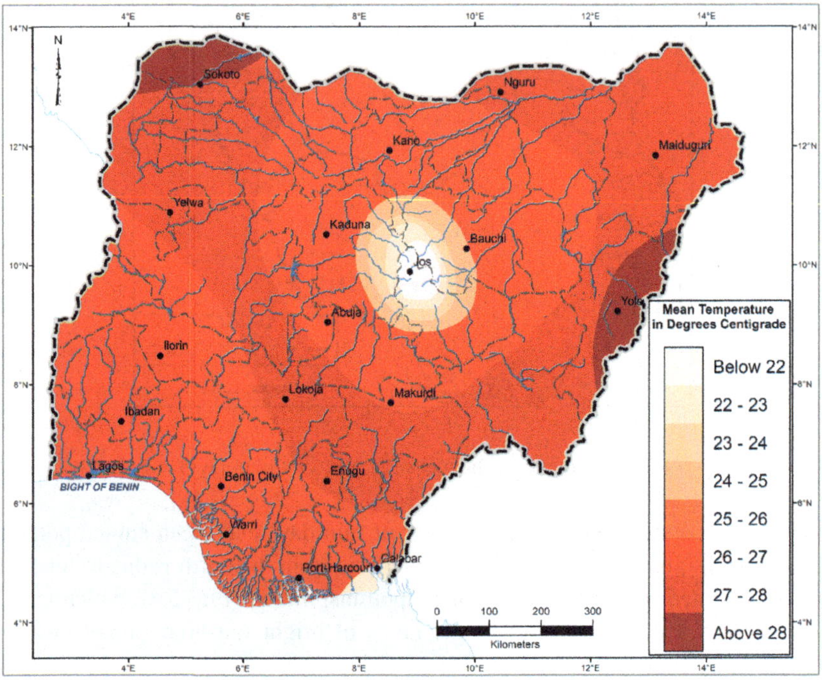

FIGURE 5.5 Mean annual air temperatures (°C) in Nigeria (Modified from Macmillan, 2010)

in the north of the country (Fig. 5.6). The surplus is due to the fact that more rain falls than can be taken up by the plants or returned to the atmosphere through evaporation from bare and vegetated surfaces (evapotranspiration); consequently, the soil remains charged with water. However, during the dry season, when little or no rain falls and evaporation and evapotranspiration processes are very high, plants are forced to draw upon soil moisture reserves to survive and this often leads to a deficit. Thus, throughout the country, there are alternating periods of water surplus and water deficiency. The period of water surplus is, as we have seen, longer in the south and the Jos Plateau area than in the far north, but it should be noted that variations in soil texture and moisture holding capacity may cause local differences.

The hinterland region of southern Nigeria beyond the equatorial coastal and deltaic zone experiences the phenomenon known as the 'little dry season' in mid-

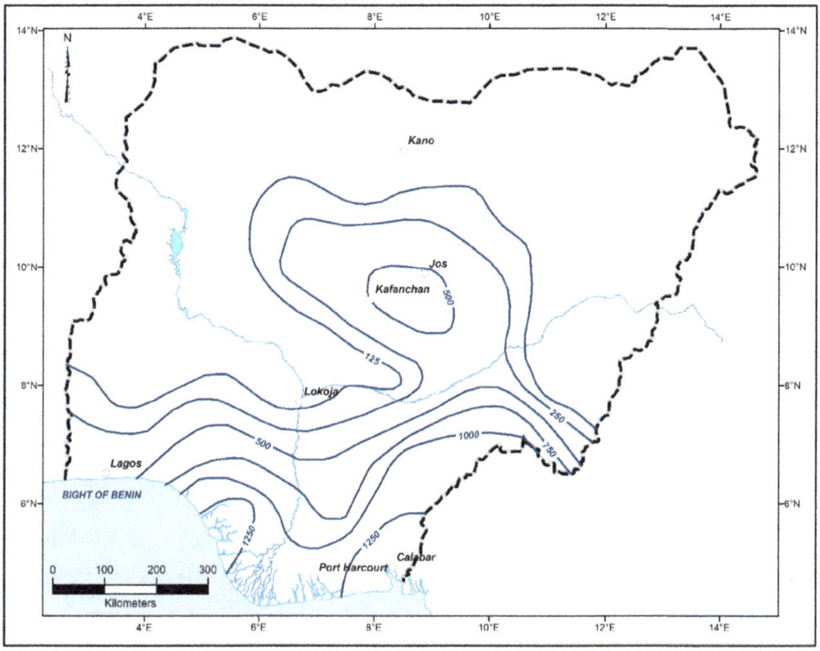

FIGURE 5.6 Mean annual water surplus (mm) in Nigeria

July to early August. This is a period of comparatively low rainfall between the two peak periods (June and September) of high rainfall in the year. The duration of the wet season decreases from over three hundred days in the coastal region, to less than ninety days in the extreme north-east (Fig. 5.3), and the number of rain days drops from an average of two hundred in the coastal and deltaic areas to less than forty in the extreme north-east.

The wet season in the extreme north-east of Nigeria is at times so short, the total rainfall so small and evaporation so high, that the soil never becomes appreciably moist. Hence, the area is always susceptible to drought.

Fig. 5.6 reveals that the mean annual water surplus is less than 508 mm over 75 per cent of the country. The areas with relatively high mean annual water surplus are the Niger delta and Imo and Cross river basins. The mean annual water surplus in these south-eastern parts of Nigeria is greater than the total annual rainfall of most of the savanna regions beyond the southern Guinea zone. Part of the country's water supply problems in agriculture and industry can be attributed to this low water surplus over most of the country. It is partly responsible also for the highly seasonal regime of Nigerian rivers.

The drainage network and surface water resources

Fig. 5.7 shows the drainage network of Nigeria. The network falls into four major

FIGURE 5.7 Nigeria: Relief and drainage network (Modified from Macmillan, 2010)

surface water basins (Table 5.1; FAO, 2016). The drainage density (as calculated on the 1:50,000 topographic maps of Nigeria) is low being about 1.5-3.5 per square kilometre on crystalline rocks and much less on sedimentary rocks. Although many rivers and streams are dry for a period in the year the drainage network gives some idea of the volume of water that is potentially available for use. At present much of the water is lost through runoff to the sea during the wet season.

Hydrology of Nigerian rivers

Table 5.2 shows the major hydrological divisions of Nigeria and the areas covered by each. There are eight (8) hydrological areas (HA-1–HA-8; FAO, 2016) in all

TABLE 5.1 Nigeria's main surface water basins (after FAO, 2016)

Surface water basins	Area (km^2)	% total area of country	Main rivers
Niger Basin	584,193	63	Niger and its tributaries (Benue, Sokoto and Kaduna
Lake Chad Basin	179,282	20	Komadugu Yobe and its tributaries (Hadejia, Jama'are and Komadougou Gena)
Southwestern littoral basins	101,802	11	Rivers from the hilly area south and west of the Niger river
Southeastern littoral basins	54,493	6	Cross, Imo

Source: FAO AQUASTAT FAO's Water Information System Nigeria: Water Resources `http://www.fao.org/nr/water/aquastat/countries_regions/Profile_segments/NGA-WR_eng.stm`

(Fig 5.8). The broad features of the hydrology of Nigerian rivers are fairly well known, in particular the high correlation between stream flow and rainfall. Nearly all the rivers, even in the humid south, have a seasonal flow regime. But the bigger rivers do not dry out completely during the dry season; this is particularly true of the estuaries which are fed by many tributaries. On the other hand, in the sub-humid areas, many rivers are reduced to a trickle, or isolated muddy pools, while in others the water in the channel is very shallow and is underlain by deposits of mud and other forms of river bed load.

The majority of Nigeria's rivers flow over Basement Complex rocks, which are comparatively poor aquifers. The underground supply of water to maintain stream flow depends very much on the thickness and hydrological characteristics of the weathered mantle in each locality. In most areas the water storage capacity of the weathered mantle is insufficient to sustain the flow of the streams during the dry season when there is no replenishment by local rainfall. The situation is compounded by man's destruction of the vegetation in many watersheds. Reduced infiltration of water into the soil and increased surface runoff resulting in reduced groundwater recharge and a lowering of the water table have caused many streams to dry up completely in the less humid parts of the country such as the Sokoto basin. Other rivers, such as those in the derived savanna zone, have become more

TABLE 5.2 Nigeria's hydrological areas (after FAO, 2016)

Hydrological Areas		Area (km^2)	Precipitation mm/year	Main related RBDAs	Main related states
HA-1	Niger North	135,100	767	Sokoto-Rima	Katsina, Zamfara, Sokoto, Kebbi
HA-2	Niger Central	154,600	1170	Upper Niger Lower Niger	Niger, Kwara, Kaduna, Kogi, Federal Capital Territory
HA-3	Upper Benue	156,500	1055	Upper Benue	Adamawa, Taraba, Gombe, Bauchi
HA-4	Lower Benue	74,500	1341	Lower Benue	Plateau, Nassarawa, Benue, Kogi
HA-5	Niger South	53,900	2132	Anambra-Imo, Niger Delta	Bayelsa, Delta, Edo, Kogi, Anambra, Rivers
HA-6	Western Littoral	99,300	1541	Ogun-Osun, Benin-Owena	Lagos, Ogun, Oyo, Osun, Ondo, Edo, Ekiti
HA-7	Eastern Littoral	57,400	2106	Cross River	Abia, Anambra, Imo, Enugu, Ebonyi, Cross River, Akwa Ibom, Rivers
HA-8	Lake Chad	178,500	610	Hadejia-Jama 'are, Chad	Kano, Jigawa, Yobe, Borno, Bauchi, Plateau, Adamawa

Source: FAO AQUASTAT FAO's Water Information System Nigeria: Water Resources `http://www.fao.org/nr/water/aquastat/countries_regions/Profile_segments/NGA-WR_eng.stm`

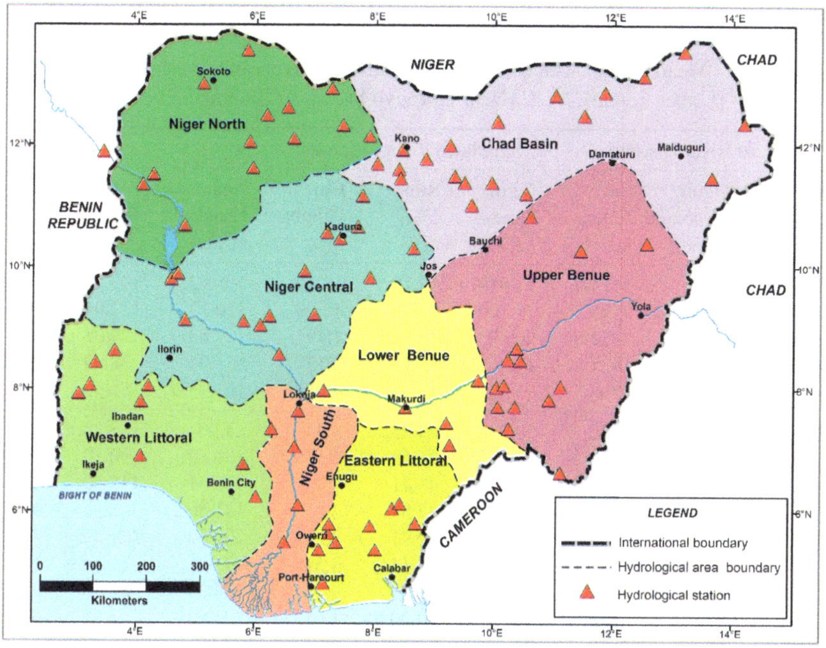

FIGURE 5.8 Nigeria: major hydrological areas and river gauging stations

markedly seasonal in flow and dependent on runoff from each local rainfall. Table 5.3 gives the discharge figures for a number of rivers in Nigeria.

Physical features of the rivers

The physical features of the rivers are just as important as the hydrological characteristics in determining the use that is made of Nigeria's surface water. Throughout the country the solid rocks are generally covered by a thick mantle of deeply weathered materials, but, except in rocky outcrops the depth of the weathered layer tends to decrease from the interfluves to the river channels. Indeed, many rivers have cut through the weathered layer down to the underlying solid rocks. Such rivers, which include the Niger in its middle course, are well known for their spectacular falls, rapids and rocky islands that impede navigation.

Also, the topography of the Basement Complex areas consists of a series of beveled erosion surfaces separated by steep, scarp slopes. Thus, a waterfall and a gorge usually occur where a river descends from a higher to a lower surface.

TABLE 5.3 Mean annual water discharge of selected rivers in Nigeria, 1990–2019 (*UB* [Upper Benue], *NGD* [Near Goronyo Dam], *AR* [Annual Runoff])

Annual River Discharge			Simulated River Discharge					
River Station	Ogun Abeokuta m^3/s	*UB* Jimeta m^3/s	Galma Zaria *AR* Mm^3	Rima *NGD* m^3/s	Gaji Gombe m^3/s	Anambra Nsukka m^3/s	Orji Enugu m^3/s	m^3/s
1990		1621.5	11,619	19.16	16.40	105.92	103.74	141.66
1991		2595.5	17,700	62.79	48.99	120.72	122.91	160.34
1992		2379.5	20,493	36.41	1.49	207.63	29.05	54.41
1993		1421	13,861	53.19	0.17	164.94	38.99	60.47
1994		3206	15,927	28.64	35.86	161.95	56.90	87.95
1995	44.97	1832	20,211	28.23	4.44	84.66	118.28	164.42
1996	33.49	2803	17,326	36.62	1.139	181.54	90.10	120.76
1997	93.77	2053	12,877	41.41	2.22	139.31	117.53	152.03
1998	88.15	2641	30,487	27.54	28.60	141.56	43.97	63.00
1999	70.85	3587	31,995	32.91	30.07	200.93	1625.42	133.38
2000	54.78	2790	24,532	21.14	4.54	97.83	89.54	118.37
2001	—	2440	25,201	52.09	9.52	204.38	68.73	91.07
2002	37.96	2670	19,287	49.30	1.01	125.92	65.75	100.45
2003	16.63	2980	26,171	72.70	49.61	380.76	14.47	21.74
2004	24.24	1930	15,331	3.09	4.52	115.73	14.01	21.30
2005	—	1680	18,076	41.91	7.30	172.10	58.88	75.00
2006	29.51	2769	25,861	33.43	42.40	180.08	150.78	190.98
2007	72.03	2825	28,440	31.38	45.56	126.48	113.15	145.08
2008	58.12	2090	22,664	74.05	8.51	269.76	79.23	109.82
2009	92.29	2580	25,431.06	37.85	1.20	130.47	78.72	108.23
2010	192.91	2775	21,079.87					
2011	147.60	1175	10,460.28					
2012	43.08	3200	29,490.05					
2013		1152.5	15,992.63					
2014		1373.5	19,671					
2015		1612.5	17,114.26					
2016		1782.5	20,567.01					
2017		1346.5	18,067					
2018		2480	22,171					
2019		1915	19,331					

Sources: Simulated river discharge data based on: Berit Arheimer, Rafael Pimentel, Kristina Isberg, Louise Crochemore, Jafet C. M. Andersson, Abdulghani Hasan, and Luis Pineda, 2020 Global catchment modelling using World-Wide HYPE (WWH), open data, and stepwise parameter estimation. *Hydrol. Earth Syst. Sci., 24, 535–559, 2020* `https://doi.org/10.5194/hess-24-535-2020`

Such falls and gorges occur on many of the rivers flowing into the Niger and Benue troughs from the Jos Plateau and the high plains of Hausa land, for example, Gurara falls on the Gurara River and the Shiroro gorge and Zungeru falls on the Kaduna River. These waterfalls offer great opportunities for the development of hydroelectric power in the country. In fact, since Independence in 1960, the federal government has engaged in the building of dams to generate hydro-electric power on the Niger River and along the streams flowing from the Jos Plateau and northern high plains of Hausa land into the Niger and Benue troughs. Examples include the Kainji and Jebba dams on the River Niger and Shiroro gorge dam on River Kaduna.

Unfortunately, stream bed or channel deterioration is now a common feature of many rivers because of increased surface runoff and sediment yield consequent upon the degradation of the vegetation in many catchment areas. In some cases, the high sediment load on the stream bed delays the flow of the river until the time the deposits are fully saturated. The deposits may be so great that long after the rains have come, or in years of low rainfall, the stream is no more than a maze of shallow channels (braided channels). Braiding is very common in streams flowing through broad valleys of low gradient but with heavy loads, for example, along the Sokoto and Hadejia rivers in northern Nigeria and the Nyaba River near Enugu.

In some cases, stream bed deterioration has increased the incidence and destructive impact of river floods. Flash floods in particular commonly arise from heavy torrential rains. In a number of cases, the inability of the river channels to contain floods is due to thick vegetal growth on the bed deposits which seriously reduces the hydraulic capacity of the channels.

The river valleys and floodplains

The physical and hydrological conditions of the river valleys also influence the potential uses of river water and may affect the scope and direction of future development efforts. Except, perhaps, in such rugged uplands as are found in the eastern borderlands, most rivers have fairly broad and shallow valleys even close to their headwaters. In the savanna region these broad valleys, known as *fadama,* usually have more fine-textured and water retentive soils than the adjacent hill slopes and are therefore of vital importance to agriculture. In the south, the major yam and rice producing areas of south-eastern Nigeria have the same types of soil and terrain conditions along the tributaries of the Imo and Cross rivers. Elsewhere such valley bottomlands have not been intensively utilized because they are mostly waterlogged and densely vegetated. But in some parts of the derived savanna zone such valley lands, referred to as *àkùrọ̀* in Ekiti and Kogi states, are cultivated to produce sugar cane, vegetables, early yams and, more recently, paddy rice. The extensive fadama lands around Agaie and Badeggi in the Niger trough have long been noted for large-scale paddy rice cultivation.

Unfortunately, except in the case of the Sokoto and lower Niger rivers, the flood plains of most of Nigeria's major rivers are swampy and thus remained largely undeveloped until the last few decades. Such, for example, was the case, with the Ogun and Osun flood plains; the lower Cross river floodplains; Enyong Creek plains; the Benue flood plains and the adjacent great plains of Muri. Although a section of the middle Niger flood plains below Jebba had been developed for sugar plantations, for example at Bacita, other sections for long remained virtually empty and undeveloped. From all indications it appeared that the major factor that militated against the use of these flood plains by the local inhabitants was the sheer difficulty of clearing the dense vegetal growth on them. Once the initial bush clearing has been done, local farmers have successfully cultivated the land as is the case in parts of the Ogun estuarine flood plains, for example, at Mokoloki. Indeed, the River Basin Development Authorities (RBDAs) have steadily opened up and facilitated access to these difficult floodplain terrains since the 1980s.

The broad and shallow nature of many river valleys poses special problems for the construction of dams particularly in the sub-humid savanna regions. Reservoirs created in any of these valleys tend to be shallow and spread over an extensive area of often valuable agricultural land. Such an extensive body of water is subject to great evaporation losses due to the bright sunshine, high temperatures, low relative humidity and strong winds. This is one of the most serious problems of water conservation in the northern parts of Nigeria; in fact, it is not restricted to reservoirs in the far north but it also affects the water supply schemes at Oyo, Ede, Osogbo, Ogbomoso and Eruwa all in the derived savanna zone.

Another problem is associated with the high permeability of the surface sediment deposits which are generally sandy and characterized by high internal drainage. A common feature of many northern rivers and streams is their tendency to lose a great deal of water to or even disappear completely into thick beds of sand. This is common especially during the dry season. The Hadejia-Yobe, for example, only flows through to Lake Chad during the wet season; during the dry season it loses a lot of water to the underlying sands and finally disappears into the swamps bordering the lake.

When a dam reservoir is created in these areas much water is lost through percolation into the ground. The solution to this problem is to excavate and remove the sandy surface layer so that the reservoir water lies on the impermeable clayey beds which form the substrata of much of the far north. Another factor which causes great losses of water from reservoirs is the dense growth of the water lettuce, hyacinth, as well as reeds and other weeds in the rivers. Large rivers such as the Ogun at Abeokuta, for example, are equally affected. The weeds absorb great quantities of water and transpire profusely. The remote Owo Water Supply Scheme on the Osse River is among the worst affected. Apart from the weeds growing on the surface, vegetal growth encroaches on the body of water from the surrounding slopes. Weeds are most prevalent at the beginning of the wet season when the first surface runoffs and seepage water bring in fresh supplies of nutrient elements from the surrounding areas. The weeds also invade rivers and reservoirs at the onset of the dry season when the water depth becomes progressively shallower and the weeds can take root in the underlying mud.

With the large-scale degradation of vegetation in the watersheds, many reservoirs receive a great deal of sediments during the wet season floods. It is true that the land areas in the immediate vicinities of the reservoirs are often protected and sometimes planted with commercial trees. But land use activities upstream and on the tributary catchments are usually beyond the area of control of the water authorities. The drying up of reservoirs, which occasionally occurs during the dry season, is as much the result of the silting up of the bed as it is of the hot dry weather.

Flooding phenomenon on Nigerian rivers

Fears have been expressed about the negative impact of climate change on Nigeria's water resources. Indeed, climate change has led to reductions in rainfall and in the discharge of rivers generally. Yet, because of reductions in the vegetation cover, runoff from rainfall events during the rainy season is still considerable. In his analysis of rainfall and discharge in the Benue River at Jimeta, Yola between 1980 and 1999, Adelalu (2012) observed that while monthly analysis of runoff in some years revealed increases in runoff above the mean, this additional runoff did not add to water availability for use. This was because the increased runoff occurred during the wet season and there was no specific storage to hold the water through to the dry season. Rather, what has been happening is that the country has been experiencing more frequent flooding along its major rivers since the onset of global warming and climate change.

BBC News in September 2018 highlighted Nigeria's losses to flooding in recent times after floods claimed nearly 200 lives and displaced thousands of people. In 2012 [*] two million Nigerians were forced out of their homes and 363 were killed by devastating floods; in 2015, more than [†] 100,000 were displaced, and 53 died;

[*] `https://www.bbc.co.uk/news/world-africa-20221451`

[†] `https://www.acaps.org/sites/acaps/files/slides/files/20180921_acaps_brie`

in 2016, [‡] 92,000 were displaced and 38 died; while in 2017, floods affected [§] 250,000 people in the eastern-central region. Apart from loss of lives, destruction of homes and other buildings and the displacement of people, the floods inundated thousands of hectares of farmland and destroyed crops and the means of livelihood of thousands of rural inhabitants.

In addition to the explanations given above for the increased frequency of floods in Nigeria, there are other equally significant factors. In spite of the efforts of the river basin development authorities established since 1976, there are still not enough dams (and reservoirs) on Nigeria's main rivers to accommodate increased runoff from rainfall events in the wet season. Hence, as noted above, flash floods have become more frequent and devastating. Just as Nigeria has been struggling with containing increased runoff and overflowing dam reservoirs so have neighbouring countries of Niger and Cameroon been struggling with the problem of overflowing reservoirs. Water releases from dams upstream of the Niger and Benue rivers in Niger Republic and Cameroon respectively compound the problem of flash floods along the two rivers during the rainy season. The coming together of Rivers Niger and Benue at Lokoja is responsible for the flooding experienced in the lower Niger River valley and delta. But, high tides and sea surge during the rainy season also contribute to the flooding in the delta and coastal areas. Flooding in urban centres is caused also by such other factors as blocked drainage ditches, river channels overgrown by dense vegetation, and dumping of solid wastes in rivers and gutters (Chapter **??**).

Groundwater

The magnitude of Nigeria's groundwater reserves is not fully known. However, well and borehole investigations carried out by the U.S. Geological Survey and other interested bodies yielded valuable data. Groundwater in Nigeria is smaller in quantity and less evenly distributed than surface water but, significantly, it is available in certain areas, such as the Chad basin, which are poorly endowed with surface water and where its exploitation could be crucial to the economic development of such regions.

fing_note_floods_in_nigeria_0.pdf

[‡]https://www.acaps.org/sites/acaps/files/slides/files/20180921_acaps_briefing_note_floods_in_nigeria_0.pdf

[§]https://www.acaps.org/sites/acaps/files/slides/files/20180921_acaps_briefing_note_floods_in_nigeria_0.pdf

Basement Complex rock areas

The Basement Complex rock areas of Nigeria are comparatively poor in groundwater reserves (Table 5.4). The rocks consist of igneous and metamorphic rocks which are neither porous (porosity ranges from 1-3 per cent) nor permeable except in areas where they are cleaved, shattered, jointed or fissured.

TABLE 5.4 Water Yield over Nigeria Based on Borehole Records

Major Rock Belt	Zone	Sub-zones	Water Yield liters per second
Basement Complex	Over 60% of country		1.0 – 2.0
Sedimentary Basins	Sokoto Basin, North West		<1.0–5.0
	Chad Basin, North East	Upper Confined Aquifer	1.2–1.6
		Middle	1.5 – 2.1
		Lower	-
	Middle Niger Basin	Sandstone	0.7 – 5.0
		Alluvium	7.5 – 37.0
	Benue Basin	Sandstone Aquifers	1.0 – 8.0
	South West		-
	South East	Cretaceous Sediments	-
	South Central (Delta)	Cretaceous and Tertiary Sediments	3.0 – 7.0

Source: FAO compiled from Jim Kundell, 2008 Encyclopaedia of Earth

These weak zones in the rocks are too localized, however, to be of any great significance as reservoirs of water. Nevertheless, there is some groundwater in the Basement Complex rock area, which has served as the major source of water supply through wells and boreholes in many rural areas and housing estates in the urban centres. The major problem is how to locate the available bodies of groundwater. Up till today, the failure rate of wells and boreholes in hitting water in any appreciable amount is still high. This is in spite of the application of geophysical techniques in locating viable sites for digging the wells and boreholes.

The complex structural characteristics of the underlying Basement Complex rocks, the complex lithology of the deep weathering profile and, especially, the irregular undulating nature of the weathering front all make the siting of wells and boreholes rather fortuitous. The weathering front is irregular, occurring at various depths in different places and within short distances. The front, therefore, resembles a dome-and-basin terrain but the rises and falls do not follow the surface relief or slope profile. Because of the locally fairly high porosity and permeability of the weathered materials, some of the basins form isolated or confined aquifers. The fact that springs in these Basement Complex rock areas flow in different directions and rise at different points even within short distances is indicative of the discontinuous nature of the aquifers. Thus, in spite of the increased use of seismic techniques and other geophysical methods to map the weathering front and locate the water bearing basins, the outcomes are not guaranteed until the actual process of drilling and pumping is concluded.

Sedimentary basins

Most of the available data on groundwater yield were provided from drillings in the 1960s. Since then, emphasis has shifted to the analysis of the quality parameters of the water derived from various sources as we shall see later in this discussion.

The sedimentary rock formations contain large reserves of groundwater. The coastal belt of Nigeria is particularly well endowed and the water lies very close to the surface in some places (Table 5.5). Fortunately, only along a narrow belt near the lagoons is there any risk of saltwater contamination of groundwater sources, for example in Delta, Bayelsa and Rivers States. But, contamination by industrial and domestic effluents and oil spill is an ever-present risk in modern times, especially as a result of over-pumping of groundwater in large cities, especially, Lagos metropolis.

The Chad basin is also well-endowed with groundwater as it is built up of thick layers of marine and continental sediments, but the regional water table is deeper than in the coastal region. Furthermore, the lateritic crusts and clay pans intercalating the Chad basin sediments created numerous perched aquifers, which are easily tapped by shallow wells and this has been of great help to the local people who would have had to dig quite deep to reach the main aquifers below. Wells dug into the main aquifers to penetrate the regional water table by a few meters usually obtain adequate supplies by seepage. Since there are clean sands and gravels in the geological sequence especially at the western part of the area, wells and boreholes are high yielding.

TABLE 5.5 Depth to water and water yield (liters per hour) in wells boreholes in the Lagos region

Location	Depth to water (m)	Yield (lph)	Depth to water in borehole(m)	Drawdown (m)
Textile factory, Ikeja	-	-	23.8	-
Asbestos, Ikeja	23.2	55,915	-	5.5
Maryland	24.7	45,460	-	2.7
Ikorodu Road, Ikeja	23.2	10,460	43.6	17.7
Cocoa Factory, Ikeja	32.0	40,505	101.5	17.5
Wahum, Ikeja	45.7	16,593	73.8	7.6
Universal Regeneration, Ikeja	23.5	45,460	86.3	11.0
Pepsi Cola, Mushin	17.4	33,186	16.8	15.5
NBC, Ikorodu	6.7	11,365	11.4	1.5
Benson Residence, Ikorodu	17.1	1,391	23.4	-
Ikorodu 1	33.5	23,003	-	11.6
Ikorodu 2	24.0	41,777	88.4	3.3
Farm Settlement, Ikorodu	26.2	37,504	82.0	17.7
Agege	35.4	37,504	68.3	13.4
Govt. Farm Agege	34.1	22,730	69.5	38.4
Dairy Farm Agege	20.7	49,778	72.2	8.7
Iganmu	0.6	78,418	89.6	10.2
Afprint, Iganmu	4.0	22,730	91.1	12.2
President Clothing, Iganmu	2.2	24,094	84.9	12.2
Tailorbird, Apapa	1.8	9,547	62.5	9.1
Apapa	3.4	54,552	73.8	21.0
Badagry	2.0	39,095	76.7	0.9
Epe	32.0	45,460	-	-
Bhojson 1, Ilupeju	19.5	45,460	41.3	4.9
Bhojson 2, Ilupeju	18.3	40,914	31.3	5.5
Aswani Textile, Isolo	12.2	40,914	146.3	3.7

Note: lph = litres per hour

A general rise in the regional water table was said to have occurred in the western part of Chad basin in the 20th century as a result of subterranean augmentation of the groundwater (Carter and Barber, 1956). This led to the rise of numerous springs in the area. This was a reversal of those processes which previously caused a lowering of the water table when hardened land surfaces reduced infiltration of water into the ground.

In the Sokoto basin the principal water-bearing beds are the surface laterites, sandstones and grits in the Gwandu formation; the limestone beds in the Kalambaina formation; the sandstones in the Wurno and Taloka formations and the grits and sandstones in the Gundumi and Illo formations. In their study of the groundwater resources of the former Northern Region du Preez and Barber (1965) also identified other areas with good groundwater (Fig. 5.9), among them the Cretaceous sediments of the Middle Niger and the Upper and Lower Benue basins; the sediments of the Idah-Ankpa Plateau in Kogi State and the Kerri Kerri formation in

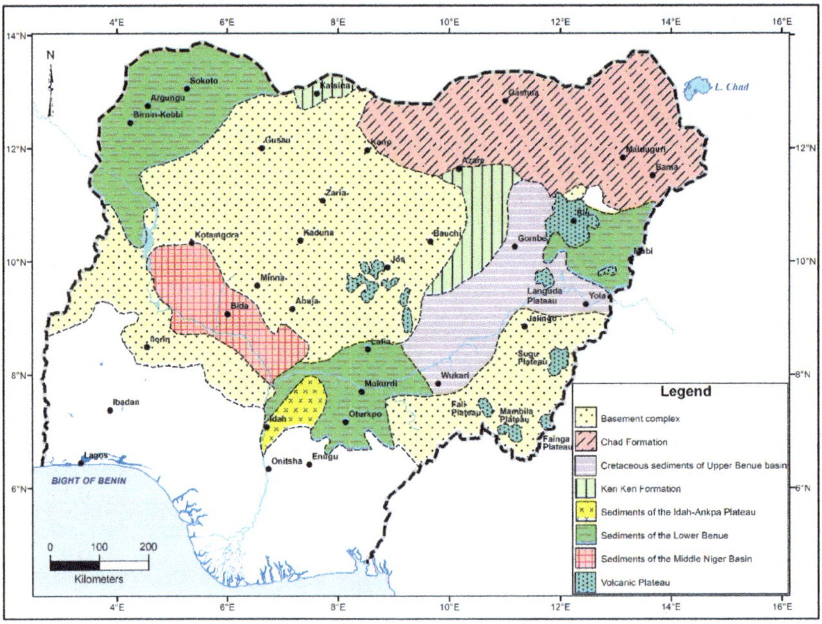

FIGURE 5.9 Groundwater areas of northern Nigeria
(Source: du Preez and Barber, 1965)

Bauchi State. The volcanic plateau areas of Biu, Jos and Mambila, among others, contain water-bearing basaltic lavas. Groundwater is also present near the surface in river alluvium. Water from alluvium is particularly important in the Basement Complex rock areas. For example, water for Kano city was pumped from the alluvium of the Challawa while Yola, Gusau, Sokoto and Zaria obtained their water from the local river alluvial deposits. Many settlements are sited near rivers because the alluvium provides easily accessible water supplies during the dry season.

Finally, springs are very important sources of water especially in the rural areas of Nigeria. Small water table springs and seeps, most of which are perennial, issue from perched or confined aquifers particularly in the Basement Complex rock areas. The seepage springs are important not only because they provide local water supplies but also because they support perennial flows in large rivers such as the Kaduna and the Gurara (du Preez and Barber, 1965).

Large springs are common in the sedimentary rock areas including Sokoto and Kerri Kerri formations. The large springs were the major sources of water in the false-bedded sand-stone areas of Idoma division of Benue State and of the Nsukka (Udi) Plateau in Enugu State. Large perennial springs also issue out of basaltic lavas. On the Jos Plateau, for example, springs with a total discharge of nearly 455,000 liters emerged from basalts near Ta Hoss station (du Preez and Barber, 1965).

The uses of water

Types of water use

Water use can conveniently be divided into two types: the *consumptive* and the *non-consumptive*. The latter includes water transport by boats and ships and the floating of logs, fishing, recreation, river sand mining and hydroelectric power generation. These activities leave the water in place and more or less unchanged, although damage to or modification of the water environment may occur sometimes. The consumptive uses of water are in agriculture, irrigation, industrial processes, and domestic use, all of which involve taking water out of the hydrologic cycle although it may be returned and recycled one way or another.

Non-consumptive uses of water

Water transport

Water transport is important on the lower Niger and the Benue rivers, in the delta region (the oil rivers) and on the lagoons, where motor-powered boats and ships operate. In numerous other rivers the use of boats may be locally important. This is especially the case on Lake Chad and many rivers in northern Nigeria. For larger ships, navigation on the rivers is periodical due to the seasonal regime of river flow, but the small dug-out boats can usually ply the major rivers and coastal creeks throughout the year. River transportation by ship seems to be suffering the worst impact of stream bed deterioration especially on the Niger, the Benue and the delta rivers. Not only are the stream channels and estuaries being silted up, but the shoals also shift about in the channels making navigation very hazardous. The formation of sand banks and islands along the rivers further restricts the shipping lanes. In the delta region the problem is being combated by dredging the water channels and outlets, as was done in the Escravos, lower Niger and Opobo channel/Imo River dredging projects. Dredging and the construction of river canals have also spread to the lagoon and creeks, a typical example being the Aiyetoro-Okitipupa canal. In the case of the Benue and the Niger the location of sand banks and shifting sands is usually charted each year, before the shipping season begins. The lagoons are navigable throughout the year for light ships and boats and provide a relatively safe, sheltered waterway for the movement of people, farm products and imported goods between the various coastal towns and states.

The use of the rivers and creeks for the rafting of logs is prevalent on the Ethiope and Cross Rivers which serve the nearby wood factories. More extensive use is made of the inland lagoons and creeks for log rafting than for shipping since these waterways are rather too shallow for big ocean ships. The waterlogged terrain of some of the timber-yielding coastal forests has made road building very difficult so that greater reliance has had to be placed on transporting logs by water. Very often the logs stay in water for quite a long time for example at the factory waterfront in Sapele and the Lagos lagoon at Ebute Metta. The tree barks get soaked and decay thereby contaminating the water.

Fishing

Fishing is, perhaps, the most widespread use made of Nigeria's surface waters. There are numerous ethnic groups whose culture and livelihoods are based on fishing. The Ijaw of the Niger delta and the Ilaje and Awori of the south-western coastal lagoons are typical. The fact that fish is estimated to account for about 60 per cent of the locally produced protein in the country (see Chapter 4) should give some idea of the magnitude and importance of this resource process. The governments of Rivers, Cross River, Akwa Ibom, Ondo, Ogun and Lagos States, for example, place great emphasis on the development of both inland and offshore fishing. The same is true of Sokoto and Borno States which control the fisheries of the Sokoto-Rima and Lake Chad basins respectively.

In Nigeria, like many coastal developing countries, fish is an important source of food for the population, which is currently estimated at 206 million people (projection based on 2006 census and annual growth rate of 2.5). As shown in Table 5.6 Nigeria now produces over 1 million tonnes of fish per year from both capture fisheries and aquaculture.

TABLE 5.6 Annual Fish Production in Nigeria, 2010 - 2018

Year	Annual Production in tonnes		
	Aquaculture	Capture	Total
2018	291 323	878 155	1,169,478
2017	296 191	916 283	1,212,474
2016	306 767	734 731	1,041,498
2015	316 727	710 331	1.027,058
2014	313 231	759 828	1,073,059
2013	278 706	721 355	1,000,061
2012	253 898	668 754	922,652
2011	200 535	635 486	836,021
2010	200 535	616 981	817,516

Source: FAO, 2007 - 2020: Fishery and Aquaculture Country Profiles: The Federal Republic of Nigeria, [accessed 19 October 2020]. `http://www.fao.org/fishery/`

However, the role of aquaculture in Nigeria is growing, as a result of increasing economic and political incentives (Bradley *et al.*, 2020). It is believed that culturing more fish species would widen the nutrient diversity to feed the growing Nigerian population. Aquaculture in general, can help meet the demand for fish domestically, as over 600,000 t of fish—primarily marine fish—are imported per year to meet the demand of Nigeria's growing population (FAO 2007-2020).

A variety of fishing gear is employed including nets, baited and unbaited hooks and traps of various types. Unfortunately, this very important and lucrative activity suffers from some human-induced environmental ailments. In some parts of the country, particularly in south-western Nigeria, local farmers use cocoa fumigants, Perenox and Gamalin 20 to catch fish. The chemicals are poured into the streams where they kill any fish that come in contact with them. The chemicals contaminate the aquatic environment and jeopardize the life, not only of the fish, but also, of other aquatic organisms. Streams so contaminated may be used by unsuspecting farmers and villagers for bathing, washing, cooking or drinking. The fish so caught are themselves contaminated and may be poisonous to eat.

In northern Nigeria, insecticides used in riverine areas to eradicate the tsetse-fly and, particularly, *simulium* (sand fly), which breed in the swift-flowing streams that descend into the Niger and Benue troughs from the north have similar adverse effects on the rivers and associated aquatic ecosystems. However, the impact may not be as devastating as in south-western Nigeria because the fumigants are not poured directly into the rivers to kill fish.

Fishing in Nigerian waters is mostly controlled at the local level where different communities have ways of enforcing the beginning and ending of the fishing season. Other than these local laws, fishing in the country is not under any strict control. There are no regulations guiding the use of fishing grounds or the quality and quantity of the catch. Small and large, young and mature, fish are caught indiscriminately; no form of management is applied to the maintenance of the fishing grounds or to the provision of optimum conditions for the spawning of the fish. However, the dumping of garbage and other biodegradable wastes can in some circumstances encourage fish spawning and feeding. Local fishermen are usually very knowledgeable about aquatic habitats and know where fish may be caught in large numbers.

In some parts of the northern states fishing is better organized under the authority of the emirs and village heads. In particular, there are fixed periods for fishing during the year. The Argungu fishing festival, for example, marks the beginning of the season on the Sokoto River in that district.

Except in a few favoured locations such as Lake Chad, Lake Kainji and, later, Lagos lagoon, fishing on the rivers, creeks and lagoons has enjoyed very little government attention in form of detailed biological surveys of the water resources and the environment or through the introduction of appropriate management techniques. However, most governments now seem to be placing greater emphasis on freshwater fisheries, especially since so many dam reservoirs have been constructed. Furthermore, the river basin development authorities and the Federal Department of Fisheries have been promoting modern approaches to fisheries and fishing.

Unfortunately, in the coastal region, fishing as a major means of livelihood among the rural grassroots inhabitants is under serious threat. The mangrove swamps that constitute the main spawning grounds for fish are now fragmented due to sand filling for housing and industrial layouts and road construction. The mangrove trees are heavily exploited for firewood by both farmers and fishermen for cooking and fish drying. Oil exploration and exploitation activities have destroyed the spawning grounds for fish and depleted biodiversity. According to Oribabhor and Udo (2011) oil spills and development projects have led to large mortalities of invertebrates and fishes, total decimation of shell fish, polychaete worms and crustaceans in the mangrove areas. In addition, as stated in Chapter 1, the aggressive exotic invasive Nypa palm has replaced the mangrove species in many areas. In spite of the rich resources of crude oil and natural gas and the once rich fishing resources both onshore and offshore, the River Niger delta region is characterized by high levels of poverty among the rural, grassroots communities. Consequently, Oribabhor and Udo (2011) have called for concerted efforts both locally and at the international level to put an end to the destruction of mangroves in order to sustain biodiversity and the livelihoods of the rural folks.

River sand mining

Sand mining in the rivers and creeks is a major economic activity throughout Nigeria but it is perhaps most organized and commercialized in the Lagos region (Faniran, Sada and Areola, 1976). The demand for river sand for the building industry is high in the Lagos area. The reasons for the popularity of river sand are many but perhaps the most important is its purity. Unlike beach sand, which may contain shells and sand grains of various sizes the river sediments being exploited in the Lagos area are pure fine sands of uniform grade. The sediments are mineralogically homogeneous being mostly quartz. Although it is not as white as beach sand, the grey to black colour is due primarily to organic matter colouration, which does not affect the quality in any way. Table 5.7 gives some idea of the quantity of sand removed from the coastal rivers and creeks near Lagos.

Sand mining in the rivers and creeks is advantageous in the sense that it helps to prevent silting in these important natural drainage lines which are crucial in any agricultural development programme on the coastal river flood plains and swamps (Tahal, 1974). On the other hand, sand mining is destroying the habitat of freshwater fish; there is no restriction whatsoever on which rivers or creeks the sand miners may operate or where along the water courses they may dig. Although interviews conducted in the field did not reveal any apparent conflict between sand miners and fishermen - indeed some fishermen are also sand miners - investigations by the Lagos State Fisheries Division as early as the mid-1970s, revealed that the fish population had been adversely affected by sand mining. There is clearly a need to strike a just balance between these two interests. Some of the water courses may have to be protected and developed as fishing grounds.

TABLE 5.7 Truckloads of river sand as counted along Lagos-Ikorodu Road

Day	Dry Season		Wet Season	
	No. of Truckloads	Tonnes	No. of Truckloads	Tonnes
1	721	3,662.7	150	762.0
2	789	4,008.1	142	721.4
3	543	2,758.4	123	624.8
4	747	3,794.8	154	782.3
5	544	2,763.5	137	696.0
6	758	3,850.6	147	764.8
7	705	3,581.4	112	569.0
8	821	4,170.7	133	675.6
9	784	3,982.7	180	914.4
10	909	4,617.7	140	711.2
Total	7,321	37,190.7	1,418	7,203.4

Source: Faniran, A., Sada P. and O. Areola (1976)

River sand mining is not restricted to the Lagos region alone. It takes place on a large scale near all the major urban centres especially since the boom in the building construction industry in the 1970s during the military era. Many valley bottom lands are dug up and the flow of the streams disrupted. Runoff from the valley sides sometimes collects in the ditches made by the sand miners instead of flowing into the stream. The problem of the quarrying of earth materials in general is discussed more fully in Chapter 8 on Mining.

Water-based tourism

It has been observed that tourism in Nigeria so far has centred on events (festivals) due to the country's diversity of ethnic groups (Wikipedia, 2020). Some of the events are water-based such as the Argungu Fishing festival on River Sokoto in Sokoto State and the boat regattas common in the coastal states of Rivers, Akwa Ibom and Cross River. But, gradually, tourism is moving more and more towards outdoor recreation based on available natural resources in the various ecological zones. Water based tourism is highly favoured by the wide availability of both surface and ground water bodies with their associated land forms and features including beaches, protected bays, sandbars, the ocean, lagoons, creeks in the coastal region; lakes, rock pools, waterfalls and other picturesque landscapes along the rivers in the hinterland; and warm springs and salt brines or springs from the ground in certain locations.

In the coastal marine and wetland zones, the available resources for water-based tourism include

(i) *beaches* (e.g. Ibeno sand beach in Akwa Ibom State, reputed to be the longest along the West Africa coast; Lagos Bar beach, Lekki beach, Badagry beach, Elegushi, Alpha, all in the Lagos region); and undeveloped, remote beaches along the coasts of Delta, Bayelsa and Rivers States;

(ii) *estuaries* where calmer waters favour recreational activities in Lagos, Benin, Escravos, Forcados, Bonny, Brass, Cross, and Bakassi Peninsular river channels;

(iii) *lagoons*, such as Lagos, Lekki, Yewa, Epe, Mahin, Ologe, Badagry, Imo, Opobo and numerous others in Rivers, Akwa Ibom and Cross River states;

(iv) *wetlands*, that is, mangrove and freshwater swamps used for bird watching and nature reserves and wildlife sanctuaries that attract both researchers and tourists

In the interior of the country, the available resources for water-based tourism include

(i) picturesque *waterfalls* along the rivers, for example, Gurara falls on the Gurara river close to the Federal Capital Territory; Ngell river falls, Farin ruwa falls and other falls along the rivers flowing down from the Jos Plateau into the Niger and Benue troughs; Kwa falls, Agbokim Falls in Cross River State along rivers flowing down from the Oban Hills; Erin Ijesha falls in Osun State; Owu falls in Kwara State

(ii) *dams* built along former waterfalls, including Kainji, Jebba, Shiroro Gorge, Zungeru and Mambila;

(iii) *wetlands* along the river valleys noted for bird watching and/or game viewing e.g. Hadejia-Nguru Birdlife Project and Nguru Lake/Marma channel complex in the Hadejia river-Nguru wetlands in Yobe/Jigawa state; the numerous Game Reserves in the northern Sudan savanna zone e.g. River Gaji in Yankari National Park; Muri Plains and other wetlands along the Benue river in Adamawa, Benue and Nassarawa States; Ngel-Nyaki Forest and the Donga River Basin Forests in Adamawa State close to the Cameroon border; and the Ebok-Kabaken Swallow Roost and the Itu Wetlands and Swallow Roost in Cross River State

(iv) *Inland lakes*, including Lake Chad, Oguta Lake in Imo State, the suspended or hilltop lake or pool at Ado-Awaye in Oyo State, and numerous man-made lakes used mainly for sight-seeing; and

(v) *groundwater-based tourist attractions* include the Ikogosi Warm Springs in Ekiti State; the Lamurde Warm Spring in Adamawa State; the Warm Pool inside Yankari Game Reserve in Bauchi State; the salt springs (brines) in the Olachor-Abachor area (lower Benue River valley) and Awe-Azara-Keana area (middle Benue River valley)

In addition to the places listed above with opportunities for water-based tourism, there are spectacular landscapes, areas of gully erosion and badland topography produced by fluvial geomorphic processes, which now attract tourists on sight-seeing. They include the Nsude – Nsukka badlands in the Enugu -Nsukka scarp lands in Enugu State, the gullies on the Jos, Biu and Mambila plateaus, the gullied landscapes in the coastal ridge lands of the country found in Ikorodu, Lagos State; Irele, Ondo State; and Uyo, Akwa Ibom State. The areas of extreme accelerated soil erosion and badland topography are underlain by Coastal Plains Sands and Sandstones.

Holiday resorts have been established in the large urban centres by both government and private entrepreneurs to take advantage of available tourist resources to promote tourism, particularly in the coastal areas. Uduma-Olugu and Onukwube (2012) argued that the desire of Nigeria to diversify the revenue base and transform into a developed economy has made government to turn to tourism. According to them, statistics indicate growing financial gains that could be realized from exploiting the untapped wealth of coastal tourism. Lagos, Eku, Yenagoa and Calabar are examples of places where holiday resorts have been built in recent years. The Lagos megalopolis, as to be expected, has more than any other region in the country. In the hinterland of the country, holiday resorts can be found in Abeokuta, Abuja, Benin City, Eku (near Jebba), Enugu, Abakaliki, Obudu Ranch, Sokoto, Yankari National Park, Kainji National Park and Ikogosi Warm Spring to name a few. The resorts vary widely in quality of service and some are more accessible than others

The environmental challenges faced in the development and use of these water-based tourism sites include: (i) contamination by oil and grease along the coastal beaches from passing ocean traffic; (ii) oil pollution of waterways and wetlands from pipe leaks and sabotage of pipelines and oil installations; (iii) water hyacinth invasion of coastal creeks and lagoons that affect water transportation, and Nypa palm invasion of mangrove swamps; (iv) sand mining, industrial pollution and waste dumping in lagoons, creeks, as well as, rivers in hinterland areas (Ujih *et al.* (2014); and (v) siltation of river channels, river estuaries and man-made lakes invaded by aquatic weeds as a result of poor management of the catchment areas.

The development of outdoor recreation in Nigeria among the people is still at a low level and it has fluctuated with the frequent changes in government at the national, state and local levels. It is not surprising then that recreational use of both natural and man-made water bodies in Nigeria is still largely rudimentary. Local inhabitants, especially the young ones, fish as a hobby; swimming also attracts young people to the streams and pools. The dam reservoirs found in different parts of the country may be regarded as visual amenities for tourists and local sightseers; but for health reasons, they have not been opened up either for boating, fishing or any other recreational activities. Most of the reservoirs supply water for domestic use in which case swimming is not allowed.

The use of these water bodies for recreation has had little impact on the resource itself: most people visit the sites on day trips; little or no camping goes on and picnicking is restricted to public holidays. Many dam reservoirs and scenic spots are rather remote from the large urban centres from which most of the internally generated tourist populations come. Rather, recreational parks, some with man-made amenity pools and water channels, are being built in metropolitan areas across the country by both governments and private establishments and entrepreneurs. These playgrounds are favorite recreation sites at weekends and during public holidays.

Water power

In 2013, there were 264 operating dams in Nigeria while another 30 were under construction (FMWR, 2013). According to FAO (2016), the larger dams in the country had a cumulated capacity estimated at 50 667 million m^3. The major dams, Jebba, Jekara and Kainji, are devoted primarily to hydro-electric power generation. There are smaller hydropower stations, for example, on the Jos Plateau which were constructed during the colonial era to service the tin mining operations. The generation of hydroelectric power is basically a non-consumptive use of water, but the construction of dams to create reservoirs usually entails intensive modification of the ecosystem involving a change from a terrestrial to an aquatic environment on the inundated lands. Beauty spots such as rapids and waterfalls are often submerged and their recreational value lost. However, the Kainji Lake and surrounding lands have been constituted a national park and a major tourist centre.

Consumptive uses of water

The consumptive uses of water are those that affect the people's health, nutritional standards and social welfare the most. Many diseases are contracted by local people through drinking, swimming or washing in contaminated water. All water bodies harbour certain insects and facilitate their breeding; many of these are disease vectors which affect both animals and human beings.

Available water resources

Table 5.8 shows long-term estimates of Nigeria's renewable water resources by the FAO (2016). The long-term average internal renewable surface water resources are estimated at 214,000 million m^3 per year. The long-term average renewable

TABLE 5.8 Nigeria's available water resources (after FAO, 2016)

Renewable freshwater resources	Year	Amount	Unit of measurement
Precipitation (Long-term average)	-	1,150	mm/year
	-	1,062,000	million m^3/year
Internal renewable water resources (Long-term average)	-	221,000	million m^3/year
Total renewable water resources	-	286,200	million m^3/year
Dependency ratio		23	%
Total renewable water resources per inhabitant	2015	1,571	m^3/year
Total dam capacity	2015	50,667	million m^3

Source: FAO AQUASTAT FAO's Water Information System Nigeria: Water Resources `http://www.fao.org/nr/water/aquastat/countries_regions/Profile_segments/NGA-WR_eng.stm`

groundwater resources stand at 87,000 million m^3 per year out of which 80,000 million m^3 per year constitute overlap between surface water and groundwater leaving 7,000 million m^3per year as net groundwater resources. Therefore, Nigeria's total long-term average internal renewable water resources amount to 221,000 million m^3 per year. However, Nigeria also receives some water from rivers flowing in from neighbouring countries while it also exports water to them through rivers flowing out of the country, including Rivers Maradi and Komadougou Yobe. Thus, the long-term average annual total renewable resources of Nigeria are 286,200 million m^3 per year or 1,571 m^3 per year per capita.

Domestic water supplies

The vast majority of Nigerians, particularly those who live in the rural areas, obtain water for drinking and other domestic purposes from springs, rivulets, streams and shallow wells. Some of these are fairly clean and sediment free, especially the perennial seepage springs, but during the late dry season as the water level falls in streams and rivulets the waters become muddy and may be heavily contaminated. The fact that many regular sources of water dry up during this season forces people to make use of less clean and less safe sources.

Piped water supply schemes mainly serve the large urban centres but at present these schemes are being extended to rural centres. Most of the water is obtained through pumping and gravity flow from rivers and from dam reservoirs into sedimentation tanks and is either fully or partially treated before being distributed. There are also a number of boreholes from which water is taken and delivered, untreated, to consumers who then devise different means of purifying the water for use. Such boreholes are in common use in industrial establishments in Lagos metropolis and Epe in Lagos State; and in several towns and rural communities in the northern states. The quality of the water from these various sources is discussed later in this chapter.

The creation of reservoirs raises some social and environmental issues. The artificial lakes inundate large areas of normally good agricultural or forest land without any protest from the local population whose indifference can usually be traced to their political naivety and low socio-economic status. For the most part, local people welcome any form of development project in their area without thinking of the long-term losses or environmental consequences so long as it brings tangible short-term benefits. Often, they are only too glad if they can persuade the government to site some development projects in their area even if this means giving up some of their best forest or agricultural lands.

It is doubtful, however, that such a carefree attitude on the part of the people will continue for long. Acute land shortages in certain areas and the fact that people living in the immediate vicinities of development projects have sometimes been totally forgotten in the distribution of the proceeds and benefits are making local populations more circumspect about releasing their land for government projects. For example, for a long time Ikire town in Osun State was not supplied with piped water from the Asejire Water Scheme built on its land. The growing level of awareness on the part of the local inhabitants is a healthy development and should ensure that officials are more conscientious in weighing the advantages and disadvantages of competing land uses before deciding on any particular type of development project. The economic benefit of flooding good land to create reservoirs for hydro-electric power generation, for example, is controversial. Such dams as the Tiga Dam in Kano State have covered up many hectares of good arable farmland.

Dam reservoirs are breeding grounds for mosquitoes and other insect pests. In this regard, the remote locations of many reservoirs are an advantage; the few that are sited close to settlements are a menace. Weeds growing on the water surfaces also facilitate the breeding of swarms of mosquitoes. There is a limit to the safe use of pesticides which can themselves contaminate the water. In the end the best protection against mosquitoes are door and window screens and the use of treated mosquito nets indoors. In addition, it is important that in future dam reservoirs are located where they will not constitute health hazards to the population.

Finally, the storage of water behind a dam often has repercussions on the hydrology and ecology of the river downstream. The reduced and irregular river flow downstream of dams disturbs fish and other aquatic life forms, especially those which need to spawn or breed in deep water. Some organisms are deprived of the silt and nutrient elements which get held up behind the dams. Fishing in the rivers downstream may eventually become very unproductive, although if the dam reservoirs themselves are developed as fisheries that may help to compensate local fishermen. Nevertheless, some people who live by fishing downstream of the dams are deprived of their livelihoods. At Kainji the problem has been solved by resettling the displaced fishing communities near the lake so that they can continue to engage in their traditional occupation.

Water use in agriculture and animal husbandry is important in many parts of the country and it is of major concern in the sub-humid parts of the country. It is discussed in detail in Chapter 6.

Water supply and sanitation

The optimism generated by the return to democratic civilian rule in 1999 after three-and-a-half decades of military dictatorship, led to the commissioning of a World Bank study of the status of water supply and sanitation in Nigeria in 2000. The World Bank (2000) noted the centrality of water supply and sanitation to Nigeria's economic development because of the linkages between them and a cluster of key sectors, including health, education, agriculture, and environment. The World Bank study carried out in conjunction with the Federal Ministry of Water Resources painted a very dismal picture of the status of water supply and sanitation in the country. Table 5.9 gives a summary of key indicators of the status of water supply and sanitation in Nigeria compiled from the World Bank study and other sources. Arguing that water supply and sanitation was the single sector investment with high and most diverse multi-sector returns, it should be the centrepiece of any lending portfolio. Thereafter the World Bank funded many water supply schemes in the country aimed at providing higher quality water and sanitation facilities to a greater proportion of the country's population in the cities and small towns.

Tables 5.10 and 5.11 give us a clear picture of the state of water supply and sanitation as at the time of the national census of 2006. The situation has not improved that much since then.

Table 5.10 shows that only about 5 per cent of households had piped water inside the homes; another 6 per cent had access to public standpipes outside. Even, in the Federal Capital Territory, Abuja, only 15.8 per cent of households had piped water inside their premises while another 5.2 per cent had access to public standpipes. The proportions of households in the states with piped water within their dwellings ranged from 1 per cent in Taraba State to 11 per cent in Kaduna State.

The proportion of households with access to public standpipes outside their premises ranged from 1 per cent in Taraba State to 15.8 per cent in Ogun State. Most households throughout the country obtain water from wells (national average 35.8%) and boreholes (national average 13.6%). Another 20 per cent obtain domestic water supply from rivers and springs. There is greater dependence on wells and boreholes for domestic water supplies in the northern savanna areas than in the humid south.

With regard to sanitation, Table 5.11 based on the census in 2006, very much reflects the situation in the country even today. The table shows clearly that the sanitation system in Nigeria is still rather primitive. Nationally, in 2006 only approximately 15 per cent of households had the water closet type toilets and most of these were connected to individual septic tanks that have to be evacuated once

TABLE 5.9 Status of Water Supply and Sanitation in Nigeria, 2000

Indicator			
Population distribution pattern	Towns >20000		38%
	Small towns 5000 -20,000		33%
	Rural <5000		29%
Access to improved water source	Overall		54% (2010)*
	Urban		50%
	Semi-urban		20%
	Overall urban average		30%
	Rural		35%
Sample water use patterns in selected small towns	Spring/stream		32%
	Hand dug well with apron		30%
	Hand dug well (w/out/apron)		27%
	Rain		20%
	River		16%
	Borehole		14%
Access to improved sanitation			35% (2010)*
Sample of Sanitation Patterns in Selected Small Towns	Existence of latrines	Private latrines	56%
		Public latrines	11%
	Disposal facility in household	Traditional pit latrine	54%
		Open field	39%
		VIP pit latrine	3%
		Flush toilet	2%
		River	1%
		Other	1%
Sewer system	Only available in Abuja & Lagos		

Source: Extracted from The World Bank 2000 ***Water Supply & Sanitation Interim Strategy***

Note: November 2000. Federal Ministry of Water Resources

*WHO / UNICEF Joint Monitoring Programme (2010) for Water Supply and Sanitation (from Wikipedia)

TABLE 5.10 Main sources of domestic water supply in households, Nigeria 2006 *DPLDP*: Dugout/Pond/Lake/Dam/Pool

	State	Total	Main sources of domestic water supply (% total no. of households)								
			Pipe-borne inside dwelling	Pipe-borne outside dwelling	Tanker supply/ Water Vendor	Well	Borehole	Rain Water	River/ Stream / Spring	*DPLDP*	Other
1	Abia	605,987	2.3	8.7	3.2	2.2	40.0	3.5	38.8	0.4	0.9
2	Adamawa	576,348	2.4	2.4	7.8	33.0	12.5	6.5	28.8	1.9	4.8
3	Akwa Ibom	857,436	3.4	5.2	1.4	4.3	27.4	4.3	48.5	1.1	4.3
4	Anambra	882,875	3.6	6.1	8.7	10.8	19.4	10.4	28.0	3.5	9.5
5	Bauchi	847,731	3.7	2.8	2.5	58.9	7.1	12.4	9.3	0.8	2.5
6	Bayelsa	352,025	3.3	9.4	4.4	13.0	7.0	3.5	53.1	4.2	2.0
7	Benue	801,833	1.3	1.2	2.9	34.7	3.1	10.8	37.0	1.5	7.5
8	Borno	787,274	4.6	5.7	12.9	45.5	16.3	6.8	5.8	1.5	1.0
9	Cross River	645,251	2.0	4.1	1.0	6.7	18.1	1.8	61.2	1.4	3.7
10	Delta	890,312	3.4	5.5	3.7	35.4	12.1	15.9	16.7	1.7	5.6
11	Ebonyi	449,709	2.2	2.4	1.4	15.8	16.8	7.3	32.7	10.4	11.0
12	Edo	701,073	5.1	8.6	9.9	18.8	21.3	8.1	24.0	1.5	2.7
13	Ekiti	493,739	7.3	10.0	1.7	43.4	4.7	5.9	23.4	0.9	2.9
14	Enugu	725,767	4.3	7.4	12.8	14.8	5.8	12.5	27.9	3.1	11.3
15	Gombe	419,226	3.3	4.2	7.3	47.7	9.7	7.0	16.3	1.7	2.7
16	Imo	837,195	3.9	6.3	5.1	10.4	20.2	13.7	29.7	3.9	6.7
17	Jigawa	810,310	6.0	5.3	3.4	29.6	29.0	16.1	9.9	0.5	0.3
18	Kaduna	1,115,974	11.1	3.9	2.3	60.2	2.8	5.7	11.8	0.4	1.8
19	Kano	1,603,335	5.6	3.7	7.1	48.1	10.1	18.8	5.0	0.9	0.7
20	Katsina	1,066,316	5.5	3.4	3.9	51.1	5.8	20.5	6.3	1.5	2.0
21	Kebbi	562,827	3.5	2.2	2.1	61.0	5.0	16.4	7.4	0.6	1.9
22	Kogi	641,556	3.2	3.8	14.7	27.4	7.5	4.1	35.2	1.0	3.2
23	Kwara	468,780	7.2	8.2	2.1	36.0	15.7	7.1	19.2	0.8	3.6
24	Lagos	2,195,842	9.8	15.3	14.9	33.8	21.9	1.6	1.6	0.1	1.0
25	Nassarawa	342,711	4.6	4.4	4.8	33.8	6.5	4.7	36.5	1.0	3.7
26	Niger	729,964	7.6	5.3	3.4	36.7	11.1	4.3	27.3	0.9	3.4
27	Ogun	880,970	5.6	15.6	8.8	26.1	23.4	2.1	16.8	0.4	1.2
28	Ondo	763,020	1.8	3.2	1.7	44.4	10.0	5.5	30.2	1.1	2.2
29	Osun	730,313	7.1	9.1	1.3	46.6	5.0	6.7	21.0	0.6	2.5
30	Oyo	1,248,105	2.8	4.1	2.3	55.7	6.9	8.3	16.4	0.8	2.7
31	Plateau	604,491	4.8	3.2	2.7	43.9	2.9	5.0	30.5	2.8	4.2
32	Rivers	1,123,998	6.2	12.1	3.9	35.8	26.9	2.7	11.1	0.7	0.6
33	Sokoto	688,710	7.0	4.5	3.1	65.9	3.5	12.4	2.7	0.4	0.5
34	Taraba	431,385	1.0	1.0	4.9	27.5	6.0	7.3	40.2	1.8	10.3
35	Yobe	418,999	3.6	5.2	6.2	55.6	12.5	10.9	4.1	1.3	0.7
36	Zamfara	592,106	3.1	2.1	3.6	38.3	10.3	16.5	19.3	1.4	5.5
37	FCT Abuja	303,592	15.8	5.2	27.0	17.9	14.8	1.5	16.2	0.3	1.3
	Nigeria	28,197,085	5.1	6.3	5.9	35.8	13.6	8.6	20.1	1.4	3.3

Source: Proportions calculated from Nigeria Census Board: *Priority Tables Volume 1 Update, 2006 Census*

TABLE 5.11 Types of Toilet Facility in households, Nigeria, 2006

State		Total	Type of toilet facility (% of total no. of households)						
			Water closet	Pit Latrine	Bucket/ Pan	Toilet Facility in another (different) Dwelling	Public Toilet	Nearby (bush/beach /field)	Other
1	Abia	605,987	27.9	52.0	3.1	0.7	12.8	3.4	0.2
2	Adamawa	576,348	5.1	51.8	2.9	1.8	8.4	29.5	0.4
3	Akwa Ibom	857,436	9.8	74.4	3.9	1.1	6.9	3.8	0.2
4	Anambra	882,875	23.8	43.4	9.4	3.8	12.1	7.3	0.2
5	Bauchi	847,731	3.6	64.5	3.3	4.0	8.9	15.3	0.3
6	Bayelsa	352,025	12.9	6.3	1.2	1.9	37.2	37.7	2.8
7	Benue	801,833	6.5	45.6	1.3	1.2	10.3	34.8	0.3
8	Borno	787,274	6.5	63.5	3.8	2.4	6.1	17.2	0.5
9	Cross River	645,251	11.4	41.9	1.5	1.0	20.1	23.6	0.5
10	Delta	890,312	19.0	44.1	5.6	2.2	10.9	16.6	1.5
11	Ebonyi	449,709	5.6	33.0	7.5	1.7	25.4	26.3	0.5
12	Edo	701,073	22.3	50.3	4.4	1.7	4.1	17.0	0.2
13	Ekiti	493,739	12.9	44.4	1.8	1.1	7.2	32.5	0.2
14	Enugu	725,767	18.7	30.7	2.2	3.2	14.8	30.1	0.3
15	Gombe	419,226	4.6	60.4	3.3	3.6	10.0	17.8	0.2
16	Imo	837,195	14.9	56.8	9.5	4.1	9.1	5.3	0.3
17	Jigawa	810,310	4.5	58.5	12.3	5.1	4.7	14.4	0.5
18	Kaduna	1,115,974	10.8	62.8	2.0	3.1	8.9	12.0	0.4
19	Kano	1,603,335	9.1	73.2	6.5	4.2	4.3	2.4	0.4
20	Katsina	1,066,316	4.7	72.7	4.9	5.3	7.5	4.5	0.4
21	Kebbi	562,827	3.9	61.1	3.4	4.0	8.1	19.1	0.5
22	Kogi	641,556	8.3	34.1	1.5	1.3	7.9	46.5	0.3
23	Kwara	468,780	13.6	29.9	3.2	1.5	9.7	41.7	0.4
24	Lagos	2,195,842	49.7	39.7	2.1	0.5	2.1	5.5	0.3
25	Nassarawa	342,711	7.5	48.4	1.8	2.0	10.1	29.8	0.3
26	Niger	729,964	8.1	41.8	2.4	1.9	8.5	36.9	0.4
27	Ogun	880,970	19.8	54.8	1.7	0.8	2.1	20.6	0.2
28	Ondo	763,020	11.6	42.8	1.5	1.0	6.3	36.5	0.3
29	Osun	730,313	14.0	41.6	2.0	1.3	7.6	33.1	0.3
30	Oyo	1,248,105	18.9	33.6	2.3	1.0	5.7	38.1	0.4
31	Plateau	604,491	9.1	37.4	0.9	1.2	6.7	44.4	0.3
32	Rivers	1,123,998	26.8	27.8	2.4	1.9	23.9	16.2	1.0
33	Sokoto	688,710	5.1	57.5	3.5	5.3	6.6	21.5	0.6
34	Taraba	431,385	3.6	44.1	2.2	2.1	14.2	33.5	0.4
35	Yobe	418,999	4.6	49.2	4.4	3.4	7.0	30.8	0.6
36	Zamfara	592,106	4.2	60.2	6.4	6.3	15.3	6.9	0.6
37	FCT Abuja	303,592	44.4	29.5	1.6	0.7	4.2	19.5	0.2
	Nigeria	**28,197,085**	**15.2**	**49.2**	**3.7**	**2.4**	**9.1**	**19.8**	**0.5**

Source: Proportions calculated from Nigeria Census Board: *Priority Tables Volume 1 Update, 2006 Census* in a while. Approximately 49 per cent of households still made use of pit latrines while nearly 20 per cent engaged in open defecation in nearby bushes and other available spaces. It was only in Lagos State (49.7%) and FCT Abuja (44.4%), that the proportion of households with water closets was over 40 per cent. The next category comprised of states with over 20 per cent coverage including Rivers (26.8%), Abia (27.9), Rivers (26.8%), Anambra (23.8%) and Edo (22.3%) states. Nineteen states, that is over half of the states in the country, had less than 10 per cent of the households provided with the water closet type toilet facility. The proportion of households making use of nearby bushes for open defecation was over 40 per cent in some states.

Water quality

Many factors determine the quality of water including its chemical composition, the presence or absence of pathogenic organisms, the sediment load and the level of contamination by pollutants from both natural and anthropogenic sources.

Sediments

Seepage springs, from which a large number of Nigerians obtain water, are largely free of sediments either in suspension or in solution. This particularly applies during the dry season. The sediment content may increase somewhat during the heavy rains because the spring pools may be contaminated by runoff from adjacent slopes which brings in not only sediments but also organic materials. The seepage water itself may have more dissolved elements in it at this period. Wells, especially shallow wells, behave in like manner. There is an increase in the sediment content after every downpour of rain so that the water turns brown or muddy. But, with time the sediments settle and the water is clear again. However, some proportion may remain in solution for a long time especially where the well is dug in clayey materials. Rivers and streams are least free of sediments as much of the water comes from surface runoff. Except in the transitional period between the end of the wet and the beginning of the dry seasons, river water is hardly ever clear. The exceptions are small rivers flowing through forests which are relatively pure because they are fed mainly by interflow.

Organic contamination

Apart from sediments, rivers and some springs and. pools, especially those issuing from swampy valley heads in the derived savanna zone, are contaminated by natural pollutants including decayed leaves, twigs and tree trunks; and plant root exudates. Some pools at spring heads are coloured reddish brown or orange by root exudates and decayed vegetation from the waterlogged surroundings. Although the organic pollutants can affect the taste of the water, people generally get used to the flavour of their local drinking water and probably only notice the difference when they drink water from other sources. The people somehow feel that their health cannot be seriously jeopardized. Indeed, in many areas, the vegetation around the springs and pools is preserved to provide shade and thereby keep the water cool and clear of dust.

Pathogens in water

Most surface water bodies harbour numerous worms, snails, insects and their eggs and larvae. These are serious health hazards. Whilst the water from a seepage spring may itself be totally free of pathogenic organisms, it gets contaminated at the pool at its outlet from where people fill their cans and pots. Shallow wells are also easily infested by disease vectors especially where a well is not properly maintained or covered when not in use. Mosses, lichens and ferns growing on the walls of some ill-kept wells provide suitable habitats for worms and insects to breed.

The most dangerous pathogenic organisms found in water include the insect group, Diptera represented by various breeds of mosquitoes and the tsetse fly; aquatic snails, which harbor the blood flukes that cause schistosomosis (bilharzia); simulium (sand fly), the insect vector of river blindness disease; and protozoans and metazoans referred to collectively as worms. Furthermore, there are numerous viruses, fungi and bacteria which thrive in or near water bodies (see Chapter 7). Suffice it to say here that many, if not the majority of the, diseases in the rural areas of Nigeria, are traceable to water supply sources, which explains the importance given by the World Bank to improved water supplies in economic development in Nigeria.

Domestic piped water supplies place a premium on the removal of pathogenic organisms before the water is distributed to the general public. Unfortunately, the containers in which some people carry and store the water from the taps are most unclean and contain coliform. Very little routine examination of public water supplies is carried out, yet this is very necessary in order to ensure that the pathogenic quality of the water is maintained. In a few areas the local health office sends water samples for analysis at the state central laboratory but this is at very irregular intervals.

Chemical characteristics

The situation with regards to routine chemical analysis of water is even worse. Very often the only records on the chemical characteristics of public water supplies which the local or state water and health authorities have for their areas are those contained in the reports of the consultants prepared prior to the establishment of the schemes. For each consumptive use of water there are standards, or tolerance limits, regarding the mineral composition of the water. Unfortunately, there is little data on the chemical composition of the untreated water which most Nigerians drink. Table 5.12 is an example of the limited range of chemical parameters that were analysed for the untreated waters of a few rivers in the country in the past. The data were deficient in that many other chemical elements which are normally considered in judging the chemical quality of water were left out. The situation improved over the years (Table 5.13) and we know more now about the physico-chemical and biological health of the water taken by the average Nigerian than in the past. At least the water supply agencies strive to ensure that public water supplies meet the minimum standards stipulated by the WHO International standards for drinking water, 1958 but not the modern updates.

The level of dissolved salts is comparatively low. This is not surprising given the nature of the underlying rocks which are either old Basement Complex crystalline rocks or sandstones. However, the salt content varies with the seasons. As to be expected, salt concentration falls during the rainy season when the rivers are diluted by heavy rains and water from tributaries. River water in Nigeria is largely

TABLE 5.12 Mean analytical data on the chemical characteristics of some rivers in northern Nigeria (1960s)

River	**Milli-equivalents per litre**						
	Na^+	K^+	Ca^{2+} plus Mg^{2+}	HCO_3	Cl^-	**pH**	**Sample size**
Oyi (near Share)	0.20	0.08	0.53	0.8	0.03	6.8	8
Kuyi (at Pategi)	0.19	0.09	0.57	0.8	0.03	6.0	9
Gbako (at Badeggi)	0.15	0.08	0.39	0.6	0.02	6.8	8
Taraba	0.14	0.03	0.26	0.4	0.02	6.8	2
Kaduna	0.11	0.08	0.30	0.5	0.01	6.7	9
Panyam	0.15	0.08	0.42	0.6	0.03	6.7	4
Kano	0.25	0.06	0.45	0.7	0.05	7.5	3
Challawa	0.20	.11	0.64	0.9	0.10	7.4	2
-Hadejia	0.15	0.10	0.49	0.7	0.05	7.0	4
Yobe	0.14	0.07	0.44	0.7	0.03	6.7	5
Yedseram	0.39	0.07	0.61	1.00	0.02	7.5	10

Source: Extracted from Bennet, A.J. (1967)

TABLE 5.13 Major chemical characteristics of partially treated water in SW States (1970s)

Characteristics	Odogbolu Borehole Ondo State	Aiyetoro Borehole Ogun State	Oke-Igbo chlorinated water Ondo State	Lalupon chlorinated water Oyo State	Ikogosi warm spring Ekiti State	WHO Limits (1958)
Temperature °C	-	28.0	-	28.0	34.0	-
pH	5.9	6.3	6.6	7.3	6.25	6.5 – 9.2
Colour (Hazen units)	0.0	0.0	110.0	27.0	0.0	50.0
Residual chlorine mg/l	0.1	0.01	0.0	0.06	0.0	-
Turbidity mg/l	0.0	0.0	43.0	13.0	0.0	25.0
Copper (total) mg/l	0.01	0.03	0.0	0.0	0.01	0.05 – 1.5
Iron (total) mg/l	0.3	0.01	2.5	0.22	0.02	0.1 – 1.0
Manganese mg/l	0.01	0.25	0.8	0.0	0.0	0.05 - 0.5
Nitrates mg/l	6.6	7.9	5.3	6.6	6.6	45.0
Fluorides mg/l	1.4	0.4	1.76	1.9	0.7	0.7 - 1.0
Sulphates mg/l	1.0	0.5	0.0	1.5	1.0	200 – 400
Phosphate mg/l	0.05	0.10	0.32	0.16	0.22	-
Acidity mg/l	20.0	35.0	20.0	13.0	25.0	-
Alkalinity mg/l	25.0	50.0	70.0	50.0	-	-
Chloride mg/l	8.0	10.0	10.0	7.5	2.0	200 – 600
Hardness (Ca) mg/l	10.0	10.0	-	35.0	40.0	75 – 200
Hardness (Mg) mg/l	4.0	1.0	-	17.0	22.0	25 – 300
Hardness (total) mg/l	55.0	20.0	-	52.0	62.0	100 – 500

deficient in fluoride and this usually requires that domestic water supplies be chlorinated before distribution. Chlorination is considered necessary for the prevention of such human diseases as dental caries and fluorosis. Moreover, the water from most Nigerian rivers is slightly acidic which, again, is to be expected judging from the nature of the underlying rocks. Therefore, lime is often added to domestic water supplies to reduce the acidity.

Whilst the water from Nigerian rivers requires a degree of chemical treatment to make it suitable for human consumption, it is good enough for irrigation and livestock watering (Bennet, 1967). Livestock can in fact tolerate comparatively high levels of dissolved solids in their drinking water. For example, cattle can drink water with solid content as high as 10,000ppm, sheep 12,900ppm, horses 6,435ppm, pigs 4,290ppm., and poultry, 2,860ppm.

Most industries are located in urban centres and depend on piped public water supplies although some make use of water from boreholes. Generally, water quality requirements for industry are as stringent as for drinking water. Piped water in Nigeria comes either fully or partially treated. In the past, partially treated water was obtained from boreholes or by direct intake from rivers. Apart from liming and chlorination, alum was added to partially treated water to precipitate the sediments. Judging by the old minimum standards laid down by the World Health Organisation these partially treated domestic water supplies were of drinking water quality. The sad truth, however, was that at that time the people cared less about the quality of the piped water supply. A study by Akintola *et al.* (1979) found that the people of south-western Nigeria cared more about the fact that the source of water (public standpipe) had been brought nearer home, thus eliminating the long trek to the springs and stream where they used to fetch water.

Fully treated water undergoes three processes before distribution to the public: (1) sedimentation to remove both dissolved and suspended solids; (2) addition of chemicals to kill all pathogenic organisms; and (3) chlorination and liming to correct the fluoride and the pH levels respectively.

New International Standards for Drinking Water

The world has changed considerably since the World Health Organization developed the first set of standards for drinking water and other uses in 1958. Many chemical elements hazardous to human health and well-being have been released into the environment and are getting into water sources. Therefore, the WHO came up with a new set of water standards that has greatly expanded the range of elements that need to be tested especially for drinking water. Some other countries, including Canada, USA, China and the European Union have come up with their own standards for water which cover more chemical elements and set more stringent tolerance limits for many elements than the WHO standards. The challenge which these new standards pose for Nigeria and other developing countries is to expand the scope and quality of the tests carried out on public water supplies to ensure that they meet at least the WHO international standards. There are now many more elements, especially, the heavy metals, which need to be analysed to ascertain if the water supplies meet the set quality limits. Furthermore, the WHO is yet to develop tolerance limits for a number of the elements. Fortunately, environmental pollution levels have not reached the level of the highly industrialized countries and the risk of heavy metal contamination of water sources is perhaps still low.

Fortunately, also, the Nigeria Standards Organization (NSO) has approved a Nigerian Standard for Drinking Water Quality (NIS-554-2015.pdf). In developing the Standard, references were made to the Nigerian Industrial Standards for Potable Water and Natural Mineral Water, various national legislations, the National Guidelines and Standards for Water Quality in Nigeria, the updated World Health Organisation (WHO) guidelines for drinking water quality, and the International Organisation for Standardization (ISO). The Nigerian Standard for Drinking Water Quality applies to: (i) drinking water supplied by State Water Agencies; (ii) drinking water supplied by community managed drinking water systems; (iii) drinking water supplied by water vendors and water tankers; (iv) drinking water used in public or privately owned establishments; (v) drinking water used in food processing by manufacturers; and (vi) drinking water from privately owned drinking water systems and used solely for the family residence

The challenge now is to ensure that all the agencies and industries involved in water supplies for domestic and various industrial uses comply with the new standards. Rapid assessment of drinking-water quality (RADWQ) had been done in selected areas in 8 states throughout the country from time to time since about year 2000. The assessments covered three categories of water supply technologies, namely, piped water, borehole, protected dug well. But, in Lagos, a fourth type of technology, tanker water supply, was included. Table 5.14 shows that in a WHO-UNICEF rapid assessment of drinking water quality exercise 2005-6, Nigerian water supplies attained 100 per cent compliance with WHO standards only in respect of arsenic level.

TABLE 5.14 Level 1 parameters included in the RADWQ survey for Nigeria, 2005-6

Parameter	Compliance with WHO guideline or suggested value (%)
Thermo tolerant coliforms	77.0
Faecal streptococci	78.0
Arsenic	100.0
Fluoride	97.1
Nitrate	89.9
Iron	91.0
Turbidity	61.0
Conductivity	98.0
pH	57.8

Source: WHO-UNICEF-RADWQ-Nigeria-Report.pdf

All sample waters performed poorly particularly in pH, turbidity (indicative of dissolved mineral content) and bacteriological parameters, which imply that the people face high health risks. The chemical quality of the sampled waters also fell short of WHO standards most especially in terms of pH and turbidity.

Groundwater quality

Groundwater in Nigeria has always been touted as being of fairly high quality; most ground waters are suitable for both human and animal consumption. Recent studies of groundwater sources of water supply in some urban areas in different parts of the country confirm the good quality of Nigeria's groundwater, but they indicate that some of the parameters of the groundwater may have serious health implications. Ground water characteristics vary with the nature of the rocks and depth of aquifer. Shallow aquifers understandably are of lower chemical quality than deep aquifers; they also are more susceptible to contamination through human activities. Table 5.15 shows the comparative data on the physico-chemical characteristics of water from deep and shallow aquifers in a part of the Sokoto river basin in southern Kebbi State (Wali *et al.*, 2019).

Statistical comparison in Table 5.16 showed that the two sources differed significantly in most of the physical and chemical parameters.

Nnaji and Omotugba (2014) also analysed the physico-chemical characteristics of both deep and shallow groundwater sources in New Bussa (a new town built to house displaced population from Kainji dam project) in the middle Niger sedimentary rock belt, and compared them with WHO and Standards Organization of Nigeria standards (Tables 5.17). They found that, except for magnesium, cadmium and lead, all the metals were within stipulated limits. The water samples had acceptable colour and odour; and, with the exception of hardness, the physico-chemical parameters were all within acceptable limits for drinking water at most sampling points. The hardness of the water samples, which is due to high calcium and magnesium contents, ranged from moderately hard to very hard at many sampling points.

The study by Foka, Yah and Bissong (2018), in Benin City confirmed the good physico-chemical quality of Nigeria's groundwater (Table 5.18), but raised major health concerns because the total coliform counts were higher than acceptable limits. The isolated bacteria were *Pseudomonas aeruginosa* (38%), *Escherichia coli* (32.3%), *Staphylococcus aureus* (20%), *Klebsiella pneumoniae* (4.2%), and *Enterococcus sp.* (5.5%). Clearly there is contamination of groundwater from the surface as a result of human activities.

Water pollution

Water pollution introduced by human activity is spreading in Nigeria, although it has hardly affected the sources from which the rural majority of the populace obtains drinking water. A certain amount of pollution occurs along some rural water courses which are used by farmers in soaking and fermenting cassava and locust

TABLE 5.15 Physicochemical parameters of groundwater samples from deep and shallow groundwater sources

Points	T^oC	pH	TDS mg/l	EC μScm^{-1}	K^+ mg/l	Na^+ mg/l	Ca^{2+} mg/l	Cu^{2+} mg/l	Fe^{3+} mg/l	Zn^{2+} mg/l	Mg^{2+} mg/l	PO_4^{3-} mg/l	Cl^- mg/l	HCO^-_3 mg/l	SO_4^{2-} mg/l	NO_3^- mg/l
A. FAKAI Deep Groundwater																
1	34.0	8.1	111.0	226.0	33.4	2.9	13.6	0.1	1.3	0.4	21.1	0.6	2.8	20.3	84.1	21.1
2	33.0	8.0	114.0	227.0	39.9	6.7	56.4	0.3	0.2	0.3	17.4	0.5	3.5	2.7	67.3	37.2
3	33.0	7.3	382.0	763.0	39.9	0.1	53.9	0.4	0.6	1.0	26.8	0.1	2.6	2.9	91.2	24.5
4	31.0	7.3	394.0	796.0	39.5	0.1	2.7	0.3	0.3	0.9	26.3	0.4	2.1	2.8	87.5	38.1
5	33.0	7.8	65.0	138.0	38.4	0.9	10.1	0.4	1.1	0.5	9.3	0.6	2.8	12.0	133.2	45.7
6	32.0	7.7	68.0	136.0	40.3	6.7	14.6	0.6	1.6	0.4	8.2	0.4	2.5	33.3	130.1	41.2
7	34.0	8.2	111.0	225.0	38.8	2.8	3.2	0.4	0.4	0.5	18.6	0.6	1.4	10.4	154.6	12.5
8	33.0	8.2	116.0	232.0	38.8	2.8	3.6	0.4	0.2	0.6	14.2	0.1	3.1	19.5	205.8	40.1
9	34.0	8.0	94.0	189.0	37.2	0.9	14.3	0.2	0.5	0.4	19.8	0.3	1.5	17.0	214.7	17.6
10	33.0	8.4	96.0	192.0	40.7	9.8	7.0	0.5	0.0	0.5	18.6	0.2	2.4	16.7	188.4	37.9
B. FAKAI Shallow Groundwater																
1	30.4	7.6	140.0	110.0	39.0	414.0	13.0	0.3	1.1	6.3	5.0	17.0	71.0	183.0	281.3	52.1
2	30.8	7.5	320.0	260.0	39.0	437.0	32.0	0.6	1.6	4.2	5.0	11.0	320.0	183.0	213.1	49.8
3	28.0	7.7	270.0	240.0	78.0	598.0	51.0	0.3	0.5	6.3	14.0	22.0	888.0	61.0	197.2	38.2
4	31.0	7.6	110.0	80.0	39.0	439.0	19.0	0.4	2.1	3.2	8.0	20.0	213.0	183.0	125.6	60.0
5	30.4	7.7	120.0	80.0	39.0	483.0	30.0	1.1	2.7	2.0	3.0	23.0	72.0	183.0	208.4	47.9
6	30.7	7.8	220.0	160.0	39.0	552.0	30.0	1.6	2.0	4.1	30.0	22.0	249.0	244.0	218.7	26.1
7	28.0	8.1	140.0	100.0	78.0	460.0	20.0	0.3	1.1	4.4	20.0	22.0	178.0	183.0	321.8	18.7
8	21.4	8.0	90.0	70.0	39.0	552.0	12.0	0.8	0.9	6.3	5.0	23.0	249.0	122.0	206.1	68.0
9	28.0	7.3	90.0	70.0	39.0	460.0	20.0	0.2	2.0	6.1	1.0	21.0	107.0	122.0	71.7	49.3
10	24.2	7.1	50.0	30.0	39.0	552.0	7.0	1.3	2.1	6.3	3.0	18.0	107.0	122.0	89.3	55.4

Source: Wali S.U., Umar K.J., Abubakar S.D., Ifabiyi I.P., Dankani I.M., Shera I.M. and S.G. Yauri; 2019 "Hydrochemical characterization of shallow and deep groundwater in Basement Complex areas of southern Kebbi State, Sokoto Basin, Nigeria." *Applied Water Science* 9:169 `https://doi.org/10.1007/s13201-019-1042-5`

TABLE 5.16 Summary and Comparison of the physical and chemical properties of shallow and deep groundwater in southern Kebbi State, Sokoto Basin

Parameter	Borehole (Deep aquifer				Dug well (shallow aquifer				Reference Guidelines		Kruskal-Wallis Non-parametric test		
Physical	Mean	Min	Max	SE	Mean	Min	Max	SE	National Standard for Drinking Water Quality (2007)	World Health Organization (2011)	*H* (*chi*^{2})	Hc (tie corrected)	*p* (same)
T °*C*	32.4	30	34	5.4	26.8	21.4	31	4.9	Ambient	Ambient	27.1	27.4	<0.001*
pH	7.9	7.3	8.4	1.3	7.4	6.8	8.1	1.3	6.5-8.5	6.5-8.5	9.8	9.9	0.002**
TDS	180.8	65	394	62.3	142	40	320	50.6	500	1000	0.9	0.9	0.330
EC	363.3	136	796	125.9	95.5	10	260	41.1	1000	1400	21.2	21.2	<0.001*
TH	76.2	19.4	184.9$^+$	29.2	134.4	63.4	402.8	63.7	150	200	9.015	9.016	0.002**
Cations													
K^+	37.7	32.3	40.7	6.4	42.9	39	78	12.3	-	-	4.0	4.4	0.036
Na^+	3.9	0	9.8	1.5	449.9$^+$	2	598$^+$	94.6	12	200	25.3	25.4	<0.001*
Ca^{2+}	21.9	2.7	129.3	20.4	19.7	1.2	51	8.1	500	75	1.3	1.3	0.256
Cu^{2+}	0.3	0.1	0.6	0.1	0.6	0.1	1.6$^+$	0.3	1	1	1.4	1.4	0.238
Fe^{3+}	0.9	0.1	3.1	0.5	1.6	0.5	2.7$^+$	0.4	2	2	8.7	8.7	0.003**
Zn^{2+}	0.5	0.2	1	0.2	5.3$^+$	2	8.4$^+$	1.3	3	3	29.3	8.7	<0.001*
Mg^{2+}	19.4$^+$	8.2$^+$	26.8$^+$	4.2	6.6$^+$	1	20$^{+3.2}$	3.2	0.2	125	22.7	22.7	<0.001*
Anions													
PO_4^{3-}	0.4$^+$	0.1	0.6$^+$	0.1	18.4$^+$	11$^+$	23$^+$	3.6	0.2	0.2	29.3	29.5	<0.001*
Cl^-	2.5	0.6	3.5	0.6	188.8	4	888$^+$	140.4	200	250	29.3	29.4	<0.001*
HCO_3^-	14	0.3	33.3	5.3	176.9$^+$	61	549$^+$	86.8	250	125-130	29.3	29.7	<0.001*
SO_4^{2-}	131.1	45.1	245.9$^+$	38.9	206.3	67.5	327.1$^+$	51.7	200	250	8.0	8.0	0.005*
NO_3^-	36	12.5	46.8	7.4	42.7	12.6	45.2	7.1	50	50	5.2	5.2	0.022

+ Do not follow WHO and/or NSDWQ reference guidelines; $*$ Significant difference between shallow and deep groundwater < 0.001; $**$ Significant difference between shallow and deep groundwater $\leqslant 0.005$.

beans. Earlier, reference was made also to the contamination of rivers with chemicals used in fishing by villagers and as fumigants on cocoa farms in southwestern Nigeria. Nevertheless, water pollution is still largely a phenomenon of the urban centres where river valleys and drainage channels are too often the dumping ground for city wastes and, in some cases, industrial effluents. The Kaduna River was once heavily polluted within Kaduna city and it spread some ten kilometres downstream as a result of oil spills and effluents from the oil refinery and other industrial establishments. The Ogunpa River in Ibadan city gained notoriety for the refuse dumps at several points along its course. The wastes thrown into the river included human excreta, garbage, car tyres, cans, scrap metals, wood, ash, leaves and other plant materials. Such wastes rendered the water absolutely useless for drinking, swimming or boating. The fish population was also adversely affected (Odunuga, 1977).

An analysis of the spatial pattern of river pollution in Ibadan city indicated a high correlation between water pollution and land use (Akintola and Nnyamah, 1978). Judging by WHO standards, the high density, low quality traditional core of the city had the most polluted surface water closely followed by the commercial and

TABLE 5.17 Physico-chemical parameters (mean ±SD) of drinking groundwater samples in New Bussa, Niger State, Nigeria

Parameters	1	2	3	4	5	6	7	8	9	10
Temp °C	28.1±1.02	29.3±0.50	28.5±0.61	28.0±0.83	28.5±0.54	28.4±0.33	28.3±0.47	28.6±0.39	29.0±0.75	28.6±0.59
pH	7.24±1.27	7.20±0.54	7.34±0.73	7.60±0.31	6.61±0.35	7.68±0.83	6.67±0.42	7.30±0.23	7.14±0.57	8.11±0.38
TDS mg/l	0.21±0.18	0.26±0.11	0.37±0.08	0.33±0.13	0.38±0.09	0.97±0.23	0.57±0.14	0.71±0.34	0.68±0.30	0.60±0.25
EC µS/cm	290±4.90	340±17.31	220±7.56	190±12.63	40±2.45	40±4.20	540±20.88	200±16.25	160±10.94	210±13.44
TA mg/l	3.19±0.19	2.84±0.26	2.20±0.40	2.53±0.38	2.20±0.80	1.10±0.62	3.30±0.63	3.30±0.19	3.51±0.23	3.20±0.51
TH mg/l as $CaCO_3$	215.70±10.49	192.41±9.32	145.96±6.41	130.75±8.33	48.17±2.30	56.35±3.44	301±20.41	255.90±15.29	200.64±25.32	208.68±18.11
DO mg/l	4.31±1.24	4.03±0.91	4.61±1.08	5.08±2.31	6.26±1.11	6.33±1.52	5.20±1.25	6.51±2.30	5.66±2.12	6.40±1.25
Nitrate mg/l	0.76±0.008	0.85±0.07	1.66±0.28	1.45±0.63	1.61±0.70	1.84±0.76	6.20±2.45	5.30±2.16	2.78±1.08	2.56±0.24
Phosphate mg/l	0.08±0.01	0.07±0.03	0.09±0.02	0.10±0.03	0.05±0.00	0.07±0.00	0.36±0.08	0.19±0.08	0.16±0.06	0.17±0.80
Chloride mg/l	24.00±3.49	21.03±4.25	46±6.12	40.31±8.33	13±1.67	9.00±4.44	208.04±30.48	160.71±30.86	100.62±18.12	98.43±15.30
Sulphate mg/l	50.11±5.16	40.33±7.25	50.30±8.38	48.66±12.20	53.20±11.50	60.11±20.63	80.13±22.80	71.20±18.33	100.6620.15	75.74±24.38
Metal content										
Na mg/l	7.560±1.038	27.598±3.22	33.163±4.29	2.980±0.69	5.512±3.75	5.223±4.31	34.776±14.48	21.128±8.36	24.961±8.71	31.574±16.89
Ca mg/l	11.250±3.15	40.555±20.15	29.861±12.11	21.527±6.97	6.250±7.12	5.139±4.33	79.167±25.30	68.472±30.44	17.638±9.21	51.250±20.77
Mg mg/l	9.408±4.37	8.968±2.48	6.930±3.12	6.993±2.07	2.002±0.98	2.181±1.25	9.210±6.58	10.036±5.61	7.316±4.24	8.259±5.16
Cd mg/l	0.020±0.02	0.013±0.01	0.010±0.00	0.009±0.00	0.007±0.01	0.003±0.00	0.017±0.01	0.013±0.01	0.007±0.00	0.013±0.01
Cr mg/l	ND	ND	ND	ND	ND	ND	ND	ND	ND	ND
Cu mg/l	0.114±0.07	ND	0.025±0.01	0.092±0.02	0.010±0.01	0.042±0.02	0.066±0.03	0.057±0.04	0.114±0.09	0.129±0.06
Pb mg/l	ND	ND	0.016±0.01	ND	0.018±0.00	ND	0.032±0.01	ND	ND	0.016±0.01
Fe mg/l	0.114±0.03	ND	0.025±0.01	0.092±0.07	0.010±01	0.043±0.02	0.066±0.03	0.057±0.02	0.114±0.10	0.129±0.11
Zn mg/l	0.178±0.13	0.089±0.08	0.170±0.09	0.029±0.01	0.034±0.03	0.059±0.02	0.356±0.18	0.380±0.25	0.095±0.04	0.288±0.11

ND = not detected Source: Nnaji J.C & S. Omotugba 2014 Physico-chemical quality of drinking water in New Bussa, Niger State, Nigeria. *International Research Journal of Pure & Applied Chemistry* 4(4), 437–446. `www.sciencedomain.org`

modern high-density housing areas. This rating was based mainly on the levels of dissolved oxygen and suspended solids, colour, turbidity and hardness of the water. In all land use zones, the levels of copper, fluoride, manganese and phosphates were higher than the WHO limits. By contrast, the chloride and nitrate levels were much lower. The major causes of pollution in residential areas are organic wastes such as food and human waste while in the commercial and industrial centres they are primarily waste paper, leaves, foodstuff wastes, plastics, cans and other metal objects. Chemical fertilizers and crop residues are the causes of water pollution in the agricultural areas.

By the 1980s in Lagos, lagoon fishermen were complaining about dwindling fish stocks in the vicinity of the city. Fishermen farther away from the city, such as at Epe, touted the high quality and good taste of their catch as compared to the 'tasteless' fish caught in the Lagos area. Human wastes were also being dumped in the lagoons in these other areas but the quantities involved were very small as compared to those of Lagos. Indeed, these limited wastes attract and promote the breeding of fish in the vicinity of several lagoon settlements.

Thus, the dumping of wastes into water bodies, though detestable, may not necessarily be harmful in the short-term; it all depends on the type of wastes; the quantities involved; how poisonous they are; and whether they are easily biodegrad-

TABLE 5.18 Physico-chemical properties of potable water in Benin City in relation to WHO, NAFDAC and NSDWQ Standards

Parameters	WHO		NAFDAC	NSDWQ	SITES					
	Highest Desirable Level	Maximum Permissible Level			Ugbowo	New Benin	GRA	Ikpoba Hill	Tap Water (GRA)	Mean ± SD
Temperature (°C)	-	40	-	ambient	27.3	28.1	27.8	27.8	27.5	27.7 ± 0.3
Colour (TCU)	6	-	15	15	-	-	-	-	-	0
Turbidity (NTU)	5	25	-	5	0.80	0.83	0.79	0.82	0.76	0.8 ± 0.03
pH	7.0 - 8.5	6.5 - 9.2	6.5 - 8.5	6.5 - 8.5	6	6.7	7.3	7.2	7	6.84 ± 0 .52
Total Dissolved Solid (TDS)	500	1500	500	500	404	335	319	313	247	323.6 ± 56.1
Nitrate (mg/L)	-	50	-	0.2	3.6	4.4	0	1	0	1.8 ± 2.07
Iron (mg/L)	-	-	-	0.3	0	0	0	0	0	0
Chromium (mg/L)	-	-	-	0.06	0	0	0	0	0	0
Copper (mg/L)	-	-	-	1	0	0	0	0	0	0
Fluoride (mg/L)	-	-	-	1.5	0	0	0	0	0	0

Source: Frank Eric Tatsing Foka, Clarence S. Yah, Marie Ebob Agbortabot Bissong (2018) Physico-chemical properties and microbiological quality of borehole water in four crowded areas of Benin City, Nigeria, during rainfalls. *Shiraz E-Medical Journal*: 19 (11); e68911 `https://dx.doi.org/10.5812/semj.68911`

able or not. Items such as plastics, tin cans and scrap metals found in the Ogunpa River in Ibadan are clearly not biodegradable, whereas human wastes dumped into the lagoon in Lagos are biodegradable but the quantity involved is so great that their biological oxygen demand greatly exceeds the amount of dissolved oxygen in the water. In the 1970s, water pollution became a serious environmental hazard on Lagos island, not only to the fishermen, boatmen and swimmers on the lagoon but, also, to people onshore who had to live with the stench coming off the polluted water. The invasion of the lagoon waters by water hyacinths from the 1970s clearly indicated an intolerable build-up of sediments and human wastes. The profuse growth of these aquatic weeds deprived many lagoon settlements of their prized fishing grounds and lagoon waterways. Some settlements were completely cut off by the weeds as the small boats used by the inhabitants could not push through them.

Traditionally shore dwellers have dumped their refuse into the rivers, oceans, creeks, lagoons or estuaries where they live. No amount of persuasion or regulation is likely to change this state of affairs. Only when they can see for themselves that the waters and adjacent lands are not just lying idle and uncared for, but are being brought into use, will the people recognize the need to find an alternative way of disposing of their domestic and other wastes. In Lagos and Ibadan, for example, the inhabitants began to respect the regulations against waste dumping in the water bodies when the state governments and municipal authorities began to clean up the water bodies. Lagos stopped using the lagoon as its sewer by putting a total ban on the dumping of human wastes. In Ibadan, the Ogunpa channelization scheme succeeded in widening the river channel; and in clearing waste dumps and thick bushes off the channel and banks of the river. It also cleared slums and traditional markets which used the river as both sewer and solid waste dump.

The future

As indicated above, the World Bank assessment of the water supply and sanitation situation in Nigeria at the beginning of the 21st century was dismal. On average, only 30% of urban dwellers and 35% of rural population had access to safe and reliable water supply and adequate sanitation facilities. Only Lagos and Abuja had a sewerage system. The water production facilities in most parts of the country were rarely operated to capacity due to broken down equipment, or lack of power or fuel for pumping. It is common knowledge that industrialists, home-owners and estate developers provide their own water through wells and boreholes; they also construct septic tanks to take care of their sewage. Open defecation is predominant in rural areas and large parts of many towns and big urban centres. Yet the state water agencies are overstaffed (though poorly remunerated) with an average staff to customer ratio of 70 as compared to a best practice ratio of 3.5 in efficient utilities (World Bank, 2000).

The World Bank determined then that for Nigeria to meet the millennium development goal for water supply by 2015, the country needed to invest approximately N215 billion (USD 1.3 billion) annually. But Nigeria could only afford to invest about N82.5 billion (USD 0.5 billion) and much of the investment was used, not for new water supply schemes, but to rehabilitate infrastructure that had not been properly maintained.

In 2003, "the Presidential Water Initiative (PWI): *Water for People, Water for Life*" was launched by the Federal Government, which obtained loans and technical assistance from several international organizations including, significantly, the World Bank. A few state governments also embarked on community-based sanitation programmes backed by UNICEF and succeeded in providing thousands of public latrines distributed over several communities, some of which attained open defecation-free status. In spite of these very welcome achievements, they were too little to meet the needs of the country in the area of water supply and sanitation. The country was not able to meet the Millennium Development Goal, nor the follow up

Sustainable Development Goal in water supply and sanitation. The financial burden of meeting these development goals is one that has proved too heavy to carry by Nigeria. The country's external debt profile continues to worsen because of unrestrained compulsive borrowing.

Many suggestions had been put forward in the past about the best way to develop and manage the nation's water resources. In the 1970s, it was decided that in order to achieve an orderly development it was necessary to assess the nature and magnitude of the available water resources; the future domestic, industrial and agricultural water requirements; and how these requirements could be met (Ayoade, 1975). The consensus of opinion then, which was also shared by the Federal Military Government, was that these objectives and assignments could best be met and accomplished by creating river basin development authorities (RBDAs). Thus, eleven (later increased to 12; Fig 5.9) river basin development authorities (RBDAs) were created in 1976. Since then, the RBDAs have been responsible for planning and achieving the integrated development, exploitation and conservation of the water resources in their respective areas of jurisdiction (Table 5.19; Fig. 5.10).

The RBDAs have certainly extended the coverage of soil, water and other environmental studies throughout the country. They have opened up many rural areas to modern development through their agricultural development schemes. But, at the same time, they seemed to have concentrated on construction activities and agricultural production to the neglect of needed land use control and management.

Land use management in the catchment areas is necessary to maintain the reservoirs and prolong their life spans. Observers were also concerned about early incidences of dam bursts (e.g. the Ojirami dam in Edo State 1981 and the Bagauda Dam near Kano 1988), which suggested that some dams had been built without adequate design specifications. The monitoring of land use in the catchment areas of reservoirs should have been given priority by the authorities. The dam bursts were caused as much by excessive sedimentation in the reservoirs as by inadequate design (Adams, 1987). Even, in agricultural production, their main area of focus, the RBDAs had performed below their capacities as depicted in Table 5.20

The RBDAs have not done much either in the area of domestic water supply as their focus has been on providing water for irrigation purposes. But it is possible for them to contribute at least to rural water supplies through some form of arrangement with community and state water agencies. The RBDAs also have neglected other important areas of their mandate that could probably have helped in furthering the cause of water supply and sanitation in the smaller towns and villages. They include the mandates on combating drought, flood control, and hydroelectric power generation particularly to serve rural communities not covered by the National Electric

TABLE 5.19 River Basin Development Authorities in Nigeria

Name of River Basin	Headquarter Location	States Covered
Anambra/Imo River	Aba-Owerri Road Agballa-Owerri	Anambra, Imo, Enugu Abia, Ebonyi
Benin-Owena Basin	Benin - Warri Road, Obayaton, Benin City	Edo, Part of Delta North, Ondo, Ekiti
Chad Basin	Maiduguri-Marte Road, Maiduguri	Borno, Yobe, and Part of Adamawa
Cross River Basin	Murtala Muhammed Way, Calabar	Cross River, Akwa Ibom
Hadejia-Jamaáre	Maiduguri Road, Hotoro, Kano	Kano, Jigawa, Parts of Bauchi
Lower Benue River	Oturkpo Road, Makurdi	Benue, Plateau, Nasarawa and part of Kogi
Lower Niger River	GRA Ilorin	Kwara, Kogi
Niger Delta Basin	Isaac Boro Park, PH	Rivers, Bayelsa, Parts of Delta
Ogun-Osun River	Abeokuta	Lagos, Oyo, Ogun, Osun
Sokoto-Rima	Guzan Sokoto Road, Sokoto	Sokoto, Kebbi, Zamfara, Katsina
Upper Benue	Yola-Fufore Road, Yola	Adamawa, Taraba, Gombe parts of Bauchi State
Upper Niger	Bosso-Kontagora Road, Minna	Niger, Kaduna and FCT

Source: Ngene B.U. 2009 Nigerian Rain Gauge Station Optimization and National Development: The importance of Head Count

FIGURE 5.10 River Basin and Rural Development Authorities in Nigeria

Power Authority (Charles Kumolu*)

For the future of water supply and sanitation, two issues that were also touched upon in the World Bank study deserve close attention: these are funding and legislation. Nigeria has borrowed so much to improve its water supply and sanitation sector. But, unfortunately, water that is so costly to provide remains largely a social service. Indeed, a common belief among the people is that 'no one pays for water'. There are water tariffs charged by state water agencies especially in the big urban centres. Since only a minute proportion of urban dwellers obtain water from municipal sources, the revenue yielding potential is low. Then, only a minute proportion of those served with municipal water supplies actually pay the tariffs. Enforcement is almost zero as most of the pressure is exerted on institutions and corporate entities. The water rates are regarded as ridiculously low by funding agencies.

Currently, the Federal Government has embarked on a programme of partial

*https://www.vanguardngr.com/2013/04/river-basins-how-unending-policy-reversals-abet-inefficiency/

commercialization of the RBDAs to achieve better performance in their mandates and operational areas. The rural areas, it is hoped, would benefit from this kind of arrangement and state water supply and sanitation agencies could also partner with the RBDAs to effect positive changes in rural water supplies and in personal hygiene and environmental sanitation.

The high cost of borrowing to fund development activities and lack of political will or managerial capacity to make the people pay more for the facilities and amenities being provided has made policy makers turn to the strategy of private sector participation in urban water supplies. This strategy would call for reforms on the part of the state water agencies to attract private sector participation (PSP). Private sector participation (PSP) would certainly bring up the issue of profit maximization from joint ventures or agreements, which in turn would call for greater attention to the policy on water tariffs.

The World Bank study rightly noted that "one of the most important pre-requisites to successful PSP is the existence of a legislative and regulatory framework to protect the potentially conflicting interests and rights of consumers, governments, operators and the environment. This is key to attracting serious private operators to the sector as it provides clarity to their roles and responsibilities". Issues to be regulated are identified as tariffs, environmental protection, water resources, and aspects related to service standards and development. Table 5.21 summarizes these issues and the level of government required to take action in producing the required legislation and regulatory measures.

Nigeria still has a long way to go in terms of harnessing and utilizing her renewable water resources and meeting the needs of the vast majority of her population. Table 5.22 shows clearly that as of 2010, estimated total water demand was only 4.4% of total available renewable water resources from both surface and groundwater sources. Yet all sectors of the economy could do with much higher supplies if public supply were more efficient and reliable. In almost every sector demand for water is satisfied from private supplies; this partly accounts for the greater reliance on groundwater sources.

TABLE 5.20 Equipped and actually irrigated areas in the River Basin Development Authorities for the year 2004

River Basin Development Authority	Equipped area (ha)	Actual irrigated area (ha)	Actual irrigated as per cent (%) of equipped area
Anambra-Imo	3,941	10	0.3
Benin-Owena	317	0	0
Chad Basin	26,180	1,000	3.8
Cross River	364	40	11.0
Hadejia Jamaáre (including areas outside of equipped area using water from main canal)	18,475	21,000	113.7
Niger Delta	187	0	0
Lower Niger	1,344	115	8.6
Upper Niger	3,697	722	19.5
Ogun-Osun	512	110	21.5
Sokoto Rima	27,580	5,290	19.2
Upper Benue	8,410	783	9.3
Total	**92,317**	**29,140**	**31.6**

Source: `http://www.eoearth.org/profile/Jim.kundell/` Jim Kundell, 2008: Water profile of Nigeria. `http://www.eoearth.org`

Highlighting the importance of groundwater to water supply in the country thus far, FAO (2016) noted as follows:

> Groundwater is available and used in the whole country for all types of uses – water supply, private irrigation, livestock, aquaculture and industries – either in the Basement Complex area, where depth of boreholes is 50m in average for a 10 to 150m^3/day yield, or from sedimentary rock area with deeper boreholes with larger yields (50 to 500 m^3/day). There are about 64,500 boreholes in Nigeria, of which 57,600 are for public water supply. Motorized and hand pumps are installed into boreholes, but only 63 per cent of boreholes are functional due to breakdown of pumps. Groundwater withdrawal is estimated at about 2.3 million m^3/year (FMWR, 2014).

TABLE 5.21 World Bank water supply and sanitation regulatory guidelines for Nigeria

Issues	Rights	Level of Legislation Required for Regulation	
		Water	Environment
Raw water (quantity)	Water balance, abstraction, environmental impact	Federal	Federal
Raw water (quality)	Treatment, environmental impact	Federal	Federal
Treated water quality	Treatment, customer service, health impact	Federal, State	
Customer levels of service	Pressures, continuity of service, problem resolution	State, Local	
Customer protection	Tariff, confidence	State	
Operator protection	Tariff, security of payment, confidence	State, Federal	
Water environment	Water balance, effluent waste control, flooding, river regime		Federal

Source: The World Bank (2000) ***Water Supply & Sanitation Interim Strategy Note November 2000*** *World Bank wss_1100.pdf, page 32*

TABLE 5.22 Estimated water demand by sector in Nigeria, 2010

Sector	Estimated Total Water Demand in million m^3/year	% Total
Municipalities	5,000	40.1
Irrigation	4,549	36.5
Industry	1,965	15.8
Agriculture	728	5.8
Livestock	233	1.9
Total Water Demand	12,475	
As % total renewable water resources	4.4	
Demand per inhabitant	78	
Expected supply from surface water sources	41%	
Expected supply from groundwater sources	59%	

Source: Adapted from FAO 2016:AQUASTAT FAO's Water Information System Nigeria: Water Resources `http://www.fao.org/nr/water/aquastat/countries_regions/Profile_segments/NGA-WR_eng.stm`

Finally, while water supply and sanitation and agricultural development are arguably the current major concerns in the country, the non-consumptive uses of water also deserve attention because of their economic and regional development potentials. Failure of succeeding administrations to tackle the problems of electricity supply has made it almost impossible to make any headway in improving the living conditions of the general populace or to increase the pace of development in such sectors as manufacturing, transportation, commerce, banking and communications. Climate change continues to challenge the country's ability to respond to increasing need for water in agriculture while oil and waste pollution continues to lower the ability of the water bodies to support the livelihoods that are based on them including fishing, water transportation and water-based recreation and tourism. Nigeria still imports more than half a million tonnes of fish annual in spite of the fact that it has the capacity to produce much more fish through aquaculture. Tourism development is still largely oriented towards foreign visitors and the local elite. Progress in this sector would depend on promoting the culture of outdoor recreation among the indigenous inhabitants and motivating them to acquire skills and the ability to cater to the needs of both local and foreign visitors who cannot afford the huge expenses of five-star hotels and resorts. Nigeria needs to develop motels as in the United States or the Bed & Breakfast home-based services as in the United Kingdom in order to internalize the culture of tourism and outdoor recreation among the general populace.

6

Soil resources and crop cultivation

As indicated in Chapter 1, the expansion of agriculture has been the dominant feature of Nigeria's land use/land cover changes in the last four decades or so. Both forest and savanna lost ground particularly to the advance of rainfed agriculture. It was estimated that by 2010, the total area of all agricultural land in Nigeria amounted to 762,000km^2. Agriculture is the mainstay of Nigeria's rural economy; it engages about two-thirds of the population and it is the dominant element in rural land use. It is characterized by a wide variety of agricultural systems and crops, in consonance with the prevailing environmental conditions in the different ecological zones. Traditional methods of crop cultivation have evolved naturally over long periods of time, and developed through the people's local knowledge and experience. Cropping systems and crop husbandry techniques evolved through trial and error, experimentation and adaptation of methods from other cultures. While many crops are indigenous to the region, others have been successfully introduced from other regions of the world through contacts made during the slave trade and colonial eras. Present agricultural systems are based, therefore, on the local people's accumulated knowledge of crops, animals, pests and diseases, soil, weather and other ecological conditions in the different ecological zones.

The environment has been the major influence in the evolution of each system of cultivation. There is a high degree of dependence on natural forces and processes for the maintenance of yield and the quality of the produce. But if their natural surroundings have given rise to the types of farming practised, in turn the cultivation techniques adopted by the people have profoundly affected the natural ecosystems in the country. Thus, a discussion of the ecology of crop cultivation in Nigeria provides a good opportunity to understand the nature and impacts of human interaction with the natural environment.

Of all the environmental factors influencing Nigeria's agricultural systems the most crucial are soil, water supply, weeds and crop and animal pests and diseases. Water supply has already been dealt with in Chapter 5. In this chapter we shall examine major soil types and their characteristics. Weeds, pests and diseases in agriculture are the subjects of Chapter 7

The agricultural properties of Nigerian soils

Nigerian soils are highly variable even within short distances; fortunately, the spatial trends, even at the local level, are relatively easy to discern and analyse since they show marked correlations with relief and the underlying rocks or the associated weathered mantle. Most soils are intensely weathered and chemically altered as a result of deep weathering a phenomenon associated with the hot, humid tropical environment for the past several millennia.

One of the major consequences of the deep weathering is that except for those developed on fine alluvium or on basic parent materials such as basalts and amphibolites, the soils are generally deficient in weatherable mineral reserves, from which soil nutrients are replenished as they get exhausted. Soils developed on sandstones, sandy alluvial and marine deposits and on the more acidic granitic and gneissic rocks are particularly poor in weatherable minerals. Those developed on amphibolites, schists and basalt usually have weatherable reserves of feldspar, biotite mica, olivine, illite and chlorites. In spite of the intensity of weathering and leaching of the soils, few are lacking in the essential trace elements. Most soils contain sufficient quantities of elements such as iron, manganese, zinc, copper and boron, which tend to be more mobile and readily available to plants in acidic than in alkaline soils. However, molybdenum is less available in these generally acidic soils. Crop response to fertilizer treatment may be hindered by the absence of these essential trace elements in the soil.

A further consequence of the intense chemical alteration of the soils is the predominance of kaolin clay (kaolinite), which has important implications for the productivity of the soils. The water retentive capacity of most soils is low and places a severe limitation on agricultural activities in many parts of the country because the lattice structure of kaolinite is so compact that it holds little water. However, it expands and contracts less than other types of clay on wetting and drying respectively. Thus, the predominance of kaolin has bestowed on most Nigerian soils a strong, stable structure. Such soils also have well developed systems of free pore spaces so that they are highly permeable to air, water and root penetration. In this way they compensate for deficiencies in their chemical nature.

In addition to its poor water retention kaolinite has a very low cation-exchange capacity to hold and exchange plant nutrients. The nutrient status or base saturation of many soils is correspondingly low. This low natural fertility is compounded by the predominance of inert sand grade particles and free oxides. The sandy nature of the vast majority of the soils is also responsible for their rapid internal drainage and relatively high susceptibility to drought when the vegetative cover is removed and they are exposed to intense insolation. Soils developed on arenaceous sedimentary rocks and deposits in the coastal zone and on broad plateau surfaces, such as the Sokoto basin, in the north are typical examples.

In these essentially sandy or sandy-loam soils the clay fraction is still important because it gives some cohesion to the soils and reduces excessive internal drainage. But of greater importance to these, as to all Nigerian soils, is the organic matter content. Organic matter is very important in maintaining the soil structure; especially during the fallow period the restoration of the stable structure of the soil destroyed during the period of repeated cultivation, depends on the build-up of soil humus. The roles and significance of humus are well understood by local farmers who are used to assessing the potential productivity of soils by the dark colour of their surface layers. Two essential plant nutrients, nitrogen and phosphorus, are present in insufficient quantities in most soils and this constitutes one of the most serious limiting factors on agricultural productivity in Nigeria. Both nutrients are rapidly lost during cultivation as a result of the attendant fall in humus levels. Moreover, the cation-exchange capacity and base saturation of most soils are more closely related to the organic matter than to the clay content. Thus, the cation-exchange capacity and base saturation of soils under fallow increase with the age of fallow because of the build-up of soil humus over time.

Reference was made above to the rapid exhaustion of nitrogen and phosphorus during the cultivation period; this also seems to be the fate of major nutrients and trace elements alike in Nigerian soils. For instance, many soils begin to show a deficiency of potassium, zinc and magnesium after two or three cropping seasons. Continuous leaching, immobilization, crop removal and a fall in the level of organic matter during the cultivation periods are the probable causes. The tendency for nutrients to be rapidly exhausted is a serious challenge in the adoption of systems of continuous or permanent cultivation without a well-managed system of application of both organic and artificial chemical fertilizers. Agboola (1970) believed that high crop yields could only be maintained in the tropics if increasing amounts of fertilizers were applied to the soils from year to year. This is in contrast to the situation in the temperate zone where high yields are maintained by using uniform yearly fertilizer applications once the soil fertility level is adequate.

Other scientists did not necessarily share Agboola's opinion about continuous cultivation of tropical soils. Writing about the soils of the West African savanna, Jones and Wild (1975, p. 216) felt that the possibility of continuous cropping under all conditions, without the use of 'resting' crops or bush fallows was a question that could not yet be answered. However, the evidence from experimental stations was that with the judicious use of fertilizers, and other inputs, high yields could be produced for many years. In other words, the cultivation period could be longer before there would be any need for a resting (fallow) period. But there are wide areas where there has been little experimental work so that studies of long-term trends of fertility under different farming systems are still much needed.

Finally, although the phenomenon of deep weathering is widespread throughout the country the effective depth of soil for plant or crop growth is limited to that portion above the stone layer, hard pan or lateritic crust. Even more important to crops is the thickness of the comparatively humus-rich surface horizon in which plant nutrients are concentrated and which is the rooting zone of most plants. Below this rich surface layer, the subsoil is relatively deficient in plant nutrients. The layer of soil above the stone or lateritic crust and the thickness of the humus-rich layer, in effect determine how much soil can be ploughed or heaped up for crops to grow in. The depth of the humus layer is greatest on the alluvial plains and in valley bottoms, a factor that has made possible, for example, the heaping of the huge yam mounds characteristic of the croplands on the Imo-Cross River plains. Elsewhere, the effective depth of soil varies with slope, the weatherability of the parent materials and the activities of worms and termites which, in some places, are responsible for bringing fine soil particles to the surface while the stones are translocated further down the soil profile.

The major soil types of Nigeria

Fig. 6.1 shows the distribution of the major soils of Nigeria. The major soil types can be related to the factors of climate, vegetation, lithology and topography. Climatic factors, particularly rainfall, influence the rates and depths of weathering and pedogenesis which generally decrease from the humid south to the sub-humid north. The soil moisture regime, which is very important in agricultural productivity in the country, is highly correlated with the incidence of rainfall in different parts of the country.

The density of the vegetation cover also conditions soil moisture because it determines the extent to which the soils are protected against intense insolation. Generally, soils become more prone to desiccation towards the north, not only because of the less humid climate, but also because of the scantier vegetation cover. With increasing desiccation hard iron concretions and layers of iron and clay pans occur in the soil profiles. However, these features also occur in the south where deforestation by man has exposed the soil to insolation. The humus content of the soils, which is so important to their productivity and structural stability, varies with the nature and density of vegetation cover.

Within the broad ecological zones, the distribution of major soil types is largely

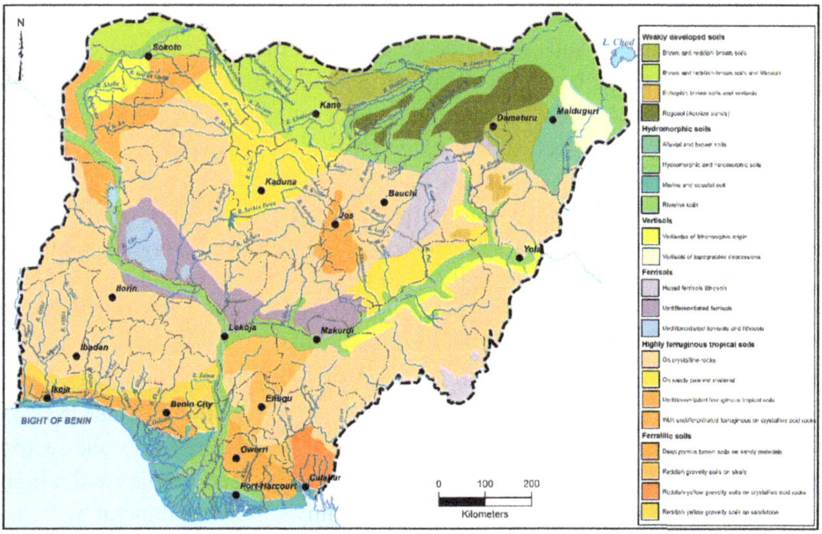

FIGURE 6.1 Major soil types of Nigeria (Modified from Macmillan, 2010)

related to parent material lithology, which influences such properties as soil depth, texture and stoniness, moisture conditions, nutrient status and the proportion of weatherable minerals. At the local level, soil types are related to slope, and it is possible to recognize a sequence of soil types from the hilltop to the valley bottom, known as a soil catena. In areas of smooth relief, the summits of the interfluves usually have sedentary clayey soils; the steep upper slopes have thinner soils; while the lower slope soils, though deeper, may be more sandy and gravelly than the summit soils because they are formed of materials washed downslope by erosion.

The major soil groups in Nigeria are described using the original terminology of Smyth and Montgomery (1962), Klinkenberg and Higgins (1968) and Areola (1978). But the soil classes indicated have been correlated with the soil classes defined in FAO (2015) World Reference Base for Soil Resources (WRB) soil classification scheme.

Hydromorphic soils (*Gleysols, fluvisols*)

These are seasonally or permanently waterlogged soils. Some (gleysols) are formed as a result of high water table or seasonally high water table, while others (fluvisols) are of fluvial origin and result from seasonal flooding by rivers. The influence of poor drainage is reflected in the greyish colours due to the reduction of the oxides in the soil. Permanently waterlogged soils show little horizon differentiation, the profile being largely uniformly coloured. The extent of profile development of the seasonally waterlogged soils is limited by the depth of the permanent water table. Brown, orange or reddish mottles occur in the zone of seasonally fluctuating water table.

Permanently waterlogged soils usually have a surface accumulation of raw organic matter which imparts a highly acidic reaction and a very dark grey colour to the soil. These are the organic hydromorphic soils. In contrast, the mineral hydromorphic soils are usually those seasonally waterlogged, where organic matter has a chance of decomposing periodically when the water level subsides. The organic types are most common along the coastal creeks, valleys, depressions and estuaries covered by mangrove swamps while the mineral types are more characteristic of the river valleys and flood plains and the Chad basin.

Among the mineral hydromorphic soils, there is a distinction between the sandy and the silty or clayey types. The sandy soils are basically arenosols and regosols while the silty or clayey types have deeper, more developed profiles and a higher natural fertility hence, classified as vertisols and fluvisols. Most mineral hydromorphic soils in Nigeria are of the sandy type; the clayey types occur in pockets within the valleys and flood plains.

Weakly developed Soils (*Regosols, Cambisols*)

The major group of immature or weakly developed soils in Nigeria comprises of the loose coarse sandy soils referred to as regosols, the other major group being the lithosols, consisting of rock fragments and usually found in isolated hilly areas.

The regosols are generally loose, excessively drained and intensely leached. Apart from the humus-rich surface layer, the regosols show little evidence of pedogenesis except for the oxidation and consequent red or brown colouring of the sands. They occur mainly on the coastal sand ridge barriers and in the Chad basin where they are developed on desert sand drift. In other parts of the extreme north the desert deposits have been subjected to greater pedogenesis giving rise to the brown and red-brown soils (cambisols) found in central Borno, and the northern parts of Kano, Katsina and Sokoto States. The sparse vegetation does not provide much litter but the plant roots which often decay *in situ* within the soil are responsible for a fairly even distribution of humus in the profile. These soils are usually clay-enriched suggesting some degree of eluviation. However, the degree of leaching is small and the soils have fairly high contents of weatherable minerals and water and nutrient holding capacities. With adequate water supply, either through rainfall or by irrigation, the soils are very productive. Most of the groundnut crop produced in Nigeria is grown on these soils.

Ferrallitic Soils (*Ferralsols*)

Deep, intensely weathered and highly leached soils with little horizon differentiation, the ferralsols (correlate with USDA *Ultisols, Oxisols*) are developed on the coastal sedimentary rocks and deposits comprising mostly sandstones or sandy deposits, with intervening clay lenses. Iron, aluminium, manganese and other oxides are diffused throughout the profile. However, the oxides form nodules and concretions in areas of fluctuating water table.

Because of the intensity of weathering and leaching the soils are highly deficient in weatherable mineral reserves and the clay content is kaolinite with low water and nutrient holding capacities. The cation exchange capacity is less than 20 μm per 100g clay and base saturation is less than 40 per cent. The productivity of these soils depends on the supply of litter from natural vegetation. Unfortunately, this source of nutrients is no longer assured because humans have greatly reduced the volume and density of the forest or woodland cover. The clayey ferralsols have a higher fertility than the sandy ones which also suffer from rapid internal drainage and susceptibility to erosion as in the scarp lands of Enugu State.

Ferrisols (*Lixisols/Nitisols*)

These have developed on sandstone formations in the Guinea savanna zone including the Nupe sandstones of the middle Niger trough, sandstones of the Lower Benue trough and the Kerri Kerri sandstones of upper Gongola basin. They are transitional soils resembling the ferralsols but having less well-developed profiles, a higher exchangeable base content, a better structure and higher biological activity. The cation-exchange capacity is usually greater than 15 μm per 100g clay and is up to 20 μm in some cases. Base saturation is less than 40 per cent. The lower annual rainfall is responsible for the reduced degree of profile development and of leaching while the relatively sparse population had left the soils protected under dense woodland vegetation, which accounts for the high humus content. The soils are usually deep red in colour with clay-enriched subsoils. Lateritic iron caps have developed on flat-topped ridges such as those found near Bida

Highly ferruginous tropical soils (*Nitisols, Luvisols*)

These constitute probably the most important group of soils in Nigeria because they cover the greater proportion of the country and support most of the important cash and food crops. The soils are derived mainly from Basement Complex and old sedimentary rocks. However, in the northern high plains they are developed on drift deposits covering these rocks. These soils are distinguished from the ferralsols by the marked differentiation of horizons, and the abundance of free iron oxides usually deposited as red or yellow mottles or concretions. In the drier or exposed sites, a distinct lateritic layer is present in the soil profile. The soils become increasingly clayey down the profile because of intense leaching and eluviation from above. The clay fraction is made up mostly of kaolinite, although there are small quantities of illite. The nutrient holding capacity is therefore very moderate but base saturation is more than 40 per cent. The weatherable mineral reserves are often quite appreciable as compared to the ferralsols. Those developed on amphibolites and related basic rocks as well as biotite and biotite-hornblende gneisses have larger reserves while they are also more clayey and water-retentive. By contrast granites, quartzites and sandstones give rise to coarser and poorer soils.

Soil productivity classes in Nigeria

The major soil types in Nigeria vary in their potential for agricultural use. Thus, they have been grouped into agricultural productivity classes by the FAO as shown in Table 6.1. None of the soils in the country was rated as Class 1 soils of *high productivity potential*. By contrast over 48 % of the Nigerian soils fall into classes 4 and 5, that is soils with *low productivity potential* due to inadequate moisture retention capacity and low organic matter. They include the vertisols, alisols, ferrasols and arenosols. Except for the ferrasols, they are the most dominant types found in the northern dry parts of the country. Fortunately, soils of *good to medium productivity potential* cover more than half (51.9%) of the country. They include the fluvisols, gleysols, regosols, luvisols, lixisols, and cambisols.

TABLE 6.1 Productivity potential of Nigerian soils

Soil Productivity grade	FAO Productivity Classes	Area km^2	% of total
High (1)		–	–
Good (2)	Fluvisols, Gleysols, Regosols	50.4	5.52
Medium (3)	Lixisols, Cambisols, Luvisols, Nitisols	423.6	46.45
Low (4)	Acrisols, Ferrasols, Alisols, Vertisols	289.2	31.72
Low (5)	Arenosols, Nitisols	148.8	16.32

Source: Originally from FAO and reported in Agboola, S. A. (1979) *An Agricultural Atlas of Nigeria*. Oxford, University Press, Oxford.

Agriculture in Nigeria

The combined influences of climate, vegetation and soils determine the spatial pattern of agricultural practice in Nigeria. It is possible to recognize distinct crop-ecological zones from the humid south to the sub-humid north (Fig.6.2).

In the southern forest zone tree crops are cultivated including cocoa (*Theobroma cacao)*, kolanut (*Cola nitida*), oil palm (*Elaeis guineensis*), rubber (*Hevea brasiliensis*), coconut (*Cocos nucifera*), plantain and banana (*Musa* spp.). This agro-ecological zone is also well suited to many food plants grown mainly for home consumption and internal trade, although a number of them have become export products in the 21st century. They include (1) root crops and tubers such as yam (*Dioscorea esculenta*), cassava (*Manihot esculenta*), cocoyam (*Colocasia esculenta*), and sweet potato (Ipomoea spp.); (2) numerous fruits such as pineapple, pawpaw, guava, citrus, mango and cashew; (3) cereals and legumes, including maize (*Zea mays*), rice (*Oryza sativa; O. glaberiima*), cowpeas (*Vigna unguiculata*), beans(*Vigna spp., Phaseolus spp.*) and groundnut (*Arachis hypogaea*); and (4) vegetables such as melon, okra (*Abelmoschus esculentus*), peppers, and a variety of greens. Carrot, cucumber and water melon are increasingly being grown on the outskirts of major cities by urban residents to augment their meagre incomes.

Nigeria is the largest producer of cassava in the world and she exports it to other countries including the United States of America where it is used primarily to feed pigs. Yam is also exported to the United States, England, European countries and other African countries where it is primarily in demand by immigrant West Africans.

The derived and southern Guinea savanna zones of the Nigerian Middle Belt combine the root crop economy of the forest region to the south and the grain crop economy of the savanna regions further north. Yam is by far the most important root crop. Others are cassava and potatoes. The major food grains are guinea corn (*Sorghum*), maize, rice and acha (*Digitaria exilis and D. iburua*), considered to

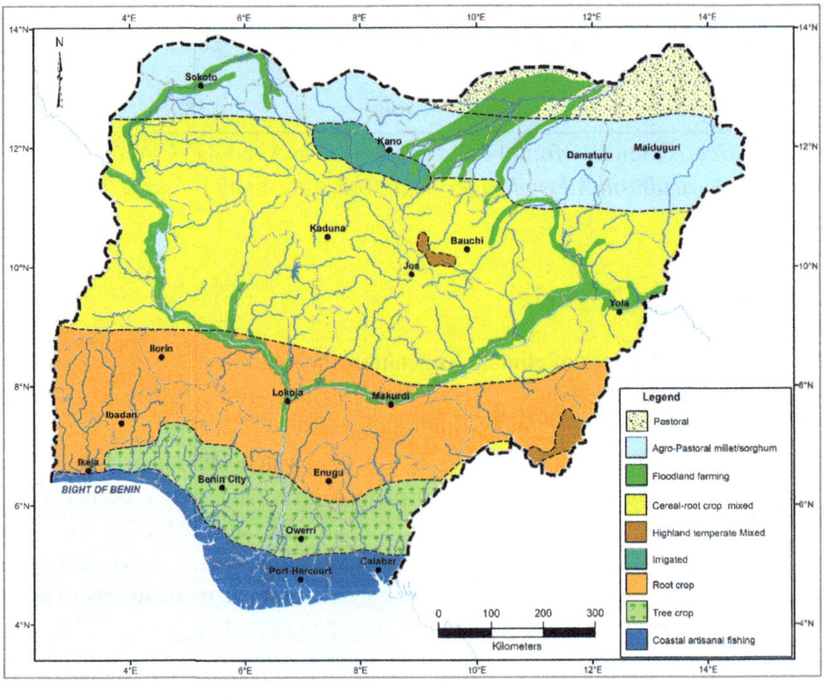

FIGURE 6.2 Major farming systems in Nigeria (Modified from Macmillan, 2010)

be the oldest African cereal. The major cash crops are soya beans (*Glycine max*), benniseed (*Sesamum indicum*), shea nuts from the shea tree (*Vitellaria paradoxa*), cotton (*Gossypium hirsupum*) and tobacco (*Nicotiana tabacum*).

In the drier, open savanna grasslands of the north, grains, legumes and livestock are the principal products. The major food crops are guinea corn, millet (*Pennisetum spp.*), groundnuts, cowpeas, tomatoes, onions and peppers. Cotton and tobacco are widely grown. Tomatoes, onions, cabbage, and peppers are the primary crops in the irrigated farming projects in the northern savannas and most of the output is sold in the big cities to the south.

Nigeria has a vast resource of land. Of the total land area of the country, approximately 923,768 km^2, only about 9 per cent is reserved as conservation areas and is unutilized as stated in Chapter 4. By the 1980s 56 per cent of the total land was devoted to forestry and permanent pastures while 35 per cent was devoted to arable and permanent crop cultivation. It was estimated then that as much as 31 per cent of the 56 per cent classified as forest or permanent pasture, constituted potential agricultural land. This meant that there was great scope for expansion in the agricultural sector of the economy. As noted in Chapters 1 and 4, there has been a considerable expansion of the area devoted to agriculture since the 1970s. By 2015 agriculture occupied as much as 40% of the country's total land area. But, as we said earlier, the overall effect of the expansion on the national economy and on the social and economic wellbeing of the people has been well below expectation. Most farming is still done by peasant farmers who cultivate relatively small parcels of land. Farm sizes are traditionally small ranging between 0.4 and 2 hectares.

Agriculture had contributed immensely to the foreign trade and internal economy of Nigeria in the past. Until the oil boom of the early 1970s, agricultural exports were the main foreign exchange earners for the country. Unfortunately, since the 1960s there has been a downward trend in the quantity of almost all agricultural exports. Part of the explanation for this was increased internal consumption of products such as cotton and groundnuts. But the real reason was the general neglect of agriculture and the failure to resolve the major constraints on progress in the sector as highlighted later in this chapter by the FAO.

Table 6.2 clearly reveals the sharp decline in both the quantity and value of Nigeria's agricultural exports by the 1980s in contrast to the steady increase in the 1960s. Palm oil, cotton, benniseed and groundnut had disappeared completely from among Nigeria's exports by the 1980s but palm kernel, palm kernel oil and cake still featured in the export trade. But, by the first decade of the 21st century the National Bureau of Statistics had greatly improved on her data collection, data analysis and database management system. Table 6.3 shows that the number of

TABLE 6.2 Quantity and value of agricultural exports, 1965, 1973, 1980-85

Crop	Quantity of Export ('000 tonnes)								Value of Exports (₦ million)							
	1965	1970	1980	1981	1982	1983	1984	1985	1965	1970	1980	1981	1982	1983	1984	1985
Cocoa beans	260	197	119	128	136	206	131	93	85.4	133.1	175.9	172.9	150.1	232.4	192.6	182.1
Cocoa butter	*	*	5	8	7	10	7	11	*	*	19.3	23.0	18.3	25.4	26.4	46.8
Palm kernel	424	186	35	91	28	54	42	32	53.1	21.7	4.4	13.8	3.7	11.9	9.2	6.2
Palm kernel oil	*	*	29	8	6	8	8	6	*	*	10.4	2.1	1.6	3.9	4.7	3.3
Palm kernel cake	*	*	13	25	16	8	1	14	*	*	0.8	2.1	1.8	0.8	0.1	0.6
Palm oil	153	8	1	-	2	-	-	-	27.2	1.13	0.2	-	0.6	-	-	-
Rubber	68,963	59,286	41	21	17	8	6	6	22.0	17.6	19.6	10.3	7.7	3.7	2.9	3.8
Groundnut oil	93	91	*	*	*	*	*	*	20.03	23.3	*	*	*	*	*	*
Groundnut	522	293	*	*	*	*	*	*	75.6	43.5	*	*	*	*	*	*
Groundnut cake	115	163	*	*	*	*	*	*	10.5	11.03	*	*	*	*	*	*
Cotton (raw)	13,746	28,288	*	*	*	*	*	*	6.6	13.13	*	*	*	*	*	*
Benniseed	20,878	12,084	*	*	*	*	*	*	2.9	2.3	*	*	*	*	*	*

agricultural products surveyed had been expanded beyond what obtained in the 20th century. Significantly, there is a disaggregation of data to show production figures for the different types of poultry while fishery is now included.

Table 6.4 gives more recent data on crop production and productivity in Nigeria as compared with the West Africa region and Africa as a whole. The internal trade in foodstuffs has grown tremendously in the last four decades. Many factors are responsible for this among which are the increasing concentration of the population in large urban centres and the tremendous improvement in road transportation which has facilitated the movement of both people and goods. But, the most important reason is because the production of food crops has not kept pace with the rate of population growth and urbanization such that Nigeria has had to import to feed her people.

The FAO (2018) explained the problem using the example of rice. It noted that

> Nigeria is the continent's leading consumer of *rice*, one of the largest producers of rice in Africa and simultaneously one of the largest rice importers in the world. As well as an important food security crop, it (rice) is an essential cash crop for it is mainly small-scale producers who commonly sell 80 per cent of total production and consume only 20 per cent. Rice generates more income for Nigerian farmers than any other cash crop in the country. In 2008, Nigeria produced approximately 2 million MT of milled rice and imported roughly 3 million metric tons.

On the other hand, Nigeria is the largest producer of cassava in the world accounting for as much as 20 percent of world cassava output. Hence, the FAO has urged the government and private sector to join efforts to develop ways to enhance cassava's competitiveness in the international market and at the same time improve the efficiency of domestic rice production and processing. Only a range of similar policies and initiatives can help Nigeria bridge the gap between population growth and food crop production.

Soil management problems

Throughout the evolution of agriculture in Nigeria farmers have had to contend with three soil-related environmental problems, namely, soil deterioration; soil desiccation; and accelerated soil erosion. A brief summary of these conditions provides a useful background to the discussion of the agricultural systems. Some land management practices in crop farming have contributed in no small measure to the emergence and intensification of these problems; but as we shall see later, some other land management practices have developed in response to the soil-related environmental problems and have, so far, been able to cope with and contain them. Most local farmers are well aware of the deleterious effects of soil deterioration, and understand its implications. But, except in the worst affected areas, the other two problems are not so well understood and have received less attention.

Soil deterioration is a marked lowering of the natural fertility and hence the productivity of the soil through a fall in soil humus and nutrient levels. Many factors including the rarity of weatherable minerals in the soil, the kaolinitic nature of the clays and the high content of inert particles predispose Nigerian soils to rapid exhaustion of fertility and structural breakdown under cultivation. *Soil desiccation* is the progressive drying out, compaction and hardening of the soil resulting, in extreme cases, in the development of clay pans and lateritic crusts in the soil profiles and the occurrence of patched bare land surfaces. Soil desiccation is most marked in the sub-humid regions of the north where prolonged crop cultivation, overgrazing and bush burning have combined to reduce the protective vegetation cover and exposed the land to the weather elements in these climatically marginal areas. But soil desiccation occurs also in the humid south particularly in some areas of sandy loams developed on sedimentary or granitic crystalline rocks. *Accelerated soil erosion* is the rapid removal of soil materials by strong winds, torrential rains, surface runoff and occasional flood waters. It has attracted attention in some parts of the country where deep, spectacular gullies have developed, notably, in Abia, Imo, Anambra and Enugu States, the Jos Plateau, Gombe and the Zaria regions. Yet the problem of soil erosion exists in other areas: sheet erosion is particularly prevalent at the beginning of the wet season when most farmlands are bare of crop or vegetation cover.

These problems are compounded by replacement of natural vegetation by crops, which neither adequately protect the soil against excessive moisture loss nor main-

TABLE 6.3 Agricultural production in Nigeria, 2006 - 2009

NBS/CBN/CCC COLABORATIVE SURVEY ON SOCIO-ECONOMIC ACTIVITIES

CROPS	Production in ('000 tons)—National			
	2006	2007	2008	2009
Millet	5940	4,388.00	4,327.61	4,884.89
Guinea corn	6474	5,429.12	5,218.40	5,270.79
Ground nut	3062	2,843.37	2,872.74	2,969.26
Beans	3770	1,921.27	2,096.76	2,369.58
Yam	28280	26,751.38	27,211.07	29,091.98
Cotton	516	819	785	772
Maize	7023	5,796.67	9,113.71	7,338.84
Cassava	38041	39,130.83	33,216.39	36,804.30
Rice	3333	3,561.55	3,369.70	3,402.59
Mellon	371		377.95	340.88
Cocoyam	2220		2,814.14	2,639.85
Soya beans				573.86
LIVESTOCK	Number of livestock			
Cattle	26,489,858	25,019,700	25,632,683	66,200,889
Goat	60,790,062	57,416,281	58,822,980	88,775,004
Sheep	37,932,521	35,827,308	36,705,077	39,825,343
Pigs	2,587,712	2,444,096	2,503,977	3,338,586
Guinea Pig	509,461	481,186	492,975	91,742
Camel	35,665	33,686	34,511	1,334,491
Donkey	639,389	603,904	618,699	2,026,866
Horse	36,839	34,794	35,647	23,591
Rabbit	65,950	62,290	63,816	353,450
Dog	1,695,571	1,601,468	1,640,704	1,677,442
Others	14,078	13,297	13,623	49,546
POULTRY	Number of birds			
Chicken	79,487,005	75,075,565	76,914,916	187,185,351
Guinea fowl	5,631,175	5,318,651	5,448,958	94,315,131
Duck	3,111,194	2,938,526	3,010,520	22,441,152
Turkey	205,861	194,436	199,199	5,737,582
Pigeon	843,208	796,411	815,923	2,354,364
Geese	338,147	319,380	327,205	210,794
Ostrich	66,004	62,340	63,868	26,960
FISHERY	Quantity (kg)			
Tilapia (kg)		333,822,775	386,900,532	410,602,014
Cat (kg)		146,932,991	216,401,911	191,012,888

Source: National Bureau of Statistics: *2009 Annual Collaborative Survey of Socio-Economic Activities in Nigeria Main Statistical Report Volume 1* August, 2010.

TABLE 6.4 Food crop production and productivity, 2011

	Crops							
	Area		Yield		Production			
	total	p.a. growth	total	p.a. growth	Total		p.a. growth	
	thousand ha	Percent	thousand hg/ha	Percent	thousand tonnes	thousand tonnes	percent	percent
	2011	2000-2011	2011	2000-2011	2010	2011	1990-2000	2000-2011
Root and tuber producers and their productivity								
Africa	25 661.7	2.1	100.7	1.8	239 052.6	258 415.0	5.5	4.3
West Africa	12 404.7	1.5	111.9	1.7	124 186.8	138 841.6	7.2	3.7
Nigeria	8 441.4	1.0	115.1	2.7	84 052.3	97 184.7	6.8	3.7
Coarse grain producers and their productivity								
Africa	90 371.3	2.6	12.5	1.6	115 310.9	112 817.9		3.9
West Africa	38 772.0	2.2	9.3	0.2	41 924.3	36 083.0	2.3	2.1
Nigeria	13 948.6	−0.1	12.5	0.2	20 073.8	17 433.4	1.8	0.1
Rice producers and their productivity								
Africa								
West Africa	6 248.6	4.1	19.1	1.1	12 397.4	11 939.5	3.5	5.6
Nigeria	2 579.5	1.5	17.7	1.5	4 472.5	4 567.3	2.8	3.0
Wheat producers and their productivity								
**Nigeria	99.0	6.0	16.7	1.6	110.0	165.0	3.9	7.7
Oil crop producers and their productivity								
Africa	34 527.9	2.8	3.0	0.8	9 981.3	10 341.5	3.7	4.6
West Africa	14 455.1	1.8	3.4	−0.3 5	402.2	4 890.9	4.2	2.1
Nigeria	7 588.0	0.7	3.7	0.7	3 036.5	2 776.0	4.9	1.3
Pulse producers and their productivity								
Africa	22 429.7	4.2	6.2	1.5	15 292.0	13 887.6		5.8
West Africa	10 559.6	3.6	4.5	0.9	6 715.2	4 728.6	4.7	6.6
Nigeria	3 320.0	−1.1	5.8	−0.3	3 424.2	1 916.8	4.7	−1.4
Vegetable producers and their productivity								
Africa	6 877.8	2.0	103.1	1.6	70 869.8	70 879.7	4.4	3.5
West Africa	2 719.9	−0.0	62.7	2.2	17 386.8	17 050.9	5.6	2.2
Nigeria	1 869.4	0.7	61.2	1.9	11 830.2	11 439.6	6.4	2.6

Notes: For Nigeria, the various classes of crops may be interpreted as follows

1. *Roots & Tubers*: Potatoes, Sweet Potatoes, Cassava, Cocoyam, Yams
2. *Coarse grains* include Maize, Millet, and Sorghum
3. *Pulses* are annual leguminous crops yielding grains or seeds e.g. cowpeas, beans, soya beans
4. *Oil crops*: coconuts, cottonseed, groundnuts, shea nuts, melon seed, Oil palm fruit, Palm kernels, Palm oil, Sesame seed, Soybeans.
5. *Vegetables*: fruit crops, melons, watermelon, chillies, peppers, tomatoes, onions, greens

* *hg/ha* = hectograms per hectare (1 hectogram = 100 grams)
** No data for Africa and West Africa
Source: Extracted from FAO *Statistical Yearbook, 2014* `http://www.fao.org/3/a-i3620e.pdf` Part 3: Feeding the World pp. 61–114

tain soil humus levels. Thus, loss of vegetation cover, repeated cultivation and bush burning all expose the soil to intense insolation leading to increased rates of evaporation, and the decomposition and oxidation of organic matter. Soil structural breakdown and accelerated erosion are, therefore, precipitated by both cultural and environmental forces.

The ecology of agricultural systems

The various systems of cultivation in Nigeria can be grouped into five broad categories: (1) shifting cultivation with bush fallowing; (2) systems of continuous cultivation with animal or farmyard manure and crop rotation but without the integration of crop and animal husbandry; and rudimentary mixed farming; (3) permanent tree crop cultivation; (4) irrigation and floodland agriculture; and (5) hill cultivation and terrace farming.

Shifting cultivation

This is by far the most widespread agricultural system in Nigeria and, in modern times, remains perhaps the most destructive of vegetation and soil fertility. But, in the past when the man/land ratio in many parts of the country was still very low, the system worked admirably being well adjusted to the tropical environment. Even today, fairly high crop yields are obtained by peasant farmers under this system of cropping (Table 6.3). But, the general decline in crop yields over time in traditional farming is also evident in the obvious downward trend in crop yields and values in Tables 6.3 & 6.4.

The system developed as a natural response to the problem of producing food in an environment where the soils are only of moderate fertility and subject to rapid exhaustion. Since the soils are not capable of sustaining continuous cultivation for an unlimited period. As there is little or no application of fertilizers or animal manure, even in the 21st century (Table 6.5), arable lands have to be rested after two or three years of cultivation. The rest or fallow period normally lasts for two to five years but sometimes it can be as long as twenty years before lost fertility can be regained. During this fallow period the land reverts to secondary bush so that soil fertility is restored under natural conditions, primarily through the build-up of organic matter. Thus, shifting cultivation is characterized by the rotation of land rather than of crops.

TABLE 6.5 Farm inputs in food crop production, 2011

	Agricultural tractors	Pesticides	Fertilizer consumption		
	Total	per ha of arable land and permanent crops	per ha of arable land and permanent crops		
	tractors 2000-12*	kg/ha 2008-12*	nitrogen kg/ha 2011	phosphate kg/ha 2011	potash kg/ha 2011
Africa			13.26	9.11	1.63
West Africa			3.74	1.72	0.92
Nigeria	24 800		2.95	0.65	0.27

Source: Extracted from FAO *Statistical Yearbook, 2014* `http://www.fao.org/3/a-i3620e.pdf` Part 1: The Setting pp. 1–30.

The role of burning

The forest is usually cleared and burnt before cultivation of the land; burning is a quick and efficient means of making available to the soil the nutrients stored in the vegetation. Except for the nitrogen and carbon contents which are volatilized and lost to the atmosphere the bulk of the nutrients such as phosphorus, potassium, calcium and magnesium is returned to the soil in the ash from the burnt litter. The nutrients concentrated on each small parcel of land substantially raise the soil pH. This is of great significance because most tropical soils are highly leached and acidic in reaction. The timing of the burning is important; if the ash remains for too long on the surface without being ploughed in, it may be carried away by winds or washed away by cloud bursts at the beginning of the rains. The fear has often been expressed that the burning of the cleared vegetation may destroy organic matter. But Nye and Greenland (1965, p.7b) had pointed out that it is only the non-humified organic litter that is burnt and there has been no evidence of any loss of humified organic matter from the soil itself. They further noted that data from Ghana, Nigeria, Liberia and Sri Lanka did not indicate any significant changes in the carbon and nitrogen contents of the soils following burning. However, repeated burning of the vegetation on the same piece of land from year to year may lead to a progressive desiccation and baking of the soil particularly since the fallow period is now drastically reduced in many parts of Nigeria.

When the land is cleared for cultivation, economically valuable trees, among them the iroko, mahogany, palms, shea butter and locust bean trees are preserved.

The stumps and roots of smaller trees are also left in place. Apart from the fact that they often provide support for crops such as yams, these are left by the famers to facilitate regeneration of the vegetation at the beginning of the following fallow period.

Soil and crop management

A lot has been written about the crude farm implements of the African husbandmen which only scratch the soil surface. Greater understanding of the system and the environmental conditions controlling it has now revealed that deep ploughing is in fact undesirable since the soil is largely infertile a few centimetres below the top humus-rich layer, and disturbance may bring up toxic elements and infertile soil materials to the rooting zone of crops. Cultivation with the hoe turns over only the top nutrient-rich layers in which crops can make full use of available resources. Shallow ploughing almost certainly reduces the susceptibility of the soils to erosion. In the savanna regions of the north and the middle belt, mulching and the ploughing in of dead organic litter are additional ameliorative soil management practices.

Ploughing to loosen heavy compact soils and improve their permeability to air and water is of course, an essential component of cultivation. On most farms mixed cropping or intercropping is practised such that the same plot may carry yams, okra, maize, pepper, gourds and melons. Thus, the crop cover reproduces to some extent the complexity of the natural plant community. Instead of aggravating the exhaustion of soil nutrients, as was once supposed, mixed cropping makes for a better ecological balance and more efficient biological activity.

In the same way the practice of gathering the soil into heaps was once believed to be a major cause of soil erosion on farmlands. The heaps were said to concentrate runoff and thereby initiate the development of gullies especially on sloping surfaces. It was therefore recommended that farmers should replace the traditional heaps with ridges. But there has been little empirical evidence to support this claim. It is true that many farms suffer from sheet erosion during the early rainy season when they are still bare of crops, but the extent of the erosion varies with the slope of the land and the type and density of crop cover, not on whether the soil is cultivated in heaps or ridges. There is no evidence that erosion is greater on farmlands with heaps and farmers in different parts of the country make heaps or ridges according to the types of crops being cultivated. Groundnuts, cowpeas, soya beans and vegetables everywhere are planted on ridges while most other crops are grown on heaps.

After two or three years of cultivation soil fertility deteriorates as a result of depletion of both organic and inorganic nutrient elements. The most rapidly exhausted elements are potassium, phosphorus and nitrogen. Nitrogen deficiency is very acute in the savanna regions especially where the natural grassland community is dominated by *Andropogoneae* as opposed to *Pennisetum and Hyparrhenia* species. Grasses tend to repress the nitrate level of the soil because of their sensitivity even to low concentrations of salt. By contrast, there is hardly any noticeable fall in the potassium content throughout the cultivation period in the savanna regions.

The fall in soil nutrient status is matched by a decline in crop yields. The rate of decline varies from one ecological zone to another. In general, it is more rapid in the forest region, where the main crops are large tubers, than in the savanna regions where grain crops are dominant. The rapid decline in yields in the forest zone is not due only to the removal of nutrients from the soil. It is also due to the more ephemeral and temporary fertility of the underlying soils which can only be maintained as long as the forest cover is left undisturbed. Once the forest is removed the source of nutrients is gone; the store of nutrients in the vegetation can only be fully utilized by the first-year crops while the amount available in subsequent growing seasons is greatly reduced. The rate of decline in yields is more moderate in the savanna regions partly because of the slower rate of nutrient uptake by grain crops and partly because there is a greater concentration of nutrients in the soil relative to that in the vegetation cleared from the land.

In all ecological zones decline in soil fertility is marked by a proliferation of weeds on the farms, although this is by no means the major factor leading farmers to abandon the land to fallow. The subject of weeds is fully covered in the following chapter.

The fallow period

The fallow period is crucial to the stability and continued productivity of the system of shifting cultivation. As secondary vegetation succession takes place on the abandoned farmland the soil humus builds up and so improves soil moisture and nutrient status. The extent to which these improvements in the soil conditions are realized depends very much on the length and totality of the fallow period. Unfortunately, in recent times the length of the fallow period has not only been shortened in most places, but, also, the fallow vegetation is frequently disturbed and secondary succession arrested by annual bush fires so that its restorative capacity is greatly reduced. Thus, the system of shifting cultivation has been weakened in many parts of Nigeria. It has already broken down completely and with disastrous consequences in Tiv and Igbo lands as a result of rapid increase in population resulting in extremely high pressure on land in those areas.

The shortening of the fallow period not only means that the soils cannot support high crop yields, but also that the fallow vegetation communities themselves are less virile and restorative. In the past farmers used to depend on the emergence of certain indicator species of plants to know that a piece of land under fallow was fully restored to its natural fertility. Nowadays it takes much longer for these indicator species to emerge, if they do so at all. The farmers become impatient and in certain areas will then try cultivating portions of the land, for a season or so, in order to find out if it is ready for cropping. Invariably, the soil is found to be still incapable of supporting high crop yields, and in the meantime the trial cultivations have further disturbed the natural restorative process of the fallows.

The efficacy of the fallow itself in restoring soil fertility, under present management conditions, is increasingly questionable. Nowhere is this more obvious than in the derived savanna zone where the farmers do not keep livestock. The establishment of a predominantly grassland type of vegetation in the 20th century has brought in its wake the practice of bush burning on a colossal scale. The soils are under fallow only in the sense that they are not being cultivated, but the land is not really at rest since the vegetation is burnt annually and when the ash is washed away by the first torrential rains the soil nutrients go with it. A comparative study of soils under savanna fallow of different ages (Areola, 1980) showed that although there had been an upward trend in soil nutrient status and although there were some improvements in soil moisture and physical conditions, the rate and magnitude of these improvements were not commensurate with the age of fallow. The phosphorus content of the soils was found to show a negative correlation with the age of fallow. This was perhaps to be expected as phosphorus is highly susceptible to leaching and erosion losses. Furthermore, sheet erosion had rendered some of the older fallows useless as the topsoil had been washed away exposing the lateritic crust below, and the ground surface was strewn with pisolitic gravel and quartz pebbles.

In the far north of Nigeria continuous cultivation and bush burning have caused the erosion of many plateau surfaces and hillside slopes. But no area has been as much eroded as the Udi-Nsukka plateau region of Enugu State. There is very little fine soil left on many of the residual hills and ridges on the plateau surface. Rather, the slopes are covered by detrital laterite from the hill summits. Crop cultivation is now restricted to the lower slopes, where cassava is the main crop, and depressions where yams and some vegetables are grown under palm groves. Ash and refuse from the dwellings are used to manure the soil.

Systems of continuous cultivation

In some parts of the Sudan zone farmers have responded to the demand for more food for larger populations by adopting one of two alternative systems of continuous cultivation The first, which operates chiefly in the densely settled Kano zone (Mortimore, 1967) but also around other large urban centres, is based on the use of animal manure, ash and other household refuse on small intensively cultivated parcels of land. The second type, mixed farming, is practised in widely scattered locations in northern Nigeria including the Jos Plateau, Kano, Borno, and Adamawa areas. The difference between the two systems is that in the former there is no integration of crop and animal husbandry while in the second a balance of crops and animals is maintained on individual holdings. Under the first system manure is obtained from the Fulani herdsmen and from households in exchange for firewood and feed for their animals. Sometimes the Fulani are allowed to camp near the farms while their cattle feed on stubble on recently harvested fields. During this process the animals deposit great quantities of dung that manure the soils. More commonly, however, the farmers obtain manure when the Fulani cart firewood to the urban centres and return with ash, dung and household refuse. The dung comes from a variety of domesticated animals, chiefly goats, sheep and donkeys. Furthermore, the farms are dotted with stands of trees, usually acacia species, which supply litter to augment soil organic matter.

Crop rotation is an essential element of this system, the main crops being groundnut, millet and sorghum. Irrigation farming is practised on a small scale in the valleys to grow such crops as sugar cane, sweet potatoes, cassava and cocoyam.

Mixed farming in Nigeria is at a very rudimentary level. The arrangement referred to above whereby Fulani cattle graze some farms for brief periods is one form, but a more integrated type of husbandry is found among the Shuwa Arabs of the Chad basin and the Mumuye of Adamawa who keep livestock on their farms and sometimes make use of farmyard manure. The animals are kept in sheds, especially during the wet season, to protect them against blood-sucking insects. Animal droppings accumulate in the sheds from where the farmers remove them daily to spread on the fields or store in a large communal heap. Farmers with cattle have a further advantage in that they can use draught oxen to pull the plough.

The essence of these systems of continuous cultivation is that the fallow period has been eliminated without necessarily making use of chemical fertilizers because animal droppings and household refuse are readily and cheaply obtained. The manures are rich in nutrients, particularly nitrogen and phosphorus which are essential to good crop growth. Ash from human dwellings also adds to the store of such inorganic elements as potassium, calcium and magnesium. One of the most important features of the system is that it is less likely to lead to environmental pollution than any type of farming based on excessive use of chemical fertilizers. For instance, the natural vegetation in Kano area has virtually disappeared, except for the widely scattered trees, yet there has been no soil erosion or general environmental deterioration. Thus, we have in the tropics an area where a stable and productive system of agriculture has been found to replace the original natural ecosystem. Unfortunately, the system is not as widespread as it should be, because there is generally a lack of integration of crop and animal husbandry in Nigeria.

Permanent tree crop cultivation

Permanent tree crop cultivation has been adopted by local farmers with greater success than with continuous food crop cultivation. This is probably because, unlike the latter, tree crop cultivation does not entail churning up the soil every year while the problem of soil exhaustion is longer in coming and does not have such dramatic effects on yields. Once fully established most tree crop communities are self-maintaining and fit well into local ecological conditions, particularly in the forest region of the humid south. The important tree crops are cocoa, rubber, kola and oil palm. Others are coffee and citrus.

Several factors have contributed to the successful establishment of permanent tree crop plantations in Nigeria as in other tropical countries. As Manshard (1974) wrote about the cocoa growers of Ghana, the Nigerian tree crop growers apply basic cultivation methods which do not entail large capital investment and thereby achieve a surplus with minimum effort. Agricultural equipment for spraying, harvesting, drying and processing is simple and easy to use. The soil, climate, and topographic conditions all favour the tree crops.

Altogether, tree crop cultivation is very destructive of forest vegetation. Mature plantations are all monocultures on which natural tree species have been more or less eliminated. Nevertheless, plantations in Nigeria and elsewhere have been established without many of the environmental problems that attend forest clearance for cereal or vegetable cultivation. The reasons for this are that the soil is hardly disturbed and the ground surface is well protected against erosion and moisture loss; litter from the trees and undergrowth helps to maintain the humus level, and the tree crops themselves are very economical in the use of soil nutrients.

But this is not to say that the soil under tree crop cultivation remains as it is under forest. On the contrary, studies by Areola (1985) and Ekanade (1988), showed that there was a decline in nutrients in soils under cocoa as compared to those under mature forests but the level of decline was less than in soils under arable cultivation (see Table 6.6). There was evidence also of the accumulation of copper in the soils under cocoa as a result of the use of cuprous fungicides, but this accumulation had not yet reached toxic levels (Areola, 1984).

Cocoa

Cocoa is about the most important of the tree crops. Until the oil boom in the years after the civil war, cocoa was one of the leading foreign exchange earners for Nigeria (see Table 6.2, 200). It grows best on well drained moisture retentive soils under conditions of moderate to high rainfall; high relative humidity with only a short dry spell during the year; high temperatures; and bright sunshine, especially for ripening. Because of its need for well-drained but moisture-retentive soils cocoa does best on clay-loams or sandy clays on hill slopes and broad interfluves. It does not do so well in the wet conditions of valley bottoms.

Cocoa plots were nearly always established on virgin land freshly cleared of the original forest vegetation, so that the seedlings were able to take advantage of the accumulated nutrient reserves of the virgin soil. The bigger trees were left standing initially to provide shade for the young cocoa trees while leaves, twigs and stems of smaller ones were allowed to decay *in situ* to add to soil organic matter. Food crops such as plantains, bananas and cocoyam were often planted to provide shade for the young seedlings while the big trees were gradually eliminated by firing at the base. As the cocoa trees matured food crops were restricted increasingly to the valley bottoms, hence the established cocoa plantations that we now see are essentially monocultures except for the kolanut trees that have been introduced on many old plantations. The mat of leaves on the floor of mature cocoa plantations to some extent helps in suppressing weeds which form the undergrowth in young plantations. The closed canopy of the trees creates favourable microclimatic conditions within the plantation such as lower temperatures and higher relative humidity. The cocoa leaves on the floor of the plantation decay easily so that the soil under cocoa characteristically has a dark humus-rich top horizon. Erosion is almost non-existent, a condition which is also partly attributable to the strong structural stability of the soils.

Rubber

Rubber plantations also bring all the ecological benefits associated with cocoa plantations but with the added advantage that they are established on lower slopes where the ground is often well-protected by undergrowth of herbaceous and woody plants

TABLE 6.6 Comparative data on soil characteristics under cocoa and forest fallow regrowth, Ibadan region, Nigeria (from Areola, 1984)

Parameters	Soil layer	Cocoa			Forest Regrowth		
		Range	Average	CV	Range	Average	CV
pH (in H_2O)	T	5.8-6.8	6.42	3.1	6.0-6.6	6.37	2.5
	S	6.1-7.5	6.66	6.1	5.8-7.1	6.37*	5.7
Org. C (%)	T	0.69-4.03	2.49	26.5	1.04-2.79	2.27	25.1
	S	0.23-1.90	1.24	29.9	0.78-1.75	1.10*	22.8
Total N (%)	T	0.03-0.36	0.18	44.4	0.03-0.4	0.21	61.9
	S	0.06-0.23	0.13	23.0	0.09-0.19	0.11	27.5
Avail. P (ppm)	T	0.64-16.08	5.23	53.2	0.64-8.36	5.56	43.9
	S	0.50-6.80	2.79	48.2	1.60-3.5	2.60	27.0
Exch. Ca^{++} mg/l	T	2.5-19.0	7.71	42.9	2.0-9.5	4.80*	56.5
	S	1.2-13.0	4.37	59.8	0.8-10.0	3.42*	89.0
Exch. Mg^{++} mg/l	T	1.15-5.60	2.79	34.4	1.25-3.04	1.97*	27.9
	S	0.22-4.26	1.71	48.0	0.17-1.78	0.84*	52.7
Exch. Na^{+} mg/l	T	0.07-2.17	0.22	136.4	0.07-4.09	0.55	227.3
	S	0.05-0.18	0.10	30.3	0.05-0.14	0.09*	31.7
Exch. K^{+} mg/l	T	0.15-1.36	0.40	65.0	0.15-0.24	0.35	57.1
	S	0.09-0.82	0.26	74.5	0.08-0.21	0.14*	28.5
Exch. Acidity	T	0.2-0.8	0.39	38.5	0.2-0.8	0.40	47.5
	S	0.40-1.0	0.50	26.0	0.2-0.6	0.46*	29.3
CEC	T	4.35-24.11	11.52	35.2	4.07-13.22	8.06*	39.8
	S	2.62-17.08	6.94	46.8	2.00-11.79	4.94*	67.8
Base Sat. %	T	899-99.2	96.00	2.5	90.2-98.5	94.17*	3.4
	S	71.8-97.7	91.30	5.6	73.2-96.7	86.55*	9.7
Extractable Cu ppm	T	0.4-12.2	2.68	75.4	0.3-4.6	1.37*	89.8
	S	Tr.-4.1	1.01	109.1	Tr.-17.0	0.90	241.2
% Sand	T	47.2-89.2	72.98	11.2	75.6-89.2	82.50*	4.8
	S	43.2-85.2	65.30	14.4	64.2-81.2	72.60*	7.5
% Silt	T	5.4-35.4	17.09	31.9	5.4-18.0	11.10*	34.1
	S	6.8-30.0	13.30	32.0	6.0-17.0	10.70*	33.7
% Clay	T	4.4-25.4	9.94	47.8	4.2-10.0	6.40*	34.2
	S	6.8-38.8	21.20	35.2	12.8-20.8	16.70*	17.9

T = topsoil 0–15 cm S = subsoil 15–39 cm CV = coefficient of variation (%)
* Statistically significant difference at 5 per cent level between cocoa and forest regrowth soils

(Aweto, 1987). Unlike cocoa, rubber is adapted to areas of sandy loam soils which are free draining but which may, in some places, suffer from poor drainage for part of the year. The vegetation cover of large parts of the Urhobo plains in Delta State is dominated by rubber which is also widespread in those other parts of the state and Edo State underlain by the Benin sands. While the trees, which are evergreen, do not produce as much litter as cocoa trees, the undergrowth, which is regularly cleared by the plantation owners, add organic matter to the soil.

Perhaps more than any other tree crop, the rubber industry had suffered from indifferent management resulting in low yields and poor-quality products. This had been attributed to the practice of share cropping, that is, the employment of itinerant tappers for a period of six or twelve months every year to tap the rubber trees. As to be expected the itinerant tappers were primarily interested in obtaining the highest possible daily output of latex in order to maximize their profit during the term of their contract. Without supervision by the farmers of their tapping or processing operations, the tappers paid little attention to correct tapping methods, with the result that the trees were tapped daily or even twice daily and cuts were too deep (often to the cambium) and too many (up to three or even more) on each tree. The poor tapping methods reduced the economic life of the rubber trees (FAO, 1966). A number of such degraded plantations became overgrown with bush, uncared for by the farmers. The rubber plantation owners seemed to have given higher priority to food crop production and were only too willing to leave itinerant workers to do the rubber tapping for them. Many rubber plantations have been rehabilitated and re-planted since the 1980s and are now better operated and managed.

Oil palm

Oil palm plantations are less frequent being restricted to a few government farms in the Edo, Delta, Ondo, Lagos and Cross River States. But there are thousands of hectares of native oil palm groves in Abia, Imo, Enugu and Anambra States and parts of Delta, Rivers, Bayelsa and Cross River States. These groves characteristically occupy the valleys and wide amphitheatre-shaped depressions between the hills and ridges characteristic of these coastal lowland regions. They are also found on low clay ridges. In general, these are areas of low relief with little danger of soil erosion. The groves are so dense that the soil is well-protected against the direct impact of rain drops. The land is usually cultivated to grow yams and vegetables often applying ash and other household refuse as soil manure.

Ecology of tree crop cultivation

The long-term effects on the soil of each of these tree crops are not yet known. There is no documented evidence so far of soil exhaustion, but there are fears that residues of the fungicides used in spraying the trees, may accumulate in the soil. The heavy rainfall in the cocoa growing areas makes this less likely in the short term as the chemicals are easily washed away both from the trees and from the soil. Cocoa, like other monocultures, has encouraged the spread of certain pests and diseases which may also affect other crops. In some areas the pests have developed resistance to the common cuprous chemicals in use (Gamalin 20 and Perenox). New chemicals have therefore been introduced. The incidence of pests and diseases is lower in rubber plantations but the trees suffer from root diseases which are treated with fungicides. The possibility of soil contamination cannot be ruled out but again the high rainfall and free soil drainage make this unlikely in the short term. However, the chemicals may end up somehow in aquifers and contaminate valuable sources of water supply.

In the 1980s chemical fertilizers were introduced on many tree crop plantations; peasant farmers freely made use of them particularly on cocoa farms. The types commonly used included the NPK fertilizers for cocoa and rubber; and sulphate of ammonia, potash and magnesium sulphate for oil palm. Their use by farmers has been irregular and it is difficult to determine the ecological impact they may have had on the environment. As with the fungicides and pesticides, the environment is not likely to suffer from overuse of chemical fertilizers for the foreseeable future. Irregular supplies and high costs of farm inputs have limited the use of these chemicals, which is to the advantage of the general environment. In order to cut down on costs, farmers cut corners by bending the rules not only on the quantities to apply but also on the number of times to apply the chemicals to the crops.

Irrigation and flood land agriculture

Since the 1960s, a number of large-scale irrigation projects had been established and, more have been established in more recent times in different parts of Nigeria notably in the sub-humid savanna regions. Oguntoyinbo (1971) grouped the earliest modern irrigation schemes into two broad types: (1) large scale commercial enterprises such as the Bacita Sugar Plantation in the Niger Valley near Jebba and the Nigerian Agricultural Development Industry (also sugar) in the Gongola Valley; and (2) enterprises organized either on a commercial basis with peasant participation, or mainly for a subsistence economy to make additional products available at local markets, for example, the Sokoto-Rima Valley Scheme, the Ta Hoss and

Ganawuri Scheme on the Jos Plateau, and the Pategi Rice Development Scheme. Other early irrigation schemes are shown in Fig. 6.3. As we discussed in Chapter 5 on water resources, many irrigation and water supply schemes have been implemented since the creation of river basin development authorities (RBDAs) in 1976. Most of the schemes by the RBDAs seem to have continued with the system of running irrigation schemes as quasi-commercial ventures but with participation by local peasant farmers. Much of the farm operations (including ploughing, fertilizer and pesticide applications, crop harvesting with mechanical harvesters) is done for the farmers by project officials. The urgent need now is to gradually cede these operations to the farmers themselves so that the innovations can take firm root and be sustainable in the long term.

Long before the establishment of these modern irrigation systems, however, there had been a tradition of simple irrigation and valley flooding in several parts of Nigeria. Floodland plain agriculture is the more prevalent up till today. It involves the cultivation of seasonally-flooded and alluvium-enriched river floodplains to grow food crops. As soon as the rains start the farmers make ridges on their arable crop fields so that when the flood water arrives it collects in the furrows. This type of cultivation is practised in almost all the major river valleys in the Sudan savanna zone including the Kano, Challawa, Hadejia, Komadugu-Yobe and Sokoto-Rima. Apart from these river valleys, floodland farming is also practised in several seasonally-inundated depressions with dark, very clayey soils (vertisols), known locally as *firki* in Borno and Yobe States. Further south there are the flood plain agricultural lands of Badeggi, Bida, Agaie and Bwari in Niger State; the Benue and Gongola valleys; the Anambra River valley and the clay vales of Abakaliki and other areas of south-eastern Nigeria.

Irrigation *per se* is usually practised along with floodland farming especially in the far north. Water storage ponds are created on the flood plains either by digging hollows or by using earth banks. The water is led to the farms through a system of shallow ditches. Shadufs are in use in many areas (even as far south as Niger State) especially where the alluvial farmland is on a terrace above the river channel. Shaduf irrigation is effectively used by peasant farmers in raising dry season crops of onions, carrots and other vegetables around the major urban centres of the Sudan zone.

The lush green vegetation which normally grows on most of the wetlands when not in cultivation promotes photosynthesis and nitrogen-fixation. Decayed vegetation which may have accumulated during the flooding decomposes towards the end of the rainy season and old rice stalks and other crop wastes are similarly broken down to add to soil organic matter. Some loss of fine soil materials may occur at high flood and many ridges may be washed away but the standing crop normally helps to stabilise the soil.

The National Fadama Development Project (NFDP)

Farming in the river valley wetlands, referred to as *fadama* in the northern savannas, *akuro* in south-western Nigeria and *aba* in Igbo land got a great boost when the military government instituted the *National Fadama Development Project* (NFDP) in 1990 to address the challenge of poverty in Nigeria. The Project significantly has targetted value chains of cassava, rice, sorghum and horticulture. Funded by the Federal Government through financial assistance from the World Bank, this agricultural development campaign encouraged many educated young men and women significantly in southern Nigeria to acquire parcels of land in the erstwhile neglected valley bottomlands to grow mostly vegetables and grains such as maize and rice along the river valleys running across urban and peri-urban areas The programme was particularly well received in the major metropolises such as Lagos, Ibadan, Port Harcourt and Warri. The programme has continued even till the present perhaps to meet the MDGs and SDGs. The government sourced for additional financing for the third phase of the project, which has the objective to increase the incomes for users of rural lands and water resources within the fadama areas in a sustainable manner. The emphasis is on improving farm productivity performance of clusters of farmers engaged in priority food staples namely rice, cassava, sorghum and horticulture in six selected states with high potential (World Bank, 2019). In addition to farmers, the Project is also mobilizing fisherfolk, hunters, pastoralists and gatherers who make use of the land and water resources. It is quite understandable then why the project has been well received in the north where fadama farming has long been firmly entrenched.

In the Benue valley Iortyom *et al.* (2019) showed that the third phase of the national fadama project had achieved its goal to increase the income of rural inhabitants making use of the land and water resources in a sustainable manner. The project succeeded in making significant economic impacts on rural livelihoods in terms of assets acquisition, provision of farm inputs and increased crop yields. Unfortunately, the age-old problem of lapses on the part of officials showed up again. According to Iortyom *et al.* (2019) the sustainability of the project in the Benue valley is threatened by poor implementation of the advisory and input support, which are key elements of the national fadama development project. Nevertheless, the World Bank (2015) is of the opinion that because of the additional funding provided in the third phase, the project is successfully turning rural farmers into *agro-*

preneurs by allowing them to turn their agricultural skills into income-generating ventures.

A significant aspect of this fadama project which is worthy of note is that it has encouraged many rural farmers to move out of familiar territory and try out another system of farming. A few examples will suffice to highlight the point being made. In Ibarapa division of Oyo State, for example, the soils are predominantly light, sandy and easy to work. But in the valley bottoms the soils are generally more clayey and darker in colour because of seasonal flooding and high organic content. These valley bottoms had largely been avoided by the local farmers who are more familiar with the lighter soils.

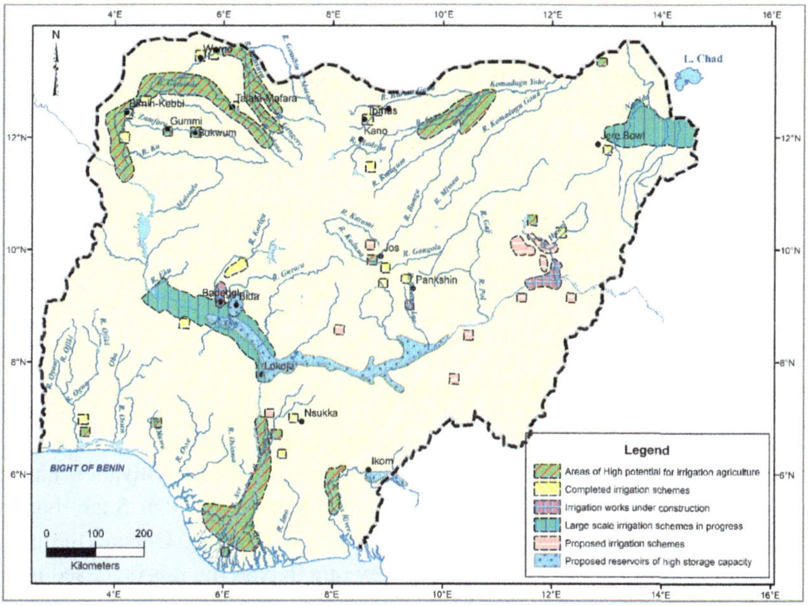

FIGURE 6.3 Early modern irrigation and land reclamation schemes in Nigeria. (Oguntoyinbo, 1971)

Yet, elsewhere, these heavier soils are put to good use. In the late 1980s, parts of the valley bottoms were brought into cultivation, not by the long-established farmers, but by newcomers who were familiar with the working of such soils or who had not been able to obtain land elsewhere. In like manner, in parts of Ekiti, lands with light loamy soils, which had been largely degraded to open grassland, had been progressively left to migrant farmers while the local people had gone after the heavier and still largely wooded areas. These arrangements were not satisfactory at all because the farmers were not learning new soil management techniques but kept to their own age-old practices in familiar environments. Real progress would only be made when local farmers themselves can learn to manage and cultivate the less familiar soils in their different localities. The National Fadama Development Project is making rural farmers do exactly that.

Compared to the flood-irrigated lands of south-east Asia, the Nigerian flood plains have a number of disadvantages, namely, (1) they are not very extensive being narrow stretches of land on both sides of the seasonal rivers; (2) the floodwater recedes very quickly as soon as the dry season sets in, so that the possibility of dry season farming is greatly reduced; (3) the alluvial deposits derived mostly from old metamorphic rocks and meta-sediments are not as rich or extensive as they are in, for example, the Deccan of India or along Rivers Ganges and Irrawaddy. Finally, the fine-textured heavy soils of the flood plains and depressions in the northern dry lands tend to harden up and crack during the dry season so that cultivation is not possible until the hard mass has been thoroughly saturated and softened by water at the onset of the following rainy season.

An important ecological phenomenon associated with flood land farming in Nigeria is the incidence of waterborne diseases in man. The most common of these is schistosomosis (bilharziasis) which is fully discussed in Chapter 7 on Weeds, Pests and Diseases.

Hill cultivation and terrace farming

Hill cultivation as a distinct system is important only in the upland areas of Jos Plateau and the Mandara and Adamawa highlands. Hillside cultivation has also been practised in other areas, such as the Udi Plateau in Enugu State, but with disastrous consequences because of the absence of terracing. Outside the upland areas mentioned above terrace farming is very rare in Nigeria; it is practised in only a few localities as for example at Maku near Awgu in Enugu State. Except on the gentler slopes attempts to cultivate hillsides without terracing have accelerated soil erosion and rendered many areas virtually useless for farming. The impact has been particularly severe because of the loose sandy nature of the soils and the underlying False-bedded Sandstones.

By contrast, in south-western Nigeria steep hillsides have been repeatedly cultivated and yet the soils have remained relatively stable. In Erinmo/Effon Alaye area of Ekiti State most of the rice and yam farms are on 30° slopes. The soils are essentially sandy clay to clay loams developed on granite gneiss, biotite gneiss and quartz schists. The reason for the structural stability of the soils is not fully understood but is thought to be related to the highly kaolinitic nature of their clay contents.

Skillful and well-developed systems of terrace farming were found among the indigenous peoples of the Jos Plateau such as the Kaleri and the Kimmuk (Gleave, 1965, 1966). The hill settlements on the fringes of the Jos Plateau were established for defensive purposes during the slave raids and internecine warfare of the pre-colonial era. Although the hillsides have thin and extremely stony soils with very low natural fertility, the hill settlers succeeded in developing a system of intensive cultivation without a fallow period but based on the judicious application of animal and household wastes and the construction of elaborate systems of terraces to prevent soil erosion and at the same time extend the area available for cultivation. The individual plots were however still very small. They were usually hedged round with cacti.

A more elaborate system of hill cultivation and terrace farming also existed among the tribes of Dikwa emirate in the Mandara highlands. Again, as with the Jos Plateau, the hill settlements arose as a result of past insecurity due to slave-raids and intertribal wars. Buchanan and Pugh (1955) listed the four major principles on which the system was based as (1) the prevention of soil erosion through an ingenious system of terraces made up of piles of stones; (2) the maintenance of the soil nutrient status through the heavy application of animal and human wastes; (3) the adoption of a system of crop rotation growing guinea corn, beans and bulrush millet in a three-year cycle; and (4) the preservation or planting of economic trees, such as *Acacia albida,* date palm, fan palm and African mahogany, to provide supplementary animal and human food, wood for fuel and building purposes and organic matter to the soil through leaf fall. An important aspect of soil management was the use of farmyard manure. Human excreta, wood ash and household wastes, weeds and cattle dung were collected and composted for use on the terraces. Each plot was usually cultivated continuously for a period of up to ten years before being left fallow for a period of not more than two or three years. Before cultivation was recommenced after each fallow period, the soil was manured with grass, wood ash and household refuse and by growing tiger nut plant on the land.

These systems of hill cultivation and terrace farming were significant in that they had been developed, not by the larger tribes and cultivators, but by the small, less sophisticated tribes who responded to their special circumstances and difficult upland environment, and evolved highly specialized systems of crop cultivation. Given the right impetus and the necessary guidance Nigerian farmers can cope with the most difficult terrains and develop sophisticated systems of crop cultivation. Richards (1975) revealed the extent of the indigenous knowledge of local people in parts of Nigeria on the ecology of crop cultivation in their local environments.

Unfortunately, with the arrival of the British in the 19th century and imposition of Pax Britannica, the hill settlers dispersed into the surrounding lower slopes and plains, abandoning the conservational systems of hill cultivation that they had practiced for so long and so skillfully. They adopted shifting cultivation with bush fallowing. On the Jos Plateau, for example, the bulk of the staple food crops, *acha, dauro, tamba* and yams, is now produced under the bush fallow system outside the boundaries of the hill settlements.

Climate change adaptations in agriculture

In the northern savanna belts where people have lived for long with drought and desertification the small holder farmers have learnt to adapt to climate change, as seen in the case of farmers in Borno State (Ahmed, 2019). First, some of the small holder farmers in Borno State have adapted through livelihood diversification. In other words, they have sought other means of income outside of farming by looking for employment primarily in construction and unskilled salaried jobs. But a greater proportion have remained in the rural economy and diversified to other rural livelihood activities such as forestry and hunting in addition to crop cultivation. However, insecurity has been responsible for outright livelihood change among many rural dwellers in Borno State. Such livelihood changes are not common in other states in the Sudano-Sahelian zone.

In other parts of the Sudano-Sahelian belt, farmers have learnt to adjust to market conditions. For instance, yam cultivation, which for long had declined in Katsina State, is staging a comeback because yam fetches more money in the market than the grain crops (Idris, 2020). Over the years, yam cultivation had been displaced with the introduction of new crops such as hybrid maize, soybeans, rice and beans. A significant aspect of the revival of yam cultivation in Katsina is that the youths are credited as being responsible for the welcome development. They source their seed yams from Zaria and other parts of Kaduna State, as well as Sarkin Pawa in Niger State. The yam farmers have attracted marketers from some big towns (Faskari, Malumfashi, Kafir) who come to buy the yams on farm. The yams find their way to the big cities including Kano, Katsina and even across the border to Niger Republic.

It is important that the eyes of the farmers have been opened to look to other crops to cope with the vagaries of climate change. They now feel that they had laid too much emphasis on maize and sorghum production while they have fertile land that could produce a lot more crops in commercial quantities including upland rice, pineapple and banana.

Climate change has made food crops to become gold in these drylands. Unfortunately, according to Idris (2020), there has been an increase in theft of yams on the farms usually at night. In other parts of the dry savanna lands, the situation is even worse. The menace of bandits in the northwest of the country has become a threat to national security. The bandits, mostly the long-neglected, uneducated and unemployable youths, engage in smuggling across the borders of Nigeria and in illegal mining of rare and precious minerals, which has devastated large tracts of savanna land. These illegal activities have become their means of survival and revenge on the society that is responsible for their plight.

The resurgence of yam cultivation in Katsina State could mean that the impact of climate change has eased somewhat in recent times. But the local people attribute it to fertile soils. Still fertile soils need moisture to support crop growth. It is quite possible that if and when the grip of climate change tightens, the famers may not be able to grow yam and other crops currently yielding high dividends. Indeed, a recent survey in Sokoto State showed that poor returns on investment had forced farmers to abandon food crop cultivation (Vanguard Newspapers, 2020). The farmers had abandoned food crops such as maize, sorghum and millet and are now growing cash crops such as cassava, sesame, tomatoes, cashew, wheat, ginger, orange, which fetch them money from manufacturing companies and industries. Thus, the vagaries of weather due to climate change are forcing farmers to be more pragmatic and rational in their choice of crops to cultivate. This is likely to continue into the future as studies and models on the future impacts of climate change in the Sudano-sahelian zone indicate that food crop yields, especially of millet, would be severely impacted (Adeyemi and Adelabu, 2020). The consequences for food security in the zone would be catastrophic. Therefore, efforts directed at combating climate change and desertification in the northern dry savanna belt must be intensified. The soil conservation methods being promoted in crop cultivation are meant to reduce soil tillage, and conserve soil moisture and organic matter content, thereby minimizing the risk of desiccation and accelerated erosion.

Nigeria's new initiative at soil erosion control and land rehabilitation

From the foregoing it is clear that traditional soil management practices have been mainly geared towards alleviating the effects of one of the three problems identified earlier in this chapter, that is, soil deterioration or soil fertility depletion. Even so, there has been great reliance on natural regeneration for the restoration of soil fertility whenever it becomes exhausted, as is clearly demonstrated by the predominance of the bush fallow system. Except in a few localities and in special circumstances farmers have had limited success in combating the problems of soil desiccation and accelerated soil erosion. The devastation of vast areas of former farmlands in south-eastern Nigeria is a rude reminder of the failure of past soil conservation and erosion control efforts.

However, in the 21st century, Nigeria has decided to confront the problem of accelerated soil erosion in a more proactive manner by establishing the Nigeria Erosion and Watershed Management Project (NEWMAP) in 2010 with the assistance of the World Bank and partner agencies. The mission is "*to address on a multi-dimensional scale the menace of gully erosion in the south east as well as land degradation in the North*" (Dahiru, 2020). Noting that states and local governments were confronted with overwhelming complex gully erosion problem, it embarked on a plan of action to reduce the vulnerability that threatened infrastructure and livelihoods in the identified watersheds initially in the seven states of Abia, Anambra, Cross River, Ebonyi, Edo, Enugu and Imo. The number of participating states increased to 22 over time. The overall objective is to reduce vulnerability of livelihoods and infrastructure to erosion in the targeted sub-watersheds. The current sustainable mitigation strategies combine natural (biological afforestation) and civil works, promotion of alternative livelihoods in partnership with local communities and a total watershed approach. Measures are geared towards (i) stabilizing severe erosion sites; (ii) preventing emerging erosion problems early when intervention costs are low; and (iii) improving landslide and flood preparedness. The erosion control plans cover entire watersheds and communities are sensitized and actively involved at every stage of plan implementation. The capacity of the local people to play their roles is assured through sensitization and training programmes. The measures taken have included resettlement of people away from hazardous locations and financial and technical assistance in watershed rehabilitation and livelihood activities.

If NEWMAP could be sustained to survive changes in government both at the federal and state levels, the prospects are high that it would positively impact the landscape and livelihoods in the project areas. Thus far, NEWMAP claims to have achieved the following in the three major areas of operation:

a. Gully Rapid Action and Slope Stabilization (GRASS)
b. Integrated watershed management
c. Adaptive livelihoods

- 76 hectares of degraded land recovered as at April 2019. Majorly from civil works in *Abia, Anambra, Enugu, Ebonyi, Edo, and Cross River*

- 11,060 lives saved across the sites. These lives cut across those living close to the gully or affected areas.
- 6,448 buildings saved due to engineering works, mainly as they are situated directly by the gullies or unstable meters from the gully.
- 5,750 jobs created which consist of food vendors, drivers, carpenters, labourers
- 60,000 people saved due to road construction works implemented in Abia State, and river channelization works currently ongoing in Ebonyi State.
- Support to 5,159 project beneficiaries: (2,346 males/2,813 females) via grants under the NEWMAP livelihood enhancement activities: beneficiaries now engaged in businesses such as livestock farming, small shops, trading, honey production etc. 3,262 Project Affected Persons (PAPs) fully compensated.

The future

In 2018, the United Nations Food and Agriculture Organization succinctly captured the problems of agriculture in Nigeria as follows:

> In spite of the oil, agriculture remains the base of the Nigerian economy, providing the main source of livelihood for most Nigerians. The sector faces many challenges, notably an out-dated land tenure system that constrains access to land (1.8 ha/farming household), a very low level of irrigation development (less than 1 per cent of cropped land under irrigation), limited adoption of research findings and technologies, high cost of farm inputs, poor access to credit, inefficient fertilizer procurement and distribution, inadequate storage facilities and poor access to markets have all combined to keep agricultural productivity low (average of 1.2 metric tons of cereals/ha) with high postharvest losses and waste.
>
> Even though agriculture still remains the largest sector of the Nigerian economy and employs two-thirds of the entire labour force, the production hurdles have significantly stifled the performance of the sector. Over the past 20 years, value-added per capita in agriculture has risen by less than 1 per cent annually. It is estimated that Nigeria has lost USD 10 billion in annual export opportunity from groundnut, palm oil, cocoa and cotton alone due to continuous decline in the production

> of those commodities. Food (crop) production increases have not kept pace with population growth, resulting in rising food imports and declining levels of national food self-sufficiency (FMARD, 2008). The main factors undermining production include reliance on rainfed agriculture, smallholder land holding, and low productivity due to poor planting material, low fertilizer application, and a weak agricultural extension system amongst others.

The persistence of these problems and issues in the agricultural economy of Nigeria is not for lack of studies and recommendations based on extensive studies and experiments carried out by both local and international researchers and organizations. As indicated in the FAO assessment above, one of the problems of agriculture in Nigeria is the "limited adoption of research findings and technologies." This is true perhaps in respect of research by indigenous researchers. But the international research institutions have found it easier to carry through many of their programmes. A case in point is the International Institute of Tropical Agriculture (IITA) in Ibadan, which succeeded in promoting adoption and spread of new crop varieties such as the high yielding yellow maize and cassava. They also pioneered the systems of minimum tillage and alley cropping that protect the soil against erosion during the cultivation period. Some of the outcomes of the agricultural research institutes have also been disseminated including high yield varieties of rice, okra, greens and fruits. But, much more could still be done to justify the efforts of researchers in Nigeria's own research institutions including the Departments of Agriculture, Agronomy, and Agricultural Biology in universities; the institutes of agricultural research, for example, at Samaru, near Zaria, Moor Plantation, Ibadan, and Umudike in Abia State; the long established research institutes on cocoa, oil palm and rubber and similar government research establishments at different locations throughout the country

Future agricultural development programmes must find effective methods of conserving and enhancing soil productivity and the natural environment if the country is to increase agricultural productivity. The declining efficiency of natural fallows in restoring soil fertility has led to experiments in alternative soil conservation measures but nowhere are they being introduced to the peasant farmers with any degree of seriousness. The long-proposed alternatives to the bush fallow system include:

1. The planting of fallow crops using fast growing plant species which help in building up soil humus and nutrient levels quickly. Forestry experts are particularly interested in this practice as a means not only of conserving the

soil and ameliorating local climate, but also of promoting forestry to provide raw materials for pulp and paper and other wood-based industries.

2. The adoption of a 'minimum tillage' technique now referred to as "conservation tillage" which does not involve clearing and ploughing the soil every season because seeds are planted into chemically sterilized soil. Under this system previous residues are left on the soil surface to prevent runoff and erosion and thereby improve soil moisture and temperature regime (Lal, 1973). This system developed at the IITA, Ibadan has given rise to the concept and system of conservation agriculture that is geared towards crop production in a sustainable environment friendly manner (Farooq & Siddique, 2015; Ayodele and Aluko, 2017). Conservation agriculture emphasizes minimal soil disturbance, diversified crop rotations, and surface crop residue retention. The goal of conservation agriculture is to preserve, mend and make judicious use of natural resources through unified management of available soil, water and other biotic constituents (Reicosky, 2003). It is being touted as a most effective way to combat the problem of weeds in crop cultivation as discussed in Chapter 7 of this book. Again, this goes to show that researchers have not relented in efforts to develop new and better systems of agriculture but the problem still remains as to how to get farmers to adopt the new systems.

3. The interplanting of food crops, especially cereals, with nitrogen-fixing legumes (alley cropping) to combat the nitrogen deficiency in many Nigerian soils particularly in the savanna regions. Again, as with the minimum tillage technique, the interplanting of crops with nitrogen-fixing legumes has become part of the system of conservation agriculture in modern times. Farmers have adopted the practice of interplanting legumes such as groundnuts and soybeans with grain crops, such as maize and guinea corn. But the practice is not yet widespread.

4. The adoption of a system of crop rotation that includes sole legume plots in the cycle; and the integration of crop and animal husbandry. Integration of crop and animal husbandry has not been adopted by traditional farmers in the villages but modern-day entrepreneurs in arable crop cultivation referred to earlier have adopted the system of crop rotation that includes sole legume plots in the cycle

As stated earlier, these measures have languished largely at the experimental stage and have yet to be adopted on a large scale by peasant farmers. By contrast, the use of chemical fertilizers has been successfully established among peasant farmers. However, for some time after they were introduced there was a tendency everywhere to restrict the fertilizer application to specific crops, usually cash crops, for example, rice in Ekiti, tobacco in northern Oyo State, soya beans on the Jos Plateau, and so on. But, today, the situation has greatly improved and government promotional efforts have encouraged farmers to apply chemical fertilizers to a wide variety of crops. Experiments in the agricultural research institutes for example at Moor Plantation, Ibadan and Samaru, Zaria had shown that crops other than those mentioned above respond very well to fertilizer applications among them maize, cowpeas, millet, cotton, and groundnuts. Weed killers have been introduced also, for example, the National Cereals Research Institute at Moor Plantation, Ibadan encouraged farmers through radio and television to make use of herbicides on their maize and rice farms. Unfortunately, the use of fertilizers, herbicides and other inputs is beset with many problems including irregular supply, high costs of procurement despite government subsidies at both federal and state levels, and corruption among government officials in the distribution of available supplies.

There is no doubt that farmers have derived immense benefits from the use of both fertilizers and weed killers. For example, a rice farmer in Ekiti could grow three successive crops of rice on the same piece of land before abandoning it for another plot. The application of chemical fertilizers has not totally removed the need for a fallow period; its main effect appears to be that the period of cultivation has been lengthened somewhat. It is conceivable, but there is as yet no concrete evidence to support this, that with the use of chemical fertilizers the fallow period could be shortened without the usual concomitant risks. Of course, there is a danger inherent in the indiscriminate use of chemical fertilizers and herbicides on the land and one only hopes that the agricultural extension officers will warn farmers of the possible environmental consequences of their injudicious use in good time. Ayodele and Aluko (2017) warned that increased reliance on herbicides could cause water contamination, weed resistance, weed flora shift, and herbicide carryover. But, given the irregular supply and low levels of application, problems associated with heavy use of fertilizers and herbicides have not yet become major concerns anywhere in the country.

With the recent efforts to improve agricultural productivity in Nigeria it is becoming increasingly clear that more technical data and detailed information on the soils and the environment are needed. For instance, the Sahelian droughts (1968-73; 1984-85) awakened scientists and government officials alike, at least in northern Nigeria, to the need for more extensive and fundamental studies of soil climate and water supplies. Technical data on soils are still very limited and reports of investigations are widely scattered in numerous files, bulletins and journals. More than ever before there is a need to collate existing technical data through a system of soil correlation. This involves not only the documentation of soil characteristics in every region but also the assessment of how different soil types are being used and managed in the localities where they occur. With the right kind of information agricultural advisory services would be able to show local farmers why they should adopt new soil management techniques or experiment with different crops. Left to themselves the people are seldom inclined to change their ways. As we have seen at Maku in the scarp lands of Enugu State terrace farming is well-established and shows very good results, but farmers in villages close by, on the same escarpment, do not seem to have noticed what their counterparts at Maku had been doing or why. These are the sort of abnormal situations that can be put right through soil correlation exercise and agricultural extension services. We also referred earlier to the positive impact that the National Fadama Development Project is making in areas where floodland farming had not previously been practised

As we can see from the foregoing discussions in this chapter, including the findings and assessments by international organizations, real progress in agricultural efficiency and productivity in Nigeria will only be achieved when major structural and technological changes take place. These include overhauling the land tenure system and reorganizing rural land use; upgrading cropping systems and introducing modern equipment and technologies and higher yielding crop and animal varieties to increase overall productivity. The political will is sorely needed to establish policies and systems that would facilitate the inculcation of research outcomes in the day-to-day farming practices of the local farmers. Much will depend also on the political will to take measures to reduce the high level of corruption in the acquisition and distribution of agricultural inputs to farmers.

Finally, it cannot be denied that the farming population has aged considerably and rural areas have lost much young labour to the cities and big towns. It is important to rectify this situation by further encouraging young, educated or literate farmers, entrepreneurs and graduates in the agricultural sciences to farming. Already, there are many people, including retired military officers, who have acquired or purchased extensive areas of land particularly in the middle belt states of Niger, Nassarawa, Benue and Taraba, and established farm projects. Still other entrepreneurs and agriculturists have acquired mostly on leasehold large estates where they practice commercial agriculture. Such farm estates are common in parts of Ogun, Oyo, Osun and Ondo States in south-western Nigeria. Many of these new entrants into farming plant exotic trees, fruit trees, cassava, maize and vegetables. Others also combine arable farming with poultry and piggery. The time has come for governments at the federal and state levels to leverage on these welcome developments in agriculture and come up with policies and strategies to mobilize university graduates in the agricultural sciences to practice agriculture or engage in agribusiness including processing of farm products (e.g. rice milling, cassava processing etc.), provision of farm inputs, farm produce storage and preservation facilities instead of running after scarce jobs in business and industry. Indeed, the National Youth Service Corps could be re-engineered to make graduates in agriculture- related disciplines spend their service year on farms or agriculture-related ventures rather than going to schools to teach. They can spend the service year laying the foundation for future enterprises in farming and agribusiness. It would then be easier to promote environment friendly sustainable agriculture in the country.

7

Weeds, pests and diseases

Weeds, pests and diseases are a fact of life in Nigeria: they reduce the efficiency of the resource processes and limit their potential in vast areas of the country. In this chapter we shall consider the major characteristics and the geographical distribution of a number of major weeds, pests and diseases in Nigeria and examine their impact on crop yield, animal health and productivity and the health and welfare of the people. Because they impair crop and animal health and reduce yields, weeds, pests and diseases greatly undermine both the quantity and quality of Nigeria's products either for export or for home consumption. The output of working people is also reduced by poor nutrition and debilitating diseases.

Control of these three indigenous problems in a developing country is, to some extent, a test of the people's will and their understanding of the environment and its resources. Since many weeds, pests and diseases are spread as a result of human mismanagement they can only be effectively controlled by a more enlightened approach to crop and animal husbandry and greater understanding of the limitations of the technological means available to deal with the weeds, pests and diseases.

The distribution of weeds, pests and diseases

In general, the geographical distribution of weeds, pests and diseases in any area is determined by (1) the natural biogeography; (2) commerce and external links; (3) the spatial distribution of crops and livestock; (4) climate; and (5) cultural patterns, including population distribution, religious beliefs and traditional ways of life (Ezcurra *et al.*,1978). Each of these factors has contributed to the incidence of weeds, pests and diseases in Nigeria.

Natural biogeography

Many weeds, pests and diseases are indigenous to Nigeria; they form part of the country's biological complex and their distribution can be explained largely in terms of differing ecological conditions from place to place. Such pests as the mosquito, tsetse fly and sand fly (simulium) are well adapted to the country's tropical environment; their proliferation has been encouraged by the warm and humid conditions in certain parts of the country. The nature of the vegetation cover is important to the spread of many of these pests. For example, the tsetse fly cannot thrive under the dense canopy of the tropical rain forest, but does well in the more open savanna woodlands, which nevertheless, provide shade against intense insolation. The tsetse fly territory reaches far to the borders of the desert along the rivers where the riverine vegetation provides a more humid and shaded environment than the adjacent semi-arid grasslands. The spatial distribution of the grasshopper also depends on the vegetation. It is not very common in the dense forest region, but thrives in open derived savanna, and invades fallow lands, farmlands and plantations.

The weeds found in each ecological zone also reflect the nature of the vegetation. In the forest and savanna woodlands, many weeds are woody shrubs or herbaceous plants while in the more open Sudan savanna, most weeds are graminoids. In general, weeds are indigenous being components of the natural flora. For example, the grass weed *Imperata cylindrica* (spear grass, *lalang*) is a native grass which invades many fallow lands and grass pastures in the savanna areas.

The incidence of certain diseases is determined by the distribution of disease vectors among the native fauna. For instance, the spread of sleeping sickness to the derived savanna zone was due to the fact that the grass vegetation that replaced the original forest vegetation attracted such animal species as antelopes, hare and bush dogs which are the natural carriers of the trypanosome pathogen. Furthermore, many native plants are hosts of crop parasites and disease agents.

Commerce and external links

Weeds, pests and diseases can also be introduced through the movements of people, crops and animal species both within a country and across international boundaries. In Nigeria, the major external contacts that have had major consequences on the geography of weeds, pests and diseases were the slave trade, European colonization, modern international trade, the Second World War, trans-Saharan trade, Arab and Indian incursion across the Indian Ocean and Holy Pilgrimage to Saudi Arabia by the Muslims. The slave trade brought Nigeria in contact with the Caribbean and the continents of North and South America. Crops such as maize, sugarcane and cocoa were introduced to Nigeria from these areas. These crops have their own associated weeds, pests and diseases.

During the Second World War, Nigerian soldiers fought in Malaysia and Myanmar from where, it is believed, they brought back seeds of the Siam weed *Chromolaena odorata* (Sheldrick, 1968). d, they Earlier, in 1936-37, some of the weed had been accidentally introduced into Nigeria with the seeds of *Gmelina arborea* (gumbar, gmelina), which were sent to Enugu from Sri Lanka for erosion control and soil conservation projects (Odukwe, 1965). In the 1970s and 80s, this weed spread rapidly throughout south-western and south-eastern Nigeria and parts of the middle belt.

Diseases of epidemic proportions occasionally spread into Nigeria across the desert from North Africa, the Mediterranean and the Middle East, following the trade routes and pilgrimage routes across the Sahara. Such diseases include influenza, cholera and small pox. The import and export trade in foodstuffs including rice, beans, yams and cowpeas has also contributed to the spread of crop pests and diseases and is responsible for the wide dispersal of weeds such as the Siam weed. In modern times, air transportation has made it easier and faster for diseases to spread, as was glaring in the coronavirus COVID-19 pandemic in 2020.

Distribution of crops and livestock

Perhaps, the most important factor in the spatial distribution of crop and animal pests and diseases is the distribution of the victims themselves. Indeed, there is a direct relationship between the intensity of cultivation and the incidence and diversity of insect pests in an area. For instance, crop weeds, pests and diseases are more diverse and prevalent in the densely cultivated parts of southwestern, southeastern and middle belt of Nigeria in the more humid forest, derived and Southern Guinea savanna ecological zones. The drier savanna regions further north have their own share of pests, particularly insects (e.g. locusts) and birds (e.g. quela quela), that thrive in arid environments

With regard to animal pests and diseases, the case of trypanosomosis has already been mentioned in Chapter 3. The spread and persistence of the disease has been attributed to the migratory movements of cattle. Fleas, which are dangerous parasites to many animals, also spread with the movements of host species such as cattle, horses, dogs, cats and birds.

Land use, particularly agriculture, creates favourable conditions for pests and disease vectors. For example, schistosomosis is common in areas where irrigation or flood agriculture is practised because of the presence of ponds or pools in which the disease vector, the water snail, thrives. Insect attacks and diseases are likely to be more devastating where monoculture is practised than in areas of intercropping because the host plants of the natural inhibitors have been removed. Toye (1969) described a banana plantation which was severely damaged by the grasshopper, *Zonocerus variegatus.*

Climate

The two most important climatic factors in the incidence of weeds, pests and diseases are heat and water supply. Some weeds cannot seed or spread in dry conditions; for example, the spread of the Siam weed is limited in areas of low annual rainfall (Hall *et al.,* 1972). Many pests and diseases thrive in hot, wet conditions. For example, mosquitoes, which are the vectors of filaria and malaria parasites, breed and thrive in the more humid southern parts of the country than in the drier northern parts.

Other pests and diseases are adapted to hot dry conditions. Human diseases such as measles, chicken-pox and small pox are more prevalent in the hot dry season than in the wet season. Poliomyelitis and cerebrospinal meningitis are more prevalent in the hot and drier northern parts of the country where the viruses move about freely on dust particles.

The restriction of certain crops to particular geographical zones results indirectly from the influence of climate on the disease agents to which those crops are susceptible. For example, large-scale production of groundnuts is limited to the sub-humid northern parts of Nigeria because the humid weather of the south favours the proliferation of rosette and other diseases of groundnuts. Black pod disease of cocoa is encouraged by the high rainfall and the long periods of humid weather characteristic of the forest ecological zone. Before the widespread application of fungicides in cocoa plantations, there was evidence of a negative correlation between total annual rainfall and cocoa yields. According to Adejuwon (1978) between 1933 and 1951, the total annual rainfall in Ibadan was below average in each of the five years when cocoa production was over 100,000 tons (101,600 tonnes). The heaviest rainfall was recorded in 1947, the year showing the poorest cocoa harvest.

Cultural patterns

As with crop and animal diseases, the distribution of human diseases reflects the geographical distribution of population. The incidence of most human diseases in Nigeria is influenced by the people's living conditions, the standard of personal hygiene and environmental sanitation, the waste disposal system, the quality of the drinking water, and the nutritive value of the food.

Disease can be transmitted through people urinating or defaecating in unsuitable places; by eating uncooked contaminated food; by drinking impure water; and generally, from poor substandard living conditions. The accumulation of waste products in the cities facilitates the breeding of flies which contaminate food. Malnutrition resulting from unbalanced diet renders many rural inhabitants highly susceptible to diseases.

Events in African countries since the 1960s up till the present have shown also that civil wars, poor governance, corruption, internal political turmoil, social injustice, banditry and other forms of civil disorder have serious consequences on the environment, food supply and the health of the people. Wars have devastating effects on both the animal and plant biodiversity of forest ecosystems; agriculture and other means of livelihood; and food security and nutritional status of the populace. In general, resistance to diseases is greatly lowered among the populace and epidemic diseases often spread more rapidly in war ravaged countries and areas. African countries including Nigeria have suffered from recurrent outbreaks of Lassa fever, Ebola fever, typhoid fever, and SARS, in addition to cholera, dysentery and other diseases

Weeds

The broadest definition of a weed is that of a plant growing out of place. It follows that almost any plant, even cultivated crop, can be a weed. Furthermore, biologists tend to regard as weeds all plants which are not commonly cultivated by man; such are regarded as unwanted plants. Yet, it has happened, in the history of cultivation, that some previously unwanted plants were later found to have uses not originally known and were removed from the category of weeds. In this book a more restricted definition is adopted, which takes a weed to be an opportunistic species of plant that appears following man's disturbance of the natural habitat but which is unwanted on agricultural land because, by prevailing considerations, it does more harm than good.

When vegetation is cleared in a given area, there is a natural tendency for ecological succession to commence in order to re-establish the climax community. Most so-called weeds, in fact, are pioneer species which would eventually give way to higher plants if the succession were allowed to continue. These pioneer species usually have limited requirements for survival as compared to higher plants or crops, and are much more efficient in making use of available resources. It follows that unless they are suppressed, the weeds soon dominate the habitat. Indeed, these weeds may predominate and persist on lands which have been greatly degraded through overgrazing or overcultivation.

Ecological characteristics

Weeds possess certain ecological characteristics which give them a competitive advantage over cultivated crops or favoured plants. Most are wind pollinated which is relatively uncommon among flowering plants and crops. They have a high reproductive capacity and produce seeds in very large numbers. These seeds can remain dormant for as long as forty years or more in some cases. Such dormant seeds are capable of responding quickly to the onset of conditions favourable to their germination. Once they germinate, they grow very fast and are capable of quickly occupying denuded areas. Many weeds are annuals but have a long flowering season, due to the fact that all the seeds do not germinate simultaneously. There are also numerous perennials which are generally deep rooted and spread both by seed and runners above and below ground. They belong to a higher stage in the ecological succession than the annuals and they are more permanent and more difficult to eradicate. *Cynodon dactylon* (Bermuda grass) is an example. Some, like *Imperata cylindrica* (spear grass, *lalang*), have underground adaptations in the form of bulbous and rhizomatous root systems, and cannot therefore be destroyed by burning, grazing or cutting. Only by tilling (with hoe, plough) can the underground systems be disrupted.

Many weeds have enormous phenotypic plasticity, in other words, they have many growth forms so that they can adapt and assume any morphological character depending on the habitat in which they germinate and develop. However, this phenotypic plasticity is reduced where environmental conditions are unfavourable.

Competition

The harmful effects of weeds lie in their competing with crops for space, light, soil water, and nutrients. Because most are fast growing, they soon outstrip the crops and deprive them of sunlight needed for photosynthesis. For example, in favourable weather conditions, the Siam weed can grow up to 25 cm in a week and in six months can form an almost impenetrable mass 2–3m tall, that can completely smother a young plantation crop of oil palms (Sheldrick, 1968). Another way in which weeds interfere with the productivity of crops is by hindering the development of branches. This is a serious problem among the grain legumes such as groundnuts, cowpeas and beans. Ishag (1971) found that on badly infested groundnut farms in the Sudan, weeds decreased the number of branches per plant and the number of pods per branch. Hand weeding of the crop fields thirty days and again sixty days after planting led to an almost five-fold increase in yield. More studies have been done since the early studies of the 1960s and 1970s. Yahaya *et al.* (2008) analysed the effects of weeds on the profitability of millet (*Pennisetum glaucum*) and groundnut (*Arachis hypogaea*) crops cultivated in mixed cropping systems (millet/sorghum/cowpea and groundnut/millet/cowpea mixtures respectively, being the most widely used biological measures of controlling weeds in some villages in Sokoto State. Most of the land was zero tilled and manually weeded. The cost of manually weeding the farms greatly reduced profits from the two mixed cropping systems. But, while the cost of weeding increased production costs by 40%, unchecked weeds depressed farm profit by about 500% and 400% respectively in millet and groundnut production.

The rapid rates of germination and development of foliage and extensive root systems by weeds also mean that they can absorb more water than cultivated crops. Indeed, many weeds have contractile roots that can be extended or contracted depending on how widely distributed water is in the soil column. Thus, at any given period weeds are able to make fuller use of the available soil water than the cultivated crops.

Many weeds are nutrient accumulators and are therefore capable of depriving neighbouring plants of much needed nutrients. The effect is very serious in poor soils especially in regard to nutrients such as nitrogen and phosphorus. Competition for nitrogen is particularly serious on maize farms as the crop soon turns yellow if deprived of nitrogen. Phosphorus too is usually in short supply in tropical soils, and is further reduced by weed uptake. The weed, *Tridax procumbens*, is well known for accumulating phosphates in its body. A beneficial beneficial effect of nutrient accumulation in weeds is that when they die the nutrients are returned to the soil in the organic litter and are once more available to the cultivated crops during the following growing season.

Finally, many weeds compete successfully against other plants by means of allelopathy. They exude toxic substances which inhibit the growth of other plants. This is true of such weeds as *Cynodon dactylon (*Bermuda grass*), Sorghum halepense (*guinea corn*)* and *Cyperus rotundus* (nut grass).

Crop-weed complexes

One of the most important adaptations of weeds is that they develop seeds that are almost similar to the seeds of crops. The seeds also mature at the same time as the crop seeds so that there is a great possibility of the farmer mixing them up at harvest time. An example is *Echinochloa colona* (barnyard grass), whose seeds are very similar to those of rice. It is difficult to distinguish between the two plants in the early stages of growth, and the weeds grow luxuriantly along with the rice, until maturity, when they can no longer be safely removed without damaging the rice plants. Another example is the weed *Rotboellia exaltata* which is very similar to maize in the early stages of growth.

Indeed, it has been observed by ecologists that crop and weed plants have evolved together. The origin of a weed may be traced to a crop and similarly the origin of a crop may be traced to a weed. Many of our present-day crops have weed equivalents or 'weed races'. The weed races are usually crossed between the wild and cultivated species of the crops. As a result, they usually compete vigorously with the cultivated varieties. The West African indigenous rice, *Oryza glaberrima,* has both a weed and a wild race. Sorghum (guinea corn) has three races: the cultivated, the truly wild which is only mildly weedy and the shatter race, a cross between the wild and cultivated species, which is a very serious weed plant. Similarly, *Zea mays* (maize), has a weed race as have other crops such as potato, tomato, pepper, millet, carrot and sunflower.

Some common weeds in Nigeria

Fig.7.1 shows a simple classification scheme for weeds. Table 7.2 shows some of the more common weeds in Nigeria, their characteristics and local uses. The narrow-leaved weeds generally are grasses which are very common in all the savanna areas and grassland patches in the forest region. Most of them provide good pasture for livestock, but some are very aggressive and troublesome, such as *Eleusine indica* (wire or goose grass), *Cynodon dactylon* (Bermuda grass) and *Imperata cylindrica* (spear grass, *lalang*).

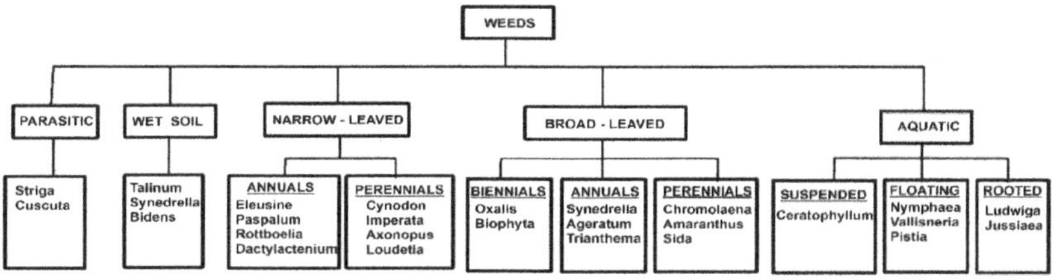

FIGURE 7.1 Classification of weeds

While *Eleusine* and *Cynodon* are most common on lawns and roadsides, *Imperata* is a major weed on grass pastures, farmlands and fallow lands, thriving in the south and in the middle belt where there is ample rainfall and where temperatures are uniformly high throughout the year. *Imperata* is also found further north because it is able to withstand long periods of dry weather. Moreover, it can survive and thrive in waterlogged areas. It spreads primarily by means of rhizomes.

The broadleaved weeds are mostly soft woody forbs and shrubs. They are among the most troublesome on arable land and tree crop plantations. In this group, the Siam weed *Chromolaena odorata* deserves more detailed discussion.

Siam weed is a soft wooded, much branched shrub reaching a height of 2-3m. It thrives in the open but cannot survive in the dense shade of a mature tree crop plantation. Although it is counted an obnoxious weed by many (Hall, *et. al.*, 1972), the rice growers of Ekiti State regard it as a good fallow plant, which supplies much organic matter to the soil (Jeje, 1977). It appears that Siam weed in the soil is more favourable to rice than to other crops. For instance, in southeastern Asia, it is used as green manure in paddy fields (Sheldrick, 1968). In Nigeria, the weed dries out rapidly during the dry season and can be cleared quickly by burning.

However, in spite of these good qualities, Siam weed is very harmful in many respects. Its rapid rate of growth makes it a danger on newly planted farmlands and young tree crop plantations because it can smother all the crops and all other weeds; and when dry it constitutes a serious fire hazard in plantations of permanent crops. The extensive woody root growth makes tilling difficult and many farmers believe it lowers the productive capacity of the soil. For example, yam and cassava growers interviewed in the Ijebu-Ode area of Ogun State claimed that it gave the soil a bad odour which is caused by its essential oil (Areola, 1977). This pungent smell also renders Siam weed unpalatable to cattle (Sheldrick, 1968).

Parasitic weeds are considerably fewer. Among those that are most troublesome is the witch weed (Striga) which attacks primarily maize and sorghum. Aquatic weeds are important in reservoirs, ponds and slow-flowing streams. They transpire profusely and are a major source of water loss in overgrown water bodies. The roots also concentrate sediments and may quicken the silting up of reservoirs. The important aquatic weeds are algae and *Eichornia crassipes* (water hyacinth), thick growths of which sometimes block the coastal lagoon waterways and fishing grounds.

TABLE 7.1 Common weeds in Nigeria

Common name	Botanical name	Weed characteristics (and uses among local people)
Grasses		
Spear grass	*Imperata cylindrical*	Useless, troublesome weed with rhizome roots
Knotgrass	*Cyperus rotundus*	Hated, hard to get rid of
Couch grass	*Digitaria abyssinica*	Roots form underground mat harmful to crops.
Bahamas grass	*Cynodon dactylon*	Good for grazing animals; medicinal
Elephant grass	*Pennisetum purpureum*	Though a weed it improves the fertility of the soil and protects against soil erosion; can be used in manufacturing.
Carpet grass	*Axonopus compressus*	Hard to get rid of; has aesthetic value
African Feather Grass	*Cenchrus macrourus*	Adaptable, thrives in cold and hot conditions, tolerant of windy areas and salty ground. Not easy to get rid of
Buffalo grass	*Stenotaphrum secundatum*	Crowd out other plants and weeds; not good in crop fields; but may be useful on ranches and pastures
Shrubs, forbs and sedges		
Siam weed	*Chromolaena odorata*	Overwhelms other plants; gives bad odour to soil; but favoured by some farmers because it adds to soil organic matter; protects soil
Billy goat-Weed (White-Weed)	Ageratum conyzoides	Invasive fast-growing weed, can crowd out other plants; have anti-inflammatory, anti-allergic, analgesic and antibacterial properties
Milkweed	*Euphorbia heterophylla*	Causes anaphylaxis and dermatitis in people who are sensitive to latex
Nuke-Noh	*Tridax procumbens*	Invasive weed, hard to eradicate; but medicinal plant
Witch weed	*Striga genus*	Evil, parasitic plant; penetrates roots and deprives cereals of nutrients
Dayflower	*Commelina benghalensis*	Causes serious harm to crops of peanuts, cassava, coffee and tea; host to a virus that can attack crops; but edible leaves
American bushmint	*Hyptis spicigera*	Medicinal mint plant; also, useful oils with potential as insecticides
Wild sunflower	*Aspilia Africana*	Multipurpose medicinal plant - with many applications
African Club Moss	*Selaginella kraussiana*	Native to tropical and southern Africa; adaptable; difficult to get rid of; attractive with aesthetic value
African Olive	*Olea europaea*	African olive shrubs are very hard to kill, and their seeds are easily spread by birds
False thistle	*Acanthus montanus*	Very prickly plant to avoid in spite of beautiful flowers
Creeping foxglove	*Asystasia gangetica*	Common roadside and waste heap weed with lovely flowers; no harm to crops; leaves useful as food or for medicine.
Yellow justicia	*Justicia flava*	Common on compound farms and waste areas; leaves can be used in making soups or for medicine (along with some other parts of the plant)

Extracted from: `https://www.legit.ng/1206477-20-common-weeds-nigeria-citizen-know.html`

Beneficial and harmful effects of weeds

Weeds have some value in the environment (Table 7.1). They protect exposed land surfaces against intense insolation and rapid runoff erosion. They can also serve as mulches in the early stages of crop growth when most of the farmland is exposed. Although being nutrient accumulators, weeds deprive growing crops of required nutrients, nevertheless the protection they provide against leaching or erosion conserves the soil nutrients. Furthermore, the decay of weed roots is responsible for the widespread and uniform distribution of organic matter in the soil from which one of the most important nutrients, nitrogen, is obtained. Weed roots also help to penetrate and break up hardpans (e.g. clay pans and compactions) in the soil. They draw water and nutrients from deep down the soil profile to supply other plants and microorganisms in the top soil layer. In some cultures, weeds have indicator value in judging the soil nutrient status or balance.

Some weeds also provide good feed for animals in the open pastures. These include species of *Cynodon, Eleusine, Chloris, Brachiaria* and *Digitaria*. The weed, shrub and leguminous components of Nigerian grasslands substantially improve their nutritive value. Some weeds are also eaten by human beings, for example, *Talinum sp., Vernonia sp., Amaranthus sp.*, and *Crassocephalum sp.* Many weeds contain vitamins and are therefore of considerable nutritional value; others have curative properties and are used by the local people in medicinal preparations.

In Nigeria, empirical studies are now revealing losses in crop yields due to weeds. Table 7.2 gives data on the results of one of the early studies on the problem from experimental plots at the International Institute of Tropical Agriculture (IITA), Ibadan. It was instructive to note that weeds could substantially reduce the effect of fertilizer applications on crop yields. In the example given in the table, crop yield on the unweeded field with fertilizer application, was only 70 per cent of that on the weeded plot, with fertilizer application. On plots where fertilizers had not been applied, the yield on the unweeded plot on Egbeda soil was only a little over half that on the weeded plot.

In addition to reducing yields, weeds lower the quality of the produce especially in the case of grain crops where weed seeds may inadvertently be mixed with the cultivated grains. The cost of harvesting also is greatly increased where weeds are present.

Table 7.3 gives more recent data on crop yield losses due to weeds. Crop yield reduction due to weeds can be very devastating. In the case of rice, it is possible to lose a whole season's crop to weeds. Losses to weeds by root crops and tubers are also high. The legumes (cowpea, soybean, groundnuts) appear to have done better against weeds than all other crop types.

TABLE 7.2 Yields of sweet potato on different soil series at different management practices

Soil type treatments	Yield (tonnes/ha)		
	Egbeda soil (IITA)	Apomu soil	Alagba soil (Ikenne)
No fertilizer, no weeding	12.69	10.73	—
No fertilizer, weeding	22.72	7.92	8.35
Fertilizer, weeding	28.79	7.38	11.76
Fertilizer, no weeding	20.17	—	—

Source: Obigbesan (1976)

TABLE 7.3 Crop losses due to weeds in selected food and cash crops in Nigeria

Crop	Percentage yield loss
Maize	18-60
Sorghum	51
Rice	28-100
Cassava	48-90
Yam	70-91
Cowpea	35-83
Soybean	40-53
Groundnut	54
Wheat	60

Source: Oerke E.C., Dehne H.W., Schonbeck F. and Weber A. (1994). *Crop production and crop protection: Estimated losses in major food and cash crops.* Elsevier p. 808

The cost of protecting farmlands against the invasion of weeds is one of the most obvious economic effects of weeds. Most crops need to be weeded at least two or three times, possibly more, during the growing season. *Cajanus cajan* (pigeon pea), which is grown in large quantities in the middle belt and in parts of Anambra, Edo and Ondo states, needs weeding between two and four weeks after planting and every two to three weeks thereafter depending on the degree of infestation. However, once the crop has developed a good canopy no further weeding is necessary. *Sorghum halepense* (guinea corn), requires two or three weedings to keep weeds in check and for large plantations it may be necessary to use herbicides on a weekly basis. *Vigna unguiculata* (cowpea) requires a minimum of two weedings during the first five weeks after sowing.

In the grazing areas, some weeds are poisonous to cattle, for example, species of *Argostemma* and *Claviceps;* others, though not poisonous, give the milk from cattle and goats, a very bad odour.

Many weeds act as alternative hosts for viral, fungal and bacterial diseases as well as nematodes which are very troublesome to crops and animals. *Physalis sp.* is a natural host for *Bacterium angulatum* which causes diseases in tobacco plants. The same *Physalis* weed and *Solanum dulcamara* harbour the virus that causes the disease tomato mosaic. The soya bean cyst nematode is similarly harboured by weeds. The golden nematode found on *Solanum dulcamara* and other weeds, causes diseases in potatoes.

With regard to human beings, the most direct adverse effect of weeds is that many of them cause allergy, particularly the graminoids (grasses). This is because they produce large quantities of pollen grains which are easily dispersed into the air where they are inhaled by human beings. They may cause incessant sneezing and interfere with the respiratory system as in hay fever. Others, such as *Dactylis glomerata*, give rise to skin rashes and red eyes.

Weed control

Ayodele and Aluko (2018) identified and described the major weed control methods in modern times, namely, (1) preventive weed management method; (2) physical weed control method; (3) chemical weed control method; (4) biological weed control method; (5) cultural weed control method; and (6) integrated weed management method.

Preventive weed management method is the most difficult to adopt with any degree of success as the outcomes are difficult to assess. The method is to curb the introduction and spread of weeds at all in the field. Several measures are applied including isolation of introduced animals that may harbor seeds in their dungs, the use of crop seeds and equipment that are free from weed seeds, and going over the fields to scout for any 'stranger' plants which may turn out to be difficult weeds, and preventing such plants from producing seeds if they succeed in escaping early detection and removal. As Ayodele and Aluko (2018 *op. cit.*) noted these measures are indirect methods of weed control with the primary objective of reducing the number of introduced plants emerging on the field plots.

The commonest method of weed control throughout Nigeria is uprooting and slashing of the weeds with the aid of hoes and machetes. This is the *physical weed control method* which involves the use of force, heat or some other physical forms of energy to break, cut off, destroy, burn or severely injure weeds (Swarbrick & Mercado, 1987). Manual weeding, mechanical weeding, and thermal weeding are examples of physical weed control. In fact, one of the greatest advantages derived from tilling the soil before planting is the control of weeds whose underground systems are thereby destroyed or damaged. Mechanical weeding involves the use of harrows, scudders and rotary weeders. As with hoeing or ploughing, the aim is to destroy the weeds by damaging or crushing the aerial as well as the underground parts. Tractor drawn implements can be used effectively only where there are no stumps or fallen trees to hinder their progress and where the crops have been planted in rows. They are of little use on Nigerian peasant farms. Finally, it is important to stress that weeds are effectively checked only when the planted crops have grown up and provided a cover for the soil surface. Most weeds cannot grow under the shade.

The *chemical weed control method* involves the use of synthetic herbicides to kill or adversely affect the growth of weeds. Herbicides could be applied to the leaves (foliar) or to the soil. Herbicides may be applied before sowing of crop seeds (pre-emergence) or after the seeds have germinated and are growing (post-emergence). Herbicides are increasingly being used in Nigeria, especially on large plantations and commercial farms. There are two categories of chemical weed killers, the selective and the non-selective. The non-selective herbicides include common salt, iron sulphate, copper sulphate and sodium chlorate which are used in high concentrations to sterilize the soil against all plant growth. These non-selective herbicides are obviously undesirable on croplands, hence the introduction of selective herbicides which are particularly effective against soft-wooded plants and grasses. The use of herbicides is an effective means of controlling weeds. However, this is associated with concerns such as weed flora shift, weed resistance, and environmental pollution. The two most commonly used chemical herbicides in cereal cultivation were *Atrazine* and *Amertryne* because they were particularly effective in controlling the witch weed (Striga), while Atrazine was also successfully used in controlling Siam weed on oil palm plantations (Sheldrick, 1968). Since then, many other chemical herbicides have been registered for use in Nigeria

It appears that in order to be effective, there must be sufficient moisture in the soil to facilitate the uptake of the herbicides by the weeds. But on the other hand, heavy rains tend to wash away the herbicides and render them less effective. Thus, if herbicides are to be used at an economic rate the weather conditions at the time of spraying become of overriding importance.

Concentrations of chemical herbicides and rates of application need to be carefully controlled. If they are too weak, they may not have a lasting effect. The need is for chemical herbicides that will stay in the soil and check weed growth for an appreciable length of time. There are other factors that determine how long a chemical herbicide will persist in the soil. These include the ease with which the organic compounds in the herbicides can be decomposed by soil micro-organisms; the rate of chemical decomposition of the herbicides through oxidation, hydration, hydrolysis and other reactions; the rate of absorption of the herbicides into soil colloids; the intensity of leaching; the rate of decomposition of the herbicides by radiant energy (photodecomposition); and the volatility of the herbicides. Extremely strong chemical concentrations are undesirable as they may damage the crops and soil micro-organisms. Some herbicides contain toxic compounds which are not easily decomposed by soil micro-organisms. Sole dependence on herbicide for weed control results in herbicide resistance, weed flora shift, soil and environmental pollution (Chhokar *et al.*, 2014).

Much research has been conducted into the *biological weed control method.* Biological weed control involves the use of other living things (organisms) and biologically based products to control weeds (Ehi-Eromosele *et al.*, 2013). There are bio-herbicides, that is, phytopathogenic microorganisms or microbial phytotoxins which are applied just as the chemical herbicides to control weeds by eating them up (Boyetchko & Peng, 2004). Among the earliest experiments were those targeted at the Siam weed (Chromolaena odorata). Wherever it occurs this plant is known to support a large and varied insect fauna which attack leaves, stems, buds and flowers (Hall *et. al.*, 1972). Certain aphids were also observed to feed on the Siam weed in Nigeria while the grasshopper is known to cat from the leaves when other plants are unavailable during the dry season. Hope of biologically controlling the weed is based on the aphids, but it has been observed in Ghana that aphids are themselves the subject of predation by the larvae of *Parous barbecues Macq.* (Syrphid: Diptera). Thus, the effective control of the Siam weed appears to depend upon the successful suppression of the *Parogus* larvae (Hall *et al.*, 1972). Other commonly used biological weed control strategies are allelopathy, animal grazing, use of resistant or tolerant crop varieties and use of phytophagous insects. *Cultural weed control method* involves the manipulation of farm practices to give crops advantage over weeds in order to grow. Farming practices that modify environmental conditions are used to suppress weed growth. The farming practices include changes to sowing time, the spatial arrangement of crops and application of fertilizers (Das, 2008). These practices are beyond the ability of peasant farmers to adopt.

Finally, in the field of conservation agriculture, Ayodele and Aluko (2018) proposed a system of *integrated weed management as an integral part of conservation agriculture* as opposed to its use in conventional farming. This is because in conservation agriculture there are limited available weed control options that can be applied in integrated weed management. Tillage and burning that destroy farm residue are strictly prohibited in integrated weed management in conservation agriculture. Nevertheless, integrated weed management in conservation agriculture is not without some faults, such as greater dependence on herbicides, reduced effectiveness of pre-emergence herbicides on the soil surface due to the interference of the crop residues present; and the fact that persistence of some herbicides affects the crop rotation component of conservation agriculture especially when the herbicide is not selective to the next crop in rotation. Still, as shown in Table 7.4, over time the integrated weed management in conservation agriculture is more effective in controlling weeds than the other methods even though it may not look so at the beginning.

In concluding this discussion on weeds, it would be wise perhaps to heed the warning by researchers at the North Carolina State University in the United States of America as reported by Brooke Boral in the Scientific American newsletter June 8, 2018, titled "Weeds are winning the war against herbicide resistance." According to the researchers, society may not be able to find its way out of the thorny problem of weed resistance to herbicides. There is a considerable chance that the evolution of weed resistance will outpace human innovation. Rather than assuming effective new herbicides will be developed, what is needed is to take a longer view of weed control that includes multiple approaches at once – like crop rotation and curbing the spread of weed seeds.

Crop and animal pests

A pest is any organism that feeds on and damages cultivated plants, attacks plant products in the field or in storage, causes a nuisance or transmits pathogenic organisms to plants, man or domestic animals and livestock (Youdeowei & Service 1993). Pests include insects and insect larvae, fungi, viruses, nematodes (worms),

TABLE 7.4 Comparison of traditional tillage (TT), conservation tillage (CT) and conservation agriculture (CA) on various issues

Issues	**Traditional tillage (TT)**	**Conservation tillage (CT)**	**Conservation agriculture (CA)**
Practice	Disturbs the soil and leave a bare surface	Reduces the soil disturbance in TT and keeps soil covered	Minimal soil disturbance and soil surface permanently covered
Erosion	Wind and runoff soil erosion maximum	Wind and runoff soil erosion reduced significantly	Wind and runoff soil erosion least of the three (less than in TT and CT)
Soil physical health	The lowest of the three	Significantly improved	The best practice of the three
Compaction	Used to reduce compaction but can also induce it by destroying biological pores	Reduced tillage is used to reduce compaction	Compaction can be a problem but use of mulch and promotion biological tillage helps reduce this problem
Soil biological health	The lowest of the three due to frequent disturbance	Moderately better soil biological health	More divers and healthy biological properties and populations
Water infiltration	Lowest after soil pores clogged	Good water infiltration	Best water infiltration
Soil organic matter	Oxidizes soil organic matter and causes its loss	Soil organic build-up possible in surface layers	Soil organic build-up in the surface layers even better than CT
Weeds	Controls weeds but also causes more weed seeds to germinate	Reduced tillage controls weed but also exposes other weed seeds for germination	Weeds are a problem especially in the early stages of adoption, but problems are reduced with time and residues can help suppress weed growth
Soil temperature	Surface temperature – most variable	Surface temperature – intermediate in variability	Surface temperature - moderated the most
Diesel use and costs	Diesel use – high	Diesel use – intermediate	Diesel use – much reduced
Production costs	Highest costs	Intermediate costs	Lowest cost
Timeliness	Operations can be delayed	Intermediate timeliness of operations	Timeliness of operations more optimal
Yield	Can be lower where planting delayed	Yields the same as TT	Yields same as TT but can be higher if planting done timelier

Source: Hobbs, P. R., Sayre, K., and Gupta, R. (2008). The role of conservation agriculture in sustainable agriculture. *Philosophical Transactions of the Royal Society B: Biological Sciences*, 363(1491), 543–555. (Quoted by Ayodele & Aluko, 2018)

weeds, rodents, birds. According to Alabi *et al.* (2006), pests damage crops at different stages of growth on the field, at harvest, during transportation and in storage and cost Nigeria the loss of 5 – 40% of her crop production annually.

Domestic livestock also suffer from infestation primarily by worms and insects, some of which transmit diseases to the animals, reduce their productivity and the quality of their products. Indeed, pests threaten the country's food security in the face of a population growth rate that has consistently outstripped the rate of food production in the last four decades or so.

Pests in the food chain

Pests are so labelled because they damage plants and livestock being tended by humans for their own use. Whether they are herbivores, carnivores or omnivores, pests occupy an important niche in the ecosystem. In both feeding and reproductive activities the pests compete with humans for their food and raw materials. Each individual pest actually consumes very small quantities of plant or animal matter. This is true particularly of the insects, ticks, mites and similar organisms. For example, a locust may consume about twice its own weight in food per day, while a caterpillar can only consume about 50 grammes of fresh leaves. The reason they cause so much damage is that they rapidly multiply and congregate in large numbers on food plants or in herds or flocks of livestock. Insects such as the grasshopper and locust which move in swarms are by far the most prolific.

Losses due to pests have been a major concern of governments and researchers since colonial times. The losses were and are still greatest in food crops especially cereals. From a survey of stem borer infestation on graminaceous crops grown in parts of Nigeria in the early 1950s, Sutherland (1952) estimated that pests were responsible for annual crop losses worth about N1,400,000 (₦1.00 =US$2.00) at the prevailing price. Booker (1963) observed that damage to pods of cowpeas by pests varied from 37 per cent in the late season to 94 per cent in the early season. Investigations carried out in the former Western Nigeria also revealed that about 20 per cent of the late maize stands were lost to stem borers (Adeyemi *et al.,* 1966). Pest control experiments revealed how much Nigeria lost to crop pests annually. For example, at a time when the average yield of cowpeas in northern Nigeria was 150-500 pounds per acre (about 27-92 kg/ha) on peasant farms, Booker (1965) achieved satisfactory control of the insect complex on experimental cowpea farms and obtained an average yield of 1,623 pounds dry seed per acre (about 298kg/ha), that is a 324 – 1104 per cent increase. Taylor and Ezedinma (1964) working in southern Nigeria, also felt that with effective pest control, yields could be raised to between 1,700 and 2,000 pounds per acre (312 – 367kg/ha) from the existing level of between 300 and 500 pounds (about 55 – 92kg/ha).

At the IITA, Ibadan, Rachie and Roberts (1974) determined percentage losses in cowpea yields due to pest damage and disease (Table 7.5). The pest problem has become more serious even in the 21st century as it threatens the country's food security. According to Zhang *et al.* (2018) insect pests and plant diseases are major yield reducing factors in Nigeria, threatening food security and farmers' incomes. For example, some other studies had shown that insect pests and diseases in yams* resulted in a 25% mean annual yield loss Tobih *et al.* (2011); Amusa *et al.* (2003) and 25–30% of yield loss of cocoa was attributed to the brown cocoa mirid bug alone Ndubuaku and Asogwa, (2006).

Many pests attack crops in the pre-flowering stage; this is to some extent an advantage because it is early enough for the damaged seedlings to be replaced. Others do more harm at the flowering and seeding stages; flower heads may be attacked so that they fall prematurely, while those that are attacked in the bud may never open at all. Damaged seeds may dry out and shrivel before reaching maturity. Fruits, seeds and tubers may be so scarred by pests that decay sets in and renders

*https://www.sciencedirect.com/topics/agricultural-and-biological-sciences/yams

TABLE 7.5 Yield reduction in cowpea (*Vigna unguiculata*) due to pests and diseases (IITA, Ibadan)

Insect	Yield reduction (%)
Thrips	50
Maruca-flower damage	20
Maruca-pod damage	20
Hemptera-seed damage	35
Laspeyresia-seed damage	50

Disease	Yield reduction (%)
Seedling blight (fungus)	75
Anthracnose	50
Leaf spots (Cercospora)	30
Bacterial pustule	12
Viruses	50
Nematodes—root knot	25

Source: Rachie, K.O. and Roberts, M (1974)

the harvest worthless.

Pest damage does not stop with the harvest, so crop yields only give us part of the picture. Further losses are incurred while the crops are in storage from attacks by a wide variety of pests, especially beetles, bugs and insect larvae. In addition, high humidity, poor ventilation, dusty and unclean storage conditions may all facilitate the breeding of numerous viruses and fungal and bacterial pathogens. Once a food store is infested by pests, a whole season's harvest may be destroyed if care is not taken.

Indeed, the impression one gets is that in Nigeria losses of food crops to pests are greater during storage than in the field. Although, government constructed granaries at strategic locations throughout the country in the 1970-80s, peasant farmers hardly patronise them. Peasant farmers still depend on their own home-based storage facilities while fumigants are not widely used by local farmers to preserve their harvests. Hence stored maize, cowpea, groundnuts, guinea corn, rice and yam tubers are all highly susceptible to attack. The only saving grace is that the middlemen who buy farm products from the peasant farmers take the pain to preserve their products for sale in the markets.

Crop pests

In this discussion of crop pests, each of Nigeria's major crops will be considered in order to present a fairly comprehensive picture of the problem. The geographical distribution of pests largely coincides with the areas where the host crops are cultivated (Fig. 7.2). A comprehensive list of insect pests in Nigeria was compiled long ago by Golding (1946) who also highlighted those which, at that time, were threats to each of Nigeria's major crops. More recently, the survey of local farmers' perceptions of pest damage carried out by Zhang *et al.* (2018) in several states across all the agro-ecological zones in the country from the coast to the Sahel, showed that a total of 54 insect pest groups were reported (Table 7.6). The most frequently reported insect pests included grasshoppers (*Caelifera spp.*), unspecified caterpillars (*Lepidoptera*: i.e. larvae of butterflies and moths), African armyworm (*Spodoptera exempta* Walker), aphids (Aphidoidea spp.), stem borers (larvae of specific Lepidoptera species that bore into plant stems), great yam beetle (*Heteroligus meles* Billb.), termites (*Isoptera spp.*), sorghum midge (*Contarinia sorghicola* Coq.), and pod borers (larvae of specific Lepidoptera species that bore into pods). While most insect groups were reported by households from all parts of the country, the occurrence of some pest insect groups corresponded with the distribution of the host crops.

A wide variety of insect pests feed on plants belonging to the same family, these

TABLE 7.6 Insect pests reported by households in various agro-ecological zones in Nigeria, 2012–2013

Groups	Reported insect pests	Frequency	%	Groups	Reported insect pests	Frequency	%
1. Aphid / mealy bug	Aphid	164	7.55	7. Stem / pod borer	stem borer	144	6.63
	maize aphid	46	2.12		pod borer	60	2.76
	cowpea aphid	39	1.79		millet stem borer	52	2.39
	groundnut aphid	24	1.10		white rice borer	21	0.97
	mealy bug	24	1.10		larger grain borer	14	0.64
	cassava mealy bug	3	0.14		large grain borer	2	0.09
	long-tailed mealy bug	2	0.09		adult borer	1	0.05
2. Diptera	sorghum midge	68	3.13	8. Other	white grub	44	2.02
	African rice gall midge	27	1.24		Grubs	2	0.09
	AFRGM	11	0.51		green stinking bug	22	1.01
	stalk eyed shoot fly	11	0.51		cowpea pod sucking bug	16	0.74
	rice whorl maggot	4	0.18		fire insect	5	0.23
3. Formicidae	Termite	104	4.79		cotton stainer	3	0.14
	Ants	29	1.33		rice leaf miner	1	0.05
4.Grasshopper	Hopper	304	13.99		cassava green spider mite	10	0.46
	Locust	43	1.98		green spider mite	1	0.05
	migratory locust	39	1.79		spider mite	1	0.05
	Cricket	20	0.92		cowpea flower thrips	5	0.23
	variegated grasshopper	9	0.41		Thrips	2	0.09
5. Leaf feeding Coleoptera	great yam beetle	140	6.44		red banded thrips	1	0.05
	blister beetle	18	0.83		Weevil	14	0.64
	cowpea leaf beetle	25	1.15		palm weevil	6	0.28
	Heleoptis bug	15	0.69		Whitefly	48	2.21
	African melon ladybird	6	0.28	**Total**		**2173**	**100**
	rhinoceros beetle	1	0.05				
6. Leaf feeding Lepidoptera	Caterpillar	219	10.08				
	African army worm	191	8.79				
	leaf roller	1	0.05				
	Worm	100	4.60				
	cotton bollworm	11	0.51				

Source: Wei Zhang, Edward Kato, Felix Bianchi, Prapti Bhandary, Gerrit Gort, Wopke van der Werf, 2018 Farmers' perceptions of crop pest severity in Nigeria are associated with landscape, agronomic and socio- economic factors. *Agriculture, Ecosystems and Environment* 259 (2018) 159–167 are the *oligophagous* pests. Those, like the banana weevil that feed exclusively on a single host plant, are described as *monophagous*. Others, like the locusts, feed on a wide variety of plants belonging to different families; they are the *polyphagous* pests and are among the most damaging.

Polyphagous pests

Perhaps the most dreaded group of polyphagous pests in Nigeria are the grasshoppers. This is because they infest large areas at the same time and damage a wide variety of crops simultaneously including those which form the main items of the Nigerian diet (Adejuwon, 1978). Thus, an outbreak of grasshopper infestation in an area leaves the population without its staple or secondary food plants, and can result in widespread famine, hunger and, in the worst cases, colossal loss of human life.

The two most important grasshopper species affecting the northern and southern parts of the country, respectively, are the locust and the variegated grasshopper. There are four species of locusts, namely, the African migratory locust, the red locust, the desert locust and the tree locust. According to Golding (1946) the African migratory locust was the only species that used to infest the whole of Nigeria. Adult swarms were always present in some parts of the country in the period from December1929 to January 1940. The only known outbreak area was the flood plains of the Middle Niger in the then French Soudan (now Mali) from where the locusts used to spread over a wide area of tropical Africa. Control measures instituted by the French Administration finally checked these outbreaks.

The red locust is found only in the Chad region where the solitary phase (a few wandering insects) is always present. Occasionally, in the past, their numbers had built up to plague proportions. For example, swarms were present in 1930, 1931 and 1932. Since the late 1970s, however, the desert and tree locusts are the species that have been menacing farmlands in the Chad region. Swarms of desert locusts move eastwards across the Sahel zone of West Africa and pass by the Chad region usually in the early growing season. The tree locust swarms in the extreme northern parts of Nigeria, usually flying at night during the late dry season. Insecurity in northeastern Nigeria as a result of insurgency by the Boko Haram and other terrorist groups has displaced thousands of people and greatly disrupted agriculture and other economic activities. The disruptions in the region have constrained the fight against locusts and weaver birds ravaging the region. According to FAO (2020) families displaced by the insecurity in the Chad region still rely on vulnerable host communities for basic needs, including food and this has put already impoverished host communities under extreme pressure, leading to increased exposure to food insecurity and malnutrition.

The variegated grasshopper became a serious pest known to cause considerable damage· to such crops as cassava, citrus, banana, plantain, cocoa seedlings, maize, cola, cow peas and various types of vegetables in the early 1970s (Toye, 1974). They are found primarily in the secondary forest, derived savanna and southern Guinea savanna zones. The insects tend to attack cassava and other perennial crops first because they carry most of the green foliage available in the dry season.

A study in three separate areas of southern Nigeria: Ikale area of Okitipupa division of Ondo State, the Kabba region of Kogi State and Alayi in Imo State (Barker *et al.*, 1977), revealed how seriously local farmers regarded the damage done by the variegated grasshopper, which they considered as their worst environmental problem. The crops attacked and consumed or destroyed by the insects included yam, cassava, cocoa, melon, maize, okra, pepper and green vegetables. The farmers interviewed felt that the spread of this grasshopper had been due partly to fluctuations in the annual incidence of rainfall, especially the lateness of the rains at the beginning of the wet season; and partly to the spread of the Siam weed (*Chromolaena*

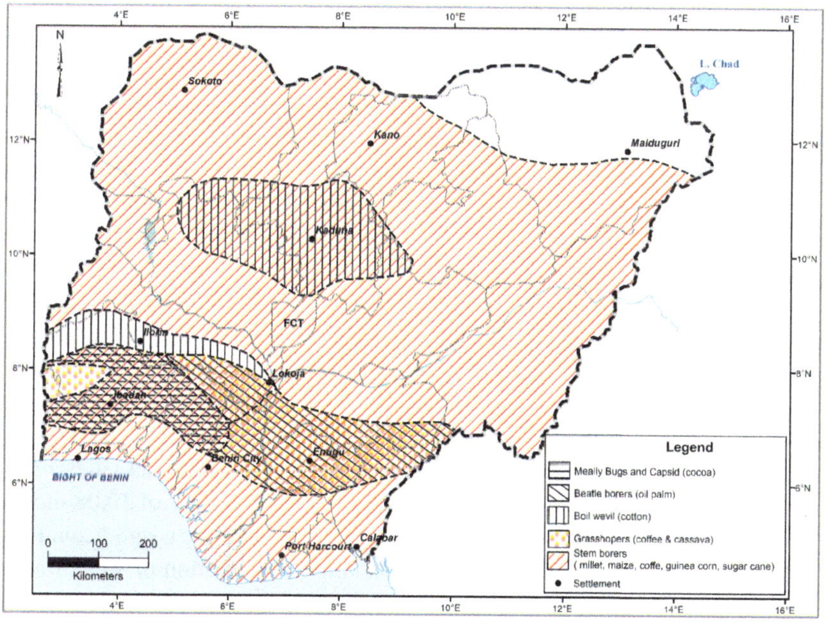

FIGURE 7.2 The distribution of some crop pests in Nigeria (Adejuwon, 1978)

*odo*rata). Entomological studies lent credence to the farmers' own observations. Since the insect breeds during the dry season, late rains prolong the period when conditions favour its survival and multiplication.

Toye (1974) suggested that there was a link between Siam weed and the survival, perpetuation and extensive distribution of the variegated grasshopper in Nigeria. He observed that in the months of October and November conspicuous feeding aggregations of the larvae were common on weeds such as *Aspilia Africana*, *Ageratum conyzoides* (goat weed), *Lantana camara* (wild sage) and, notably, *Chromolaena odo*rata or Siam weed:

> Apart from the fact that the early development stages of *Zonocerus variegatus* commonly feed on *Chromolaena,* observations showed that the weed formed adequate shelter for ovipositing adults during the egg-laying season from late March to June. It was therefore suggested that the spread of *Eupatorium* in Nigeria could partly account for the increasing abundance of *Zonocerus variegatus* in the country. It is worth mentioning, however, that the other weeds *Aspilia Africana, Ageratum conyzoides, Lantana camara* are also common in most parts of southern Nigeria but the case of *Chromolaena* odorata is significantly different in that unlike the other weeds, its introduction is more recent (Toye, 1974).

Eelworms and termites which inhabit the soil are also general polyphagous pests of crops. Termites are most prevalent in the savanna regions while eelworms exist in almost all soils, often in large numbers. Due to their adaptable mouth parts and pointed form, eelworms find it easy to penetrate plant tissue and are capable of infesting practically any plant. Similarly, termites are capable of destroying not only the underground but also the aerial parts of plants. Bandiya *et al.* (2013) noted that termites are pests to several crops including maize, groundnut, sugar cane, coconut, yams, cocoa as well as trees in several ecological zones. Termites are also notorious for destroying roofing, ceiling and window planks wooden furniture and paper and particle boards in dwellings in the settlements. Moreover, termites render the soil unproductive over extensive areas of land through their burrowing activities and the throwing up of anthills. Bandiya *et al.* (2012) reported a density of 10.08 mounds per hectrare in Sokoto State. In many savanna areas termites are a significant factor in determining soil profile morphology especially the formation of stone lines in the subsoils.

The other major group of polyphagous pests is that of the vertebrates including birds and mammals. The birds are principally bush fowls, doves and weaver birds

(*Quela quela*) which seriously damage grain crops, such as, rice, maize and guinea corn. In 1977 and 1978 the Chad region suffered its worst invasion by weaver birds ever witnessed till then. A German firm had to be called in to help eradicate the pests by spraying the fields with chemicals from helicopters and light aircraft.

The mammals that damage food crops include the squirrel, hare, rabbit, small rat, mouse and grasscutter. The threat posed by these vertebrate pests is magnified by the fact that they attack crops at all stages from seed sowing to harvest. Small rats, mice, bush fowls and doves commonly dig up and consume freshly sown or germinating seeds. Field rats, giant rats and scare rats gnaw, cut and chew up the stems of growing plants while seeds and fruits are pecked and eaten by birds and squirrels. Grasscutters feed on cassava tubers, sugarcane and the stems of most grain crops, while herds of bush pig have been known to destroy entire fields of yams just before they were ready to be harvested.

Major crops and associated pests

Cereals:

The major cereals grown in Nigeria are maize, rice, guinea corn (or sorghum) and millet. The major pests of these cereals are the stem boring insects and their larvae. The most widespread stem borers are the larvae of moths and butterflies. The two species attack all grain crops but most especially maize. According to Golding (1964), damage by these stem borers in Ibadan region in the past, used to be particularly severe in late maize.

Cereal crops are also subject to attack by another type of larvae known as army worms. The moths migrate in swarms at night, the females depositing their eggs on grass. On hatching the larvae migrate into the grain fields where they may destroy most of the crop if not checked. Army worms are widespread in Nigeria; they occur usually in May and occasionally in September. They consume the leaves of plants.

Finally, the cereal crops are also subject to attack by bugs which destroy seeds and cobs. The rice bugs also attack guinea corn. The bugs feed on grass seeds and such weed grasses may be the source of infestation of grain crops.

Legumes:

The major leguminous crops grown in Nigeria are groundnut, cowpea, lima bean (*Phaseolus lunatus*), soya bean (*Soja max*) and pigeon pea (*Cajanus cajan*). They are subject to attack by a wide variety of insect pests. The larvae of certain moths are important pests of cowpeas attacking the buds, pods and seeds. The larvae of other species of moth similarly attack groundnuts.

But by far the most important pests of leguminous crops are the beetles which attack the roots and leaves of cowpeas and transmit virus diseases to the crop. Groundnut leaves are attacked by the weevil while the leaves of soya beans and the pods of pigeon peas are damaged by beetles.

Apart from the moths and beetles, leguminous crops in Nigeria are attacked by a variety of bugs and thrips (*Thysanoptera spp.*). The bugs cause the shrivelling of fresh pods and their feeding lesions are often invaded by fungi. Thrips (minute, slender insects with fringed wings and unique asymmetrical mouthparts) are known generally for causing flowers to drop off prematurely. Alabi *et al.* (2006) noted that the groundnut rosette epidemic that occurred sporadically in 1975 and 1985 led to the disappearance of the famous groundnut pyramids in northern Nigeria, especially Kano. The widespread losses due to this disease discouraged farmers from growing groundnut; such has been the devastating impact of pests and diseases on Nigeria's food supply and security.

Roots and tubers:

The three most important root crop in Nigeria is cassava *(Manihot esculenta)* while yam *(Dioscoraea spp.*), and sweet potato *(Ipomoea batatas)* are the major tubers. The major pests of yam are the beetles of which the most damaging are those which feed on the tubers and yam seeds. They are widely distributed in Nigeria, being found in Edo, Delta, Anambra, Ebonyi, Adamawa, Benue and Rivers States.

The most important pests of cassava are the grasshopper as already mentioned; the larvae of certain moths which damage the leaves; and the white fly which transmits virus diseases and retards the growth of the crop. The leaves of potatoes are also attacked by the larvae of moths and the tubers by weevils.

Cotton:

The most important fibre crop grown in Nigeria is cotton, which is beset by a plethora of pests, the most important of which are the bollworms and bugs. The pink bollworm was observed for the first time in West Africa in 1944 (Golding, 1946). It is a major pest in nearly all the cotton-growing countries of the world and it soon became the most important pest of the cotton crop in south-western Nigeria (present-day Oyo Osun and Ogun states). The cotton stainer pierces the green cotton bolls and introduces fungi which cause staining and rot. These bugs were said to be the greatest factor that inhibited the growth of American cotton in southern

Nigeria (Golding, 1946). The drier weather conditions in northern Nigeria rid the fields of stainers before the cotton crops begin to set. The Jassid bug was at one time an important pest of cotton throughout the country; however, resistant strains of both American and indigenous cotton were later developed.

Oil palm:

The weevil, which attacks oil palm leaves, has been known to kill newly transplanted seedlings at the Nigerian Institute for Oil Palm Research (NIFOR) plantations near Benin. The larvae of the bruchid also attack the nuts.

Cacao:

The major pest of cocoa trees is the capsid bug, which destroys young shoots. It thrives in areas of high rainfall and humidity. One of the factors that favoured the development of cocoa cultivation in Ibadan area at the transitional zone between forest and savanna was the relatively drier weather conditions which minimized the capsid bug problem.

Stored food crops:

The pests of food crops in storage include birds and mammals, especially rats and bats. But by far the most important storage pests are weevils, beetles and bugs. The bean weevil is perhaps the most common storage pest in many homes. The rice weevil is responsible for heavy losses of most grains in storage, particularly maize. The storage pests of groundnut include the groundnut beetle, the Trogoderma beetle, and the groundnut bruchid, which attack groundnuts in the shell.

Crop pest control

Experience has shown that it is extremely difficult to destroy any species of pests completely, especially the insects. Thus, pest control only aims at reducing pest numbers to very low levels and to do this with minimum disturbance or damage to the environment. A prerequisite to the successful control of any pest is long term observation of the life cycles of the pest and the ecological conditions and cultural practices which favour its propagation. These studies provide a basis for assessing the nature and magnitude of the pest problem and should lead to the selection of the most suitable pest management method. The strategies usually adopted in crop pest control include (1) the elimination of those characteristics of the crop which make it susceptible to pest attack; (2) evasion of the consequences of pest activity; (3)

suppression of those features of the pest which make it injurious; and (4) reduction of the pest numbers to levels which are not injurious to the crop.

The first strategy involves developing crop varieties that are resistant to pests. The example of cotton has already been mentioned. In the 1970s, new varieties of maize were introduced which are not only more nutritive and early maturing, but which are also more resistant to pests and diseases. The International Institute for Tropical Agriculture, at Ibadan, also developed two varieties of cassava which are more resistant to pests and diseases.

Evasion of pest activities means growing crops only where and when the pests do not thrive. This has meant the elimination of certain crops from areas where, otherwise, they would have thrived. Cotton and groundnuts are crops which would have done well throughout Nigeria but for the numerous pests in the humid south, which not only physically damage the crops but also transmit diseases. Also, many farmers have been discouraged from growing late maize because of the problems associated with pests.

Very little success has been achieved in the field of suppressing those features of pests which make them harmful to crops. This can only be achieved by interfering with the life cycles of the pests which is extremely difficult. The most widely adopted strategy of pest control in Nigeria is that aimed at keeping the population levels of the pests very low. This is pursued in several ways. First, farms are kept clear of weeds known to be alternative hosts for crop pests. At the same time crop stands that have been infested are removed to avoid the spread of the pests to other stands. In this regard it is essential to destroy the crop residue of the preceding growing season before planting new ones. Infestation can be much reduced by uprooting and burning plant residues after harvest. Unfortunately, some farmers still hold to traditional practices and do not clear the land right away. For instance, in the middle belt they leave the maize and guinea corn stalks standing to serve as stakes for the yams. It is worth noting that in the modern system of conservation agriculture referred to earlier in connection with weed control, destruction of crop residue is also discouraged; rather it is managed and used to protect the soil and enhance higher crop yields (Table 7.4).

Picking bugs off grain and other crops by hand may be beneficial but can only be effective where infestation is very slight. Vertebrate pests are killed or kept out of farms by various devices. Weeding round the boundaries of the farms may discourage some mammals from crossing into cultivated areas. Snap traps are set to catch small rats and mice while leg-holding traps are used for squirrels, giant rats and scare rats. Fencing round the farms with split wood and palm fronds or with wire gauze is an effective method of keeping out the larger animals. Scarecrows are quite effective in deterring bush fowls and weaver birds from coming onto arable fields.

Today the most actively pursued method of pest control in Nigeria, especially on experimental farms, large plantations and co-operative farms, is the use of insecticides, pesticides and rodenticides. There are many varieties of these chemicals (Sence Agric 2020). Rodenticides are used primarily in poisoning rats and mice especially where grains are stored. Unfortunately, peasant farmers took some time accepting pesticides except in respect of cash crops, like cocoa, which are not consumed by the people. Peasant farmers were wary of applying what they perceived as "poison" to food crops. It is worth recalling the fact that the mere application of chemical fertilizers and the colour of the resulting harvest initially made peasant farmers and local inhabitants in southwestern Nigeria to refuse to consume the introduced American yellow maize preferring to sell them off to city dwellers. Local maize is white; the fear of the people was that the chemical fertilizers were responsible for 'poisoning' the maize and turning it yellow. It took a lot of public enlightenment and agricultural extension services to change the rural people's perception of the yellow maize.

In modern times, the application of fungicides, pesticides and other fumigants for pest control faces many problems. High costs and irregular supplies discourage poor farmers from using them on a continuing basis. The effectiveness of the pesticides also fades with time because the pests develop resistance to them or develop new strains that are resistant to the prevailing brands. This is of serious concern to agricultural biologists who are conscious of the potential impacts of these synthetic chemicals on the environment, biodiversity and water resources.

Livestock pests

Livestock pests include internal worms, maggots and parasites, which inhabit various parts of the body; and external pests such as lice, ticks, flea and mange. We are concerned here with the external pests referred to generally, as ectoparasites. Apart from transmitting various diseases, these pests damage the skins or hides of animals and inflict wounds on them, sometimes producing abscesses. Some of them are bloodsuckers which cause anaemia in the host animals. Furthermore, livestock are troubled and maddened by many pests that they cannot feed adequately and their growth and productivity are thereby impaired.

Some pests, like the tsetse fly, are found in particular ecological zones and attack migrating animals when they pass through such locations. However, others are not restricted to any particular ecological zones but arise as a result of keeping the animals in unhygienic surroundings. Because the herds and flocks live in close proximity the pests can quickly spread from beast to beast.

However, in certain respects, great strides were made in the control of human diseases and pests and diseases harmful to domestic animals and livestock since the Second World War. Rinderpest was virtually eradicated in Nigeria as a result of intensive vaccination campaigns; anthrax, foot rot and foot-and-mouth disease were also largely contained through intensive veterinary activity especially in the northern states of Nigeria. The training of more veterinary doctors and assistants has been a major priority of the state governments in their efforts to overcome animal diseases. It is perhaps, pertinent to remark at this point that insurgent activities by terrorists and counter-terrorist operations by the military have impaired veterinary operations and pests and diseases that had long been contained have resurfaced sporadically in recent times.

In contrast to the successful campaign which eliminated rinderpest, trypanosomosis was still a serious threat to domesticated grazers until recently. It is not long since the problem of sleeping sickness in man was effectively contained let alone trypanosomal infection of livestock. However, the ecology and the geographical distribution of the tsetse fly are now fairly well-known so that it has been possible to identify the zones where the infection of animals takes place as they move from one part of the country to another (Fig. 7.3).

The disease is transmitted by various species of the tsetse fly. Overall, thirty-one different tsetse species and subspecies are recognised, of which eleven are found in Nigeria, namely, *Glossina palpalis palpalis*, *G. tachinoides*, *G. pallicera pallicera* and *G. calliginea* (palpalis group), *G. morsitans submorsitans* and *G. longipalpis* (morsitans group), and *G. fusca congolensis*, *G. tabaniformis*, *G. nigrofusca nitrous*, *G. medicorum* and *G. haningtoni* (fusca group) (Moloo, 1993; Onyiah, 1995). These tsetse species occur widely over Nigeria (Fig. 7.4) except the fusca and nigrofusca groups that seem to be restricted to the humid forest zone in the south (Edo, Delta, Rivers, Bayelsa, Akwa Ibom States).

In the past intensive inoculation campaigns were carried out by the veterinary services of the northern states to immunize cattle against trypanosomal infection. The tsetse fly eradication campaigns, however, were very costly and less successful than inoculating the cattle. The problem in the Guinea savanna was that the flies were more dispersed so that spraying operations had to cover wider areas than was the case with the operation against *Glossina* species that were restricted to riverine areas. Fortunately, greater success was achieved when it was found that the flies tended to concentrate in three main ecotones: (1) along drainage channels; (2) on round flat-topped hills; and (3) in islands of Isoberlinia woodlands.

Clearing the land of the tsetse fly is only half the problem: continuous efforts

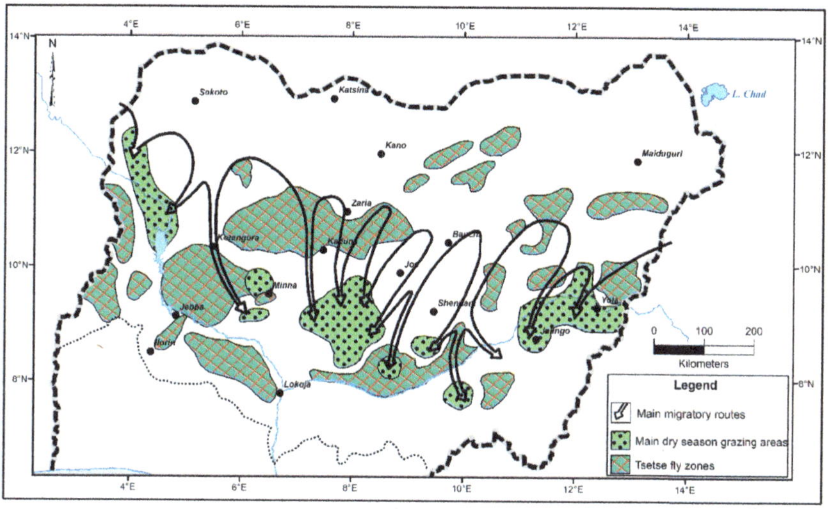

FIGURE 7.3 Cattle migration routes showing tsetse fly zones

have to be made to prevent the return of the flies. Thus, tsetse fly eradication campaigns have often been followed by programmes for the protection and utilization of the reclaimed areas. The Nigerian Institute for Trypanosomosis Research in Kaduna was largely concerned with basic research into trypanosome infection in man and beast while the state ministries of agriculture and natural resources looked after the reclamation programmes. While, there are studies that show that active human trypanosomal infection is now rare (although some doubt the efficiency of the surveillance), livestock trypanosomal infection persists in many grazing areas (Table 7.6; Gier *et al.*, 2020; Onyekwelu *et al.* (2017). Furthermore, Weber *et al.* (2019) point to protected conservation areas like national parks and game reserves as reservoirs of tsetse flies and livestock trypanosomes.

Fleas are similar to the tsetse fly in being blood-sucking insects which cause irritation and scratching. Lice puncture the skin and suck the blood of their victims. Each species of animal has its own species of lice including cattle, sheep, goats, pigs, dogs, cats, horses, and poultry. Pig lice are found usually on the head, neck, ears, on the belly and between the hind legs; while in horses, they usually infect the hair of the tail, under the chin and the mane.

The mange mite is a common pest of most domestic animals and poultry and is the worst ectoparasite of pigs. Unfortunately, early infestations are not easy to detect unless skin scrapings are examined under the microscope. The helminths

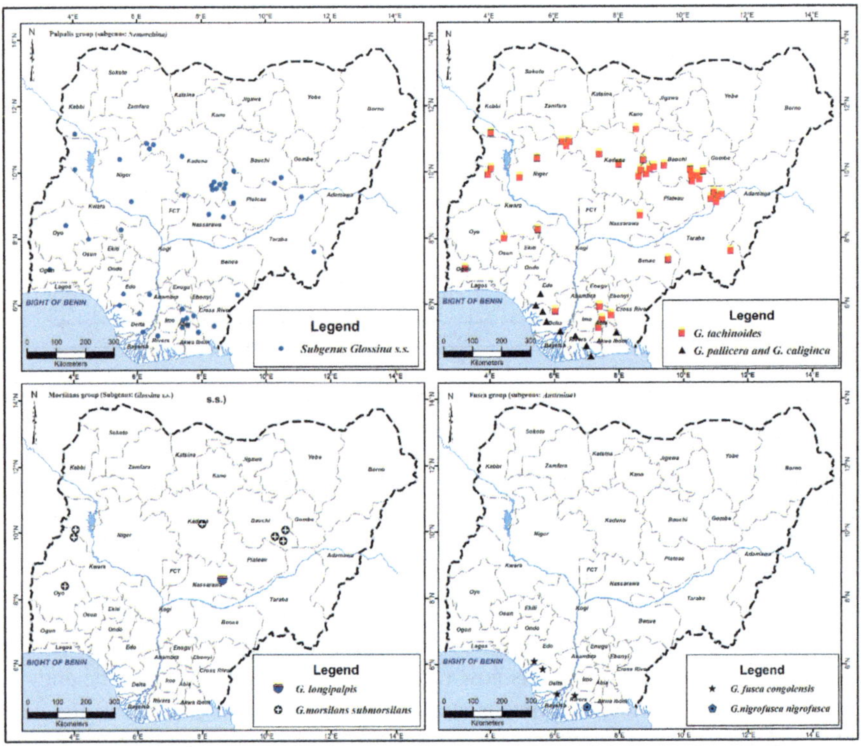

FIGURE 7.4 Distribution of major Glossina species in Nigeria
Redrawn from: de Gier J., Cecchi G., Paonea M., Dede P. and W. Zhao 2020 *The continental Atlas of tsetse and African animal trypanosomosis in Nigeria*. Food and Agriculture Organization of the United Nations (FAO), Viale delle Terme di Caracalla, 00153, Rome, Italy.

also cause skin irritation and cellular damage in a number of animals.

Ticks damage the skin of many animals including poultry often producing wounds and abscesses, while most are also afflicted by flies. For example, the sheep nostril fly, which attacks the noses of sheep and goats, lays its eggs in the nostrils. The infected animals have watery eyes and a nasal discharge; and sneeze frequently.

Control of livestock pests

The two most important factors in the spread of ectoparasites are poor sanitation and overcrowding. They are therefore best controlled by improved herd and flock management, including keeping the animal sheds clean and disinfected.

The most intractable animal pests are the blood suckers such as the tsetse fly whose survival tends to be encouraged by certain environmental factors. They occur over very large areas of land, and usually require a combination of measures to keep them under control, including the breeding of livestock varieties that are resistant to the disease pathogens carried by the flies; the control of wild game which are the alternative hosts to the pathogens; the modification of the habitats of the flies to make them less favourable, for example, through bush clearing; and the extensive use of insecticides to eradicate the pests.

Crop and animal diseases

Crop and animal diseases are caused by several agents most of which may be grouped into four categories, namely, *parasites* including many of the pests already mentioned; *viruses*, *bacteria* and *fungi*. Diseases cause great losses in the quantity and quality of Nigeria's agricultural products. Lana and Adeniji (1976) estimated losses in food crops in Nigeria in the 1970s were of the order of thirty to fifty per cent. As indicated earlier, more recent estimates put the losses as ranging from five to forty per cent (Alabi *et al.*, 2006), reflecting wide variations from place to place. The country's domestic livestock are also plagued by numerous wasting diseases some of which result in fatalities.

Crop diseases

It is not possible to list all the known diseases of crops and their agents in Nigeria. In this section, attention is drawn only to the most important (Fig. 7.5a, b and c). The majority of disease agents are present in the environment all the time: when conditions are favourable there is a sudden proliferation leading to infestation of the victim crops. The vagaries of the weather may be a factor; for example, the incidence of some diseases increases if the rains come earlier than usual or if the total amount is above the average for the area. On the other hand, prolonged dry weather encourages diseases whose agents are carried on dust particles. Alternative hosts on, or close to, the farm may cause an outbreak, so may intensive cultivation of a susceptible crop which may give rise to increased outbreaks of a particular disease. Poor soil conditions sometimes reduce the vitality of crops and render them more susceptible to disease agents. Even the planting of unhealthy seeds may have the same result.

Diseases of grain crops

The viral diseases of grain crops primarily affect the leaves, destroying green matter and thereby reducing the ability of the plants to photosynthesize. The most common of these viral diseases are brown and red leaf spot and leaf mosaics. In general, losses due to these viral diseases are not as great as those due to other diseases. One of the most virulent diseases of grain crops is rust. Spores are formed which break through the epidermis of the leaves and form circular brown powdery pustules all over both surfaces of the leaves. Rust is particularly common in maize and millet.

Smut, in maize and millet, is a common disease of grains, characterized by swellings, often called 'boils', on the leaves and ears. The boils grow to a large size and individual grains grow to enormous sizes. The boils, which are formed of abnormal plant tissue mixed with fungus tissue, are then filled with a blackish-brown dust which is released when the boil bursts. In false smut disease of rice some grains are replaced by a large olive-green fungus which may contaminate other grains giving them a sour taste.

A much-dreaded disease of rice is blast. The fungus attacks the base of the panicle and cuts the ears off from the food supply in the stalks. Thus, little or no grain will be formed in the ears.

Bacterial diseases have been identified in guinea corn. For example, bacterial blight causes the leaves to turn yellow and wilt, while the stems are also discoloured and weakened internally and may eventually collapse.

Diseases of leguminous crops

The most important diseases of leguminous crops are caused by fungi and result in wilt, leaf fall and dieback. For example, the two most important diseases of pigeon pea are wilt caused by the soil fungus *Fusarium udium* which gains entrance into the plants through wounds on the roots; and dieback or collar crack, caused by the fungus *Diplodia cajani* which results in the stem surfaces splitting at ground level and exuding gum. Tikka disease is a fungal disorder which causes spotting of the leaves of groundnuts followed by general chlorosis and leaf fall. The disease is only serious if it occurs early in the life of the crop. It also affects cowpeas. Wilt, caused by *Sclerotium rolfsii,* is the most serious fungal disease of groundnuts. It attacks the stem and collar of the spreading groundnut and contaminates the seeds, turning them brown. The fungus, which grows best in wet weather, is a serious limiting factor on the cultivation of groundnuts in humid southern Nigeria.

Viral diseases of leguminous crops include mosaic and rosette disease. Mosaic causes spotting, wrinkling and distortion of the leaves and stems of cowpea. Rosette disease attacks groundnut plants which then become stunted in growth and turn yellow. The virus is transmitted from plant to plant by aphids. In addition, parasitic nematodes cause irregular swelling on the roots of groundnut plants which therefore become stunted in growth (Table 7.3).

Diseases of root crops

The viral diseases of root crops are numerous. They include mosaic, lanceolate mottled leaf, and brown and red leaf spot. Mosaic disease of cassava is capable of reducing yields by as much as thirty per cent while also, reducing the starch content of infected plants (Obigbesan, 1976). It is transmitted by a white fly, B*emisia nigeriensis.* It causes chlorosis and distortion of the leaves and stunts the growth of the crop. Lanceolate mottled leaf affects yams in a similar fashion. Leaf spot and brown and red leaf spot lead to the formation of spots and blotches on the leaves of yam and cassava.

Witch broom is a fungal disease of yam caused by *Phyllentypa dioscoraea.* It causes chlorosis of the leaves and stunts the growth of the branches of white yam. The major fungal disease of sweet potato is stem rot caused by *Fusarium oxysporum* but it is not yet a very serious disease in Nigeria.

Bacteria wilt is a disease of cassava caused by *Xanthomonas manihotis.* It is characterized by leaf wilt, defoliation and dieback accompanied by the exudation of a yellow substance from the infected parts.

Cotton diseases

The growth of cotton in Nigeria has been greatly limited by its susceptibility to fungal diseases which thrive in the humid conditions of southern Nigeria and the

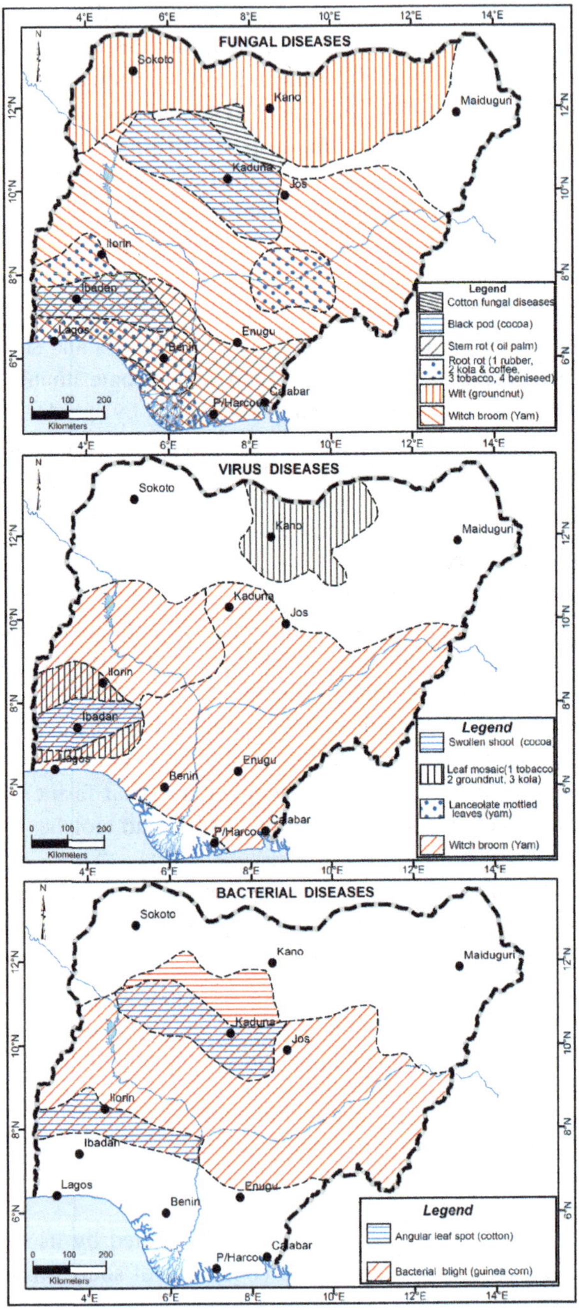

FIGURE 7.5 The distribution of some crop diseases in Nigeria (a) Fungal diseases (b) Virus diseases (c) Bacterial diseases (*Source:* Adejuwon, 1978)

middle belt. Grey mildew (*Ramularia gossypii*) gives rise to irregular whitish spots on the underside of the leaves while anthracnose affects the stems and the leaves. Another fungal disease which sometimes infects cotton is the black-pod of cocoa.

The bacterial diseases of cotton are bacterial blight, and angular leaf spot.

Cocoa diseases

Cocoa is susceptible to a range of fungal and viral diseases. Swollen shoot disease is caused by a virus which leads to swelling on the nodes, internodes and tips of quick growing shoots. The important fungal diseases are black pod, brown pod and mealy pod. These diseases are most prevalent in years when the rainfall is excessively high. They are found in all cocoa growing areas. The pathogens causing the fungal diseases originate in the soil and spread to the cocoa pods by means of spores. Observations made by Okaisabor (1965) in a cocoa plantation at Agodi, Ibadan showed that over 84 per cent of black pod occurred below a height of 50 cm on the tree; 11per cent between 50 and 75cm.; and only 4 per cent above 75cm. He noted, therefore, that suppressing normal pod development at the lower parts of cocoa trees could prevent the pathogen from making contact with the cocoa pods and so help in controlling the black pod disease.

Control of crop diseases

The principles and methods of disease control in crops are more or less similar to those of pest control. It begins with good farm and crop management, including planting healthy seeds, good spacing of the stands to avoid overcrowding, regular weeding and removal of diseased plants wholly or only the diseased parts. Crop rotation is another control measure as continuous growing of the same crop on the same plot may encourage the build-up of disease agents.

Good drainage is important; many viral and fungal diseases are encouraged by damp soil. Weeding some crops when the foliage is wet is inadvisable because fungal spores are more easily spread in such circumstances. Also, it is important to avoid damaging any part of the growing crop because wounds offer easy points of entry for disease agents. Many Nigerian farmers, of course, are aware of the precautions they can take in order to protect their crops against diseases. Unfortunately, such protection is not enough to check major outbreaks, with the result that chemical pesticides and fungicides had to be introduced. These chemicals were first used on a large scale on the large cocoa farms and were later extended to other tree crops and arable crops. Viral swollen shoot disease of cocoa is still largely controlled by the destruction of the affected plants as no chemical compound has yet been found capable of controlling it.

Animal diseases

Disease can affect domestic animals and livestock in four ways: through the skin; in the digestive system; in the respiratory system; and in the reproductive system. The signs and symptoms of sickness in domestic animals have certain features in common, such as reduced productivity which may be reflected in weight loss or reduced milk production in dairy animals; reduced appetite; a dull lethargic appearance and a tendency to wander away from the herd or flock. There are also symptoms related to particular diseases as shown, for example, in the case of trypanosomal infection.

Diseases of ruminants

Skin diseases are very common among Nigeria's grazing animals. They cause itching and scratching, thickening, folding or cracking of the skin and the development of sores and lumps. In severe cases, skin diseases may cause dehydration and emaciation of the animal. Skin diseases are caused by parasites such as lice, ticks, flies, fleas, mange mites and helminths. Viruses are responsible for pox diseases and the lumpy skin disease of cattle while bacteria and fungi cause wounds and abscesses, ringworm and foot rot diseases. Most of these diseases are not fatal but they do lower the quality of the hides and skin. By contrast, *Streptothricosis* is one of the most serious diseases affecting cattle and it can be fatal. It occurs in the rainy season when it may spread all over the body of the affected cow.

Respiratory diseases are also very common among the ruminants. They result in coughing, laboured or fast, shallow breathing, nasal discharges and fever. The causative agents are (1) parasites, the most important being lung worms which infect the bronchi and the alveolar space; (2) bacteria, which cause such ailments as bacterial pneumonia, bronchitis, tuberculosis and lung abscesses; and (3) viruses which cause viral pneumonia such as catarrh in sheep and goats.

These respiratory diseases, particularly pneumonia and tuberculosis, have afflicted cattle in Zaria, Kano, Adamawa, Eastern Borno and Enugu environs (Fig. 7.6). The agents normally inhabit the respiratory system and become harmful when they are activated by other factors such as exposure to cold or stress due to overloading or long trekking.

Diseases of the digestive system are, along with skin diseases, the commonest in Nigerian ruminants. The feeding habits of the animals expose them to many dis-

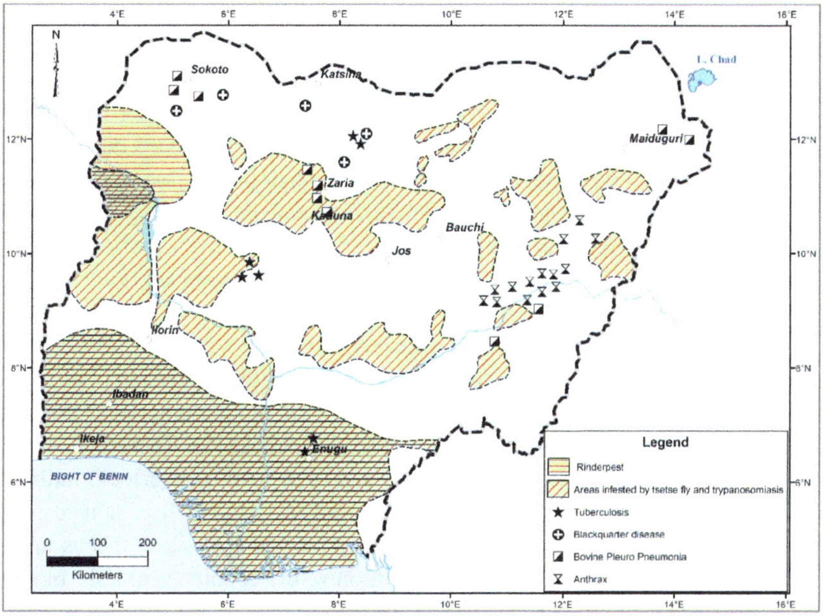

FIGURE 7.6 The distribution of some animal diseases in Nigeria (Source: Adejuwon, 1978)

ease agents. The diseases may give rise to excessive salivation, sore mouths, diarrhoea, anaemia, dehydration, constipation and abdominal pain. The agents include parasites such as helminths, roundworms, flatworms, liver fluke and protozoans which suck blood, destroy the cells of the gut walls, consume half-digested food in the gut and secrete waste products which irritate the walls of the gut. Other agents are bacteria which are responsible for such diseases as Enterotoxaemia, Salmonellosis and Collibacillosis, especially in young animals.

The viral diseases affect the mouth and other parts of the digestive system causing mouth diseases and pox diseases; catarrh in sheep and goats; and rinderpest in cattle only. Rinderpest is a killer plague of cattle. The most susceptible breeds of cattle are the *n'dama* and *muturu* found in the humid south. The disease had been largely checked in the north during the colonial era but cases were noted in Sokoto and Adamawa in the 1970s, when it was believed to have been introduced from neighbouring countries (Adejuwon, 1978). It is a sad commentary on the lapses in the management of livestock and the environment in the grazing areas that, in the 21st century, rinderpest and other diseases that had long been put in check have suddenly re-occurred, though sporadically, in some years. As stated earlier, the insecurity in the northeastern parts of the country is thought to be responsible for the lapses.

Diseases of the reproductive system, particularly those affecting the females, are the most dreaded by the Fulani cattle rearers. This is because the diseases often result in abortions, infertility, failure to produce milk or production of poor-quality milk. The female parts affected are the genital tract and the mammary glands which are most susceptible to diseases during lactation. The causative agents include parasites, bacteria, fungi and viruses. Two major bacterial diseases are brucellosis and tuberculosis.

African animal tsetse-transmitted trypanosomosis is a severe vector-borne disease. Infected animals suffer variously from intermittent fever, anaemia, decreased appetite, progressive loss of condition, decreased milk production, reduced fertility and increased mortality in affected herds (Desquesnes, 2018). Therefore, directly trypanosomosis causes reduced productivity and even death in livestock. Indirectly it reduces the efficiency of draught animals used in farming. It is ranked as the second most important livestock disease in Nigeria (Dede *et al.*, 2007) as up to a third of the cattle population of over 20 million in the country is at risk. In ruminants, particularly cattle, African animal trypanosomosis is predominantly caused by *Trypanosoma vivax*, *T. congolense* and *T. brucei spp* (Table 7.7). The disease is also transmitted to horses, pigs, sheep and goats. African animal trypanosome has also been fingered as the cause of massive utilization of veterinary drugs in Nigeria, with dire issues of substandard medicines and drug resistance (Kingsley, 2015).

Control of animal diseases

Most animal diseases are spread by contagious agents; the air expired by an animal at each breath contains numerous disease agents particularly viruses and bacteria. The incidence of disease can therefore be minimized by avoiding overcrowding and by ensuring clean living conditions. Whenever the livestock congregate diseases may be spread by infections in the urine, faeces or mud wallows. Frequent washing or cleaning of the animals is therefore important and is the most effective method of preventing skin diseases.

Internal parasites, viruses and bacteria are best controlled through good nutrition, inoculation and vaccination. Animals easily fall victim to disease agents when their natural resistance has been reduced by poor nutrition. This is a major problem where livestock depend on natural pasturage which has not been supplemented with sown pasture, fodder crops or concentrates. Many viruses and bacteria are destroyed by treating the animals with antibiotics. Vaccines are also effective against

TABLE 7.7 Availability of published surveys on, and reported presence of, African animal trypanosomosis, tsetse flies and tsetse trypanosomal infection in Nigeria. January 1990 – March 2019.

Geopolitical zone	State	African animal trypanosomosis		Tsetse flies		Tsetse trypanosomal infection	
		Published surveys	Reported presence	Published surveys	Reported presence	Published surveys	Reported presence
North West	Jigawa	x	x	x			
	Kaduna	x	x	x	x	X	x
	Kano	x	x	x	x	X	
	Katsina	x	x				
	Kebbi	x	x	x	x	X	x
	Sokoto	x	x				
	Zamfara						
North East	Adamawa	x	x				
	Bauchi	x	x	x	x	X	x
	Borno	x	x	x			
	Gombe	x	x	x			
	Taraba	x	x	x			
	Yobe						
North Central	Benue	x	x	x	x	X	
	Kogi	x	x	x	x	X	
	Kwara	x	x	x	x		
	Nassarawa	x	x	x	x	X	X
	Niger	x	x	x	x	X	X
	Plateau	x	x	x	x	X	X
	FCT						
South West	Ekiti	x	x				
	Lagos	x	x				
	Ogun	x	x	x	x	X	X
	Ondo	x	x	x			
	Osun	x	x				
	Oyo	x	x	x	x	X	X
South East	Abia	x	x	x	x	X	X
	Anambra	x	x				
	Ebonyi	x	x				
	Enugu	x	x				
	Imo	x	x				
South South	Akwa Ibom	x	x	x	x		
	Bayelsa			x	x		
	Cross River	x	x	x	x		
	Delta			x	x	X	X
	Edo			x			
	Rivers			x	x		

Note: Published surveys, and the related reported presence, refer to peer-reviewed scientific papers Source: de Gier J., Cecchi G., Paonea M., Dede P. and W Zhao 2020. *The continental Atlas of tsetse and African animal trypanosomosis in Nigeria*. Food and Agriculture Organization of the United Nations (FAO), Viale delle Terme di Caracalla, 00153, Rome, Italy.

such diseases as trypanosomosis, tuberculosis and rinderpest.

Human diseases

The major human diseases in Nigeria may be grouped into four main categories: (1) those which are caused by insect pests, namely, malaria, yellow fever, river blindness and sleeping sickness; (2) water-borne diseases such as guinea worm, bilharzia, and hook worm, which infect human beings through using impure or contaminated water; (3) those diseases such as cholera, dysentery, typhoid, leprosy, and small pox which arise as a result of poor personal hygiene and sub-standard living conditions; and (4) novel diseases caused by viruses, for example, Lassa fever, Ebola, SARS and coronavirus.

Diseases caused by insect vectors

Brown (1955) rightly observed that the ecology of human diseases in Nigeria was closely related with the ecology of insects. This is because some of the most common diseases are transmitted through insect bites. Tables 7.8 and 7.9 show that malaria is the most prevalent illness in Nigeria.

Mosquitoes are probably the most widespread insect vectors of human diseases in Nigeria. They breed wherever there is a standing body of water or a wet surface.

However, their population decreases with increasing aridity towards the extreme North. They transmit about forty different types of human diseases the most important of which are malaria, yellow fever and filariasis. Malaria is caused by protozoans *(Plasmodium spp.)* carried by Anopheles mosquitoes, which are injected into the blood stream whenever the mosquito bites a person and sucks blood. Malaria is a debilitating disease which is often fatal to children under the age of three. However, older children and adults tend to develop a partial immunity to the disease.

Efforts to eradicate malaria from Africa and other tropical lands have been championed by both local and international non-governmental organizations including, in recent times, the Bill and Melinda Gates Foundation, which is funding the development and distribution of anti-malaria vaccines. The Bill and Melinda Gates Foundation like other NGOs work in collaboration with the World Health Organization. Unfortunately, the campaigns by these international NGOs have been the subject of much misunderstanding and suspicion. There is deep suspicion among the populace and rumours widely circulated that the vaccines are laced with substances that would sterilize the recipients and reduce fertility and population growth

TABLE 7.8 Reported cases and deaths from notifiable human diseases, 2007-2011

Type of Disease	Reported Cases				Reported Deaths			
	2007	2008	2010	2011	2007	2008	2010	2011
Cholera	12,194	17,854	7,152	16,119	185	-	382	347
Meningitis (CSM)	3,537	7,199	2,402	730	301		267	47
Viral Influenza	-	-	205	5,170	-	-	8	1
Diarrhoea (Non-bloody)	1,069,133	893,482	560,052	431,852	1,038	-	935	636
Diarrhoea (Bloody)	424,668	395,443	209,039	194,714	2,500	-	245	195
Guinea worm	-	38	-	-	-	-	-	-
Pertussis	374,191	364,331	9,071	9,439	-	-	-	-
Measles	12,049	65,804	26,290	53,522	230	1,670	189	336
Hepatitis B	5,222	11,184	8,135	7,825	51	55	59	19
Leprosy	1,846	1,478	1,651	670	2	12	-	3
Lympathatic Filariasis	3,726	1,478	893	14,561	-	-	-	-
Rabies (Human)	-	-	81	82	-	-	-	-
Lassa fever	43	50	49	920	20	19	18	31
Malaria	4,456,976	5,307,553	3,837,215	5,085,533	10,506	-	2620	3,222
Pneumonia	374,191	364,331	198,409	153,759	1,141	1,428	852	440
Tuberculosis	31,264	32,882	29,949	24,752	245	1,562	521	122
Trachoma	-	-	397	1,480	-	-	-	-
Filariasis	3,726	1,478	1,651	670	2	8	-	4
Onchocerciasis (River Blindness	9,407	7,050	11,423	9,101	-	1	5	1
Vaginal discharge	56,063	55,667	54,659	54,261	-	-	-	-
Schistosoma	-	-	1,591	2,824	-	-	-	1
HIV/AIDS	44,010	47,365	44,290	41,961	740	925	634	467
Tetanus (N)					155	69	114	52
Total	**6,882,246**	**7,574,667**	**5,004,604**	**6,109,945**	**17,116**	**5,749**	**6,853**	**5,924**

Note: * No records for 2009

Source: Federal Ministry of Health, Abuja (National Bureau of Statistics: *Social Statistics in Nigeria 2012 – Part III (health, employment, public safety, population & vital registration*)

TABLE 7.9 Reported cases and deaths from Malaria, 2007 – 2011

Years	2007	2008	2009	2010	2011
Malaria (General population)					
Reported cases	510,406	478,936	673,771	380,061	277,393
Reported deaths	2,620	4,690	3,033	1,463	1,608
% deaths	0.51	0.98	0.45	0.38	0.58
Malaria (Pregnant women 15+ years)					
Reported cases	352,942	347,996	475,414	330,249	359,612
Reported deaths	1,642	477	387	193	257
% deaths	0.47	0.14	0.08	0.06	0.07

Source: Federal Ministry of Health, Abuja. (Extracted from *National Bureau of Statistics in Nigeria, 2012 Part III*) in Africa. During the coronavirus pandemic in 2020, African countries were panic stricken by news that the coronavirus vaccine being developed was part of a ploy to wipe out the black race! So much talk about reducing world population by 10-20 percent further fuelled such suspicions and rumours.

Yellow fever is more dangerous a disease than malaria and takes much longer to cure. It is transmitted by the mosquito *Aedea aegypti* and other species of the same genus. It is controlled by means of inoculation which is expected to give immunity for a period of about ten years. Outbreaks of yellow fever and Lassa fever since the 1980s point to a serious lowering of the people's immune system due mainly to lapses in the immunization efforts of the medical services, increasing poverty among the populace and general insecurity in the northern parts of the country.

Dengue fever is another mosquito-borne (*Aedea aegypti*) viral disease associated with urban environments and manifests as sudden onset of fever and severe headache. Filariasis is less of a hazard than malaria (Table 7.8) but it can be quite debilitating. It is transmitted by some species of anopheles and most especially, by the mosquito, *Culex pipiens fatigans.*

Trypanosomal infection in man results in the disease, sleeping sickness, which causes drowsiness and sometimes results in death. The most severely affected parts of the country were southern Kaduna State, Nassarawa, Benue, Taraba and Plateau States. Fortunately, the incidence has been reduced considerably with the more effective control of the tsetse fly over the years. As Table 7.8 shows, the number of reported cases of sleeping sickness throughout the country is now relatively small compared to malaria and other types of fever. But the number of deaths, though small in absolute terms, was until recently proportionally large compared with deaths from other diseases with much higher rates of infection.

River blindness, like filariasis is caused by parasitic nematodes. Except in rare cases, it is not fatal but causes much discomfort and leads to the development of skin lesions and eye cataract which blurs or completely inhibits vision. The pathogens are carried by the simulium fly (sand fly), which breeds in fast flowing streams. The gallery forests or woodlands provided shade and a comparatively humid environment for the insects which were found in large numbers in eastern Niger State, the area now constituting the Federal Capital Territory (Abuja), and in the Benue lowlands where many swift flowing streams descend from the Jos Plateau to the Niger and Benue rivers. The landscape has been greatly modified and built up since the establishment of the Federal Capital Territory in eastern Niger State and concerns about river blindness have disappeared over time.

Water-borne diseases

Guinea worm is widespread in the rural areas of Nigeria. The local pattern of occurrence is very complex. One village may be heavily infested while another, only a few kilometers away, is quite free of it. The most important factor determining the incidence of the disease appears to be the quality of the drinking water.

Communities with good, clear spring water are little affected by guinea worm. In order to prevent the occurrence of the disease, many communities forbid the keeping of pigs which, they believe, carry the germs of this and other diseases. Guinea worm has been largely eliminated from towns and villages with piped water supply. International non-governmental organizations have worked hard at total eradication of guinea worm in Nigeria, particularly in the northern parts of the country. President Jimmy Carter of the United States of America, personally and through his foundation, committed much time, money and effort to the cause.

Typhoid fever is a dangerous and sometimes fatal disease which is contracted by drinking contaminated water or eating raw food or vegetables contaminated by *Salmonella typhae*. Infection from the disease is one of the health hazards faced by urban dwellers during periods of water shortage when they have to resort to fetching water from streams and unprotected wells. Unfortunately, outbreaks of the disease still occur sporadically in urban centres even in the 21st century.

Sleeping sickness (Bilharzia or schistosomosis) is another disease contracted from contaminated water. The blood fluke which causes the disease enters through the skin and, depending on the species, subsequently invades the bloodstream of either the urinary system or the intestine. However, it rarely results in death.

Diseases due to poor living conditions

Overcrowding of people and their belongings in poorly kept homes and slum areas has attracted or bred many pests which are a constant irritation to people. These include bed bugs, head and body lice, fleas and cockroaches. The first three are blood suckers which can transmit disease pathogens into the human body.

Overcrowding also encourages the spread of such contagious diseases as measles, chicken-pox and small pox. Measles is among the commonest diseases in Nigeria. In the period 2007 - 2011 reported cases of measles ranged from 12,049 to 65,804 (Table 7.8). Measles is particularly severe on children with reported number of deaths in 2007–2011 ranging from 230–1,670. By contrast chicken-pox rarely kills.

Small pox has been virtually eliminated in Nigeria. In fact, the World Health Organisation declared the country free of the disease in 1978, although it does still occur sporadically from time to time. The almost complete elimination of small pox from the country was due to the massive vaccination campaigns launched since the colonial era. In most places, new born babies are now vaccinated against small pox as a matter of routine.

Dysentery and cholera are contracted by human beings eating contaminated food and both result in severe diarrhoea. Food is easily contaminated by flies which breed in urban environments where solid wastes accumulate in large volumes, where gutters are filthy and where living conditions are squalid. As Table 7.8 clearly shows, diarrhoea is second only to malaria as the commonest human illness in Nigeria. Complicated cases often result in death.

Cholera is a latter-day introduction; it spread to Nigeria around 1970-71. Poor sanitation in Nigeria's big urban centres has been responsible for the persistence of the disease, which has now joined the ranks of endemic diseases in the country.

Virus diseases

Apart from the mosquito-borne virus diseases, yellow fever and dengue fever, which have long existed in the country, novel virus diseases have emerged in the last five decades. Since the first case of Lassa fever was reported in Lassa village, northern Nigeria, the country has witnessed outbreaks of deadly zoonotic diseases caused by viruses. The viruses are transmitted by animals that infest areas with poor environmental sanitation. Other novel virus diseases have joined in including monkey pox, Ebola, SARS, and, lately, coronavirus. The country responded well to the Ebola and SARS outbreaks and succeeded in keeping them out of the country. But it has failed woefully in stemming the persistent outbreaks of Lassa fever and yellow fever. Tomori (2019) offered the following reasons for this failure: (1) the weak system of disease control in the country; (2) the fact that Nigeria has become too dependent on foreign aid for its surveillance needs; and (3) the country's inability to clean up its streets and manage its garbage, which has led to a spike in the rodent population. Rodents (rats) are considered[†] possible carriers of the Lassa fever virus.

Control of human diseases

According to the Institute for Health Metrics and Evaluation's (IHME's) Global Burden of Disease (GBD, 2018), the top 10 causes of death in Nigeria now are:

1. Lower respiratory infections
2. Neonatal disorders

[†] `https://theconversation.com/lassa-fever-will-keep-ravaging-nigeria-unless-better-surveillance-is-put-in-place-83847`

3. HIV/AIDS
4. Malaria
5. Diarrheal diseases
6. Tuberculosis
7. Meningitis
8. Ischemic heart disease
9. Stroke
10. Cirrhosis

They are a mix of diseases of poverty and poor living conditions (neonatal disorders, malaria, diarrheal diseases, tuberculosis, meningitis) and of modern self-indulgent and stressed society (lower respiratory infections, HIV/AID, heart disease, stroke and cirrhosis). Unfortunately, until the establishment of the Nigeria Centre for Disease Control and Prevention (NCDC) in 2011, the health sector had lacked an effective health metric and evaluation system to be able to fully plan and combat both the age old and the new causes of disease and death in the country. The system inherited from the colonial administration had collapsed and the new arrangements put in place had been beset with myriads of problems of poor funding, maladministration, corruption and poor remuneration of health personnel. Fortunately, the NCDC is learning from its links with the IHME in developing its own health metric and evaluation system.

One of the greatest achievements of the British Colonial Administration was the improvement of public health throughout the country. This was done not only by providing health facilities and medical personnel but also by improving the quality of drinking water, personal hygiene and environmental sanitation. Public health inspectors were appointed in each district to ensure that the people kept their homes and surroundings clean. Immunization campaigns against such diseases as small pox and yellow fever were also initiated during the colonial era.

Since Independence, emphasis has been on the training of health personnel, nurses, doctors, and pharmacists, and increasing the number of hospitals and dispensaries. In order to cater for the needs of rural communities, district or cottage hospitals were established in non-urban areas especially in south-western Nigeria. Unfortunately, the re-occurrence of outbreaks of yellow fever, typhoid and guinea worm in several parts of Nigeria since 1984is an indication that the health status of the populace has fallen drastically over time (Tables 7.8 & 7.10). Apart from a decline in medical services, the nutritional status of the people has fallen also. Hence, their resistance to disease is rather low. The fact that most of these diseases emanate from drinking impure water also indicates that many Nigerians still depend on doubtful sources of drinking water (Chapter 5).

TABLE 7.10 Health Statistics for Nigeria (World Health Organization, 2018)

Total population (2017)	190,886,311*
Gross national income per capita (PPP international $,)	5,680*
Life expectancy at birth male/female (years, 2016)	55/56
Probability of dying under five (per 1,000 live births, 2018)	120
Probability of dying between 15- and 60-years male/female (per 1,000 population, 2016	272/333
Total expenditure on health per capita (International ($, 2014)	217
Total expenditure on health as % of GDP (2014)	3.7

*From GBD Compare (World Bank)
Source: World Health Organization
https://www.who.int/countries/nga/en/

But the truth is that the health sector has been seriously poorly funded and managed since the era of military administration. Tomori (2019 op cit) rightly noted that to remedy the situation, "*Nigeria must recognise and understand that it makes economic sense to prevent rather than control disease outbreaks. Prevention results in a significant reduction in disabilities and deaths from diseases and can be achieved through reliable and responsive disease surveillance, prevention, control and response. Therefore, providing adequate funding to maintain and sustain an effective national disease surveillance system backed by a reliable laboratory network service is the first step in ensuring national health security and protection of Nigerians from the ravages and death caused by disease outbreaks. This calls not for dependency on foreign aid and assistance, but on national ownership of disease surveillance and control systems.*"

The Nigerian governments must summon up courage and the political will to formulate policies and programmes for improving disease prevention and control with financial commitment to ensure the sustainability of the health services and initiatives. It is commendable that Nigeria took the first step towards improving the health situation in the country by establishing the Nigeria Centre for Disease Control[‡] (NCDC) in 2011 to enhance the country's preparedness and response to epidemics through prevention, detection, and control of communicable diseases. Laboratory diagnosis of virus diseases like Lassa fever has improved with the establishment of NCDC. The NCDC is also at the forefront of present efforts to identify,

[‡]https://www.ncdc.gov.ng/

test and isolate people infected with the novel COVID-19 virus. But its effectiveness has been seriously hampered by lack of preparation due to years of negligence and poor funding by successive governments at both the federal and state levels.

Nothing portrays the standing of Nigeria's healthcare facilities more than the low priority given to them by rural communities and low-income neighbourhoods in urban centres when it comes to seeking medical care when they fall sick. Auta *et al.* (2012) household survey in a rural community near Jos, Plateau State revealed that only about a quarter of the people consult physicians when they fall sick. By contrast, approximately 35 per cent resort to self-medication (Table 7.11)

TABLE 7.11 Healthcare providers from whom household members sought treatment in a rural community near Jos, Plateau State.

Healthcare provider	Percentage response
Self-medication	34.6
Physician	24.7
Pharmacist	13.6
Patent medicine vendors	11.1
Traditional healer	7.4
Nurse	4.9
No treatment	3.7

Source: Auta A., Banwat SB., Dayom DW., Shalkur D. and MO Avu 2012 Occurrence and treatment of common health problems in a Nigerian community.*Journal of Young Pharmacists* (2012) Jan–Mar; 4(1): 49–53. doi: 10.4103/0975-1483.93571

The future

What is most glaring in the foregoing discussions is that people have been pouring more and more chemicals into the environment in order to tackle the problem of weeds, pests and diseases. But it is important to heed the warning by researchers at the North Carolina State University referred to earlier in regards to weed resistance to herbicides, that there is high chance that the evolution of weed and pest resistance will outpace human innovation. Rather than approach the issue by developing more and more lethal chemicals as weed, pest and disease resistance grows, society should take a longer view of the problem of weeds, pests and diseases control that includes multiple approaches at once. In the case of weed control the multiple approaches might include crop rotation and curbing the spread of weed seeds. In the case of crop and animal pest and disease control the multiple approaches may have in the mix crop rotation, planting of forage crops, integration of crop and animal husbandry, better hygiene on the farms, and in the animal sheds and crop storage facilities. Disease control among the populace might lay emphasis on preventive measures including better personal hygiene and environmental sanitation in the homes and work places.

8

Mineral resources and mining

After political independence from British rule in 1960, the Nigerian economy rapidly grew into a crude oil dependent economy with oil exports accounting for about 96 per cent of foreign earnings. After the structural adjustment programme of the 1980s, however, government policy began to focus on the diversification of the economy. It was during this time that government began to promote the exploitation of Nigeria's mineral resources other than crude oil, that is, the many mineral ores widely distributed throughout the country. A Ministry of Solid Minerals was established to focus on the exploitation of these minerals in commercial quantities to diversify the export and revenue base of the country. By the first decade of the 21st century, the solid minerals sector had become a major area of growth in Nigeria's national economy (Table 8.1a).

The growth and geographical spread of solid mineral exploitation have impacted the landscape in many areas and exacerbated the overall environmental consequences of mining in Nigeria. The devastation of the Niger delta ecological zone by crude oil exploration and mining operations with its social and political repercussions remains a thorny issue in the Nigerian state.

TABLE 8.1A Sectoral Growth Rate of the Nigerian Economy (%) 2005–2019

Activity Sector	2005	2006	2007	2008	2009	2010	2011	2012	2013	2014	2015	2016	2017	2018	2019
A. AGRICULTURE	**7.07**	**7.41**	**7.20**	**6.27**	**5.88**	**5.83**	**2.92**	**6.70**	**2.94**	**4.27**	**3.72**	**4.11**	**3.45**	**2.12**	**2.36**
1. Crop Production	7.13	7.49	7.25	6.22	5.83	5.77	2.85	7.51	2.54	4.12	3.49	4.34	3.64	2.26	2.51
2. Livestock	6.76	6.90	6.93	6.80	6.48	6.45	2.03	-2.67	5.98	5.42	5.93	2.94	1.61	0.33	0.16
3. Forestry	5.92	6.02	6.12	6.10	5.85	5.77	4.96	2.55	5.63	4.55	3.67	2.62	3.31	3.06	2.59
4. Fishing	6.02	6.55	6.58	6.57	6.17	5.96	8.25	7.76	8.98	6.70	5.89	-0.72	1.34	1.64	3.33
B. INDUSTRY	**2.93**	**-0.55**	**-0.09**	**-0.87**	**3.54**	**6.59**	**8.39**	**2.43**	**2.16**	**6.76**	**-2.24**	**-8.85**	**2.15**	**1.87**	**2.31**
1. Mining and Quarrying	0.52	-4.46	-4.48	-6.11	0.51	5.29	2.41	-4.78	-12.81	-1.13	-5.27	-14.45	4.62	1.11	4.43
a. Crude Petroleum and Natural Gas	0.50	-4.51	-4.54	-6.19	0.45	5.25	2.33	-4.95	-13.07	-1.32	-5.45	-14.45	4.69	0.97	4.59
b. Coal Mining	10.00	10.00	10.00	8.78	8.85	8.80	20.40	18.12	20.10	19.85	10.39	1.00	-1.45	-5.81	13.15
c. Metal Ores	10.90	12.21	12.00	11.98	11.51	11.59	14.95	6.75	15.13	16.60	7.27	20.74	21.37	26.34	-14.38
d. Quarrying and Other Minerals	9.53	10.27	12.75	12.77	12.09	12.08	14.11	20.49	16.31	14.46	7.51	-17.42	-1.12	10.33	-5.63
2. Manufacturing	9.68	9.50	9.68	9.05	7.94	7.68	17.82	13.46	21.80	14.72	-1.46	-4.32	-0.21	2.09	0.77
a. Oil Refining	10.00	10.02	10.08	10.00	6.95	7.28	6.21	-17.52	54.22	-9.67	-35.49	2.53	-27.70	-3.97	-31.39
b. Cement	11.07	11.53	12.00	11.65	10.83	10.56	7.74	13.49	39.25	29.71	22.10	-5.36	-2.20	4.50	3.11
c. Food, Beverage and Tobacco	9.57	9.33	9.50	8.80	7.82	7.51	7.31	6.56	11.81	5.63	-5.38	-6.27	2.35	2.93	2.17
d. Textile, Apparel and Footwear	9.57	9.33	9.50	8.80	7.82	7.51	62.21	42.57	34.48	31.19	-1.07	-1.09	0.82	1.69	-0.09
e. Wood and Wood Products	9.57	9.33	9.50	8.80	7.82	7.51	5.58	20.79	8.88	12.70	6.29	-4.04	0.53	1.70	1.93
f. Pulp, Paper and Paper Products	9.57	9.33	9.50	8.80	7.82	7.51	17.08	6.42	45.04	14.15	6.82	-4.17	0.12	3.44	1.25
g. Chemical and Pharmaceutical Products	9.57	9.33	9.50	8.80	7.82	7.51	54.73	58.95	49.66	37.93	18.17	1.19	0.79	0.61	0.35
h. Non-Metallic Products	9.57	9.33	9.50	8.80	7.82	7.51	66.31	13.15	32.26	34.24	14.21	3.20	1.96	-0.48	1.68
i. Plastic and Rubber products	9.57	9.33	9.50	8.80	7.82	7.51	124.79	39.83	30.15	30.22	17.88	3.59	0.99	1.54	2.69
j. Electrical and Electronics	9.57	9.33	9.50	8.80	7.82	7.51	82.26	-0.95	5.17	6.47	1.32	-8.13	-2.79	3.75	1.50
k. Basic metal, Iron and Steel	9.57	9.33	9.50	8.80	7.82	7.51	131.67	20.83	13.35	15.59	3.11	0.72	0.17	-0.75	-1.33
l. Motor vehicles & assembly	9.57	9.33	9.50	8.80	7.82	7.51	20.11	34.32	25.72	25.61	-5.55	-29.01	-21.52	-2.54	2.31
m. Other Manufacturing	9.57	9.33	9.50	8.80	7.82	7.51	39.37	32.07	33.43	28.95	-3.83	-14.00	-6.99	-0.11	0.34
3. Electricity, Gas, Steam & Air conditioner	6.50	4.65	4.70	3.45	2.94	2.96	39.51	14.61	14.56	-8.69	-9.25	-15.00	16.43	7.30	-4.86
4. Water supply, sewage, waste Management	10.50	10.91	10.92	10.89	10.19	10.20	3.19	4.10	45.34	23.08	15.38	9.27	4.16	7.21	5.47
5. Construction	12.10	12.99	13.03	13.07	11.97	11.85	15.71	9.44	14.22	13.03	4.35	-5.95	1.00	2.33	1.81

continued on page 281

TABLE 8.1B Sectoral Growth Rate of the Nigerian Economy (%) 2005–2019 , *continued.*

C. SERVICES	**10.63**	**12.37**	**12.86**	**13.00**	**12.43**	**12.96**	**4.90**	**3.97**	**8.38**	**6.85**	**4.78**	**-0.82**	**-0.91**	**1.83**	**2.22**
1. Trade	13.51	15.26	15.20	14.02	11.48	11.22	7.21	2.21	6.64	5.88	5.14	-0.24	-1.05	-0.63	-0.38
2. Accommodation and Food Services	10.45	12.91	12.95	12.94	11.89	11.95	9.22	15.85	73.86	18.33	2.27	-5.32	-1.61	1.76	2.85
3. Transportation and Storage	6.44	7.04	7.08	7.10	6.98	6.85	5.97	-3.42	3.80	4.42	4.51	0.39	3.86	13.91	10.73
a. Road Transport	6.40	6.97	7.01	7.03	6.91	6.81	2.89	-5.52	2.37	3.76	4.46	1.72	4.84	14.47	11.24
b. Rail Transport & Pipelines	6.30	6.31	6.32	6.08	5.75	5.73	13.04	20.56	8.72	7.60	2.67	-0.65	0.80	2.86	1.74
c. Water Transport	5.90	6.03	6.04	6.05	5.66	5.37	-9.74	-1.67	4.50	8.74	8.45	1.40	1.25	2.87	0.56
d. Air Transport	5.80	7.54	7.56	7.59	7.92	7.35	58.81	4.26	9.32	2.92	3.70	-4.86	1.83	20.70	13.17
e. Transport Services	5.90	6.12	6.14	6.18	5.47	5.45	19.05	23.86	17.20	14.20	5.41	-1.12	0.22	1.96	2.66
f. Post and Courier Services	10.85	10.88	10.89	10.91	10.24	10.17	2.95	8.38	9.90	9.16	5.72	-21.33	-16.16	0.09	-0.16
4. Information and Communication	21.74	25.34	26.53	27.66	28.57	29.55	2.15	3.05	8.21	6.99	6.22	1.95	-1.04	9.65	9.17
a. Telecommunications and Information Services	30.52	34.59	34.62	34.67	34.73	34.83	1.23	3.69	4.72	4.75	4.49	2.03	-2.04	11.33	11.41
b. Publishing.	10.50	10.61	10.62	10.70	9.89	9.76	38.07	4.25	12.66	12.80	10.29	2.40	2.29	6.03	2.60
c. Motion Pictures, Sound recording and Music production	10.50	10.61	10.62	10.70	9.89	9.76	0.49	2.15	24.19	20.45	4.06	-4.08	-0.57	-0.44	0.20
d. Broadcasting	8.20	8.70	8.72	8.75	8.48	8.28	11.56	-1.60	25.52	12.21	19.86	6.13	4.32	7.38	2.29
5. Arts, Entertainment & Recreation	10.50	10.61	10.62	10.70	9.89	9.76	148.31	27.36	14.93	14.89	9.40	3.72	4.13	2.53	4.12
6. Financial and Insurance	3.42	5.46	5.54	5.35	4.51	4.42	-26.93	21.02	8.63	8.13	7.12	-4.54	1.26	2.03	2.56
a. Financial Institutions	2.60	4.77	4.81	4.59	3.79	3.73	-31.49	29.40	8.92	8.27	7.43	-5.57	1.92	1.41	2.40
b. Insurance	10.85	11.30	11.33	10.99	9.60	9.05	1.95	-14.68	6.77	7.19	5.09	2.46	-2.90	6.12	3.59
7. Real Estate	11.09	11.68	11.72	11.79	10.94	10.75	0.43	5.65	11.98	5.12	2.11	-6.86	-4.27	-4.74	-2.36
8. Professional, Scientific & Technical Serv.	5.30	6.69	6.70	6.68	6.19	5.98	18.68	7.81	3.43	5.53	5.26	0.80	-0.26	0.57	0.12
9. Administrative and Support Services	5.30	6.69	6.70	6.68	6.19	5.98	5.21	-3.29	2.62	1.91	3.52	-0.69	0.64	-0.18	1.96
10. Public Administration	4.00	4.45	4.46	4.44	4.41	4.20	15.46	-20.31	-0.54	2.52	-12.28	-4.58	-0.38	-2.05	-4.01
11. Education	10.55	10.77	10.78	10.77	10.01	9.81	31.57	1.68	15.60	8.88	7.67	1.35	-0.72	-0.03	0.80
12. Human Health & Social Services	10.00	10.33	10.35	10.34	10.01	9.96	13.04	4.33	9.59	10.50	2.48	-1.79	-0.31	-0.32	0.31
13. Other Services	10.50	10.61	10.62	10.70	9.89	9.76	3.84	49.95	10.71	17.65	17.86	4.93	2.35	2.17	1.90
Total Real GDP	**7.01**	**6.73**	**7.32**	**7.20**	**8.35**	**9.54**	**5.31**	**4.21**	**5.49**	**6.22**	**2.79**	**-1.58**	**0.82**	**1.91**	**2.27**

Source: Computed from CBN Statistical Bulletin (2019)

Mining is indeed a very old industry in Nigeria although it remained a small and technologically rather simple operation until the establishment of British colonial administration. In fact, the British were led to the sources of valuable minerals by the local inhabitants who had already developed trade in metal ores with traders from other European countries before the British came into the picture. This was how the colonial geologists and district officers 'discovered' the lead and zinc deposits of the Benue valley, the tin ores on the Jos plateau, the gold at Ilesha and the coal in the Udi Hills.

However, in spite of its long history and the increased scale of operation since the introduction of modern mining technology by the British, mining was not a prominent feature of the Nigerian landscape, largely because it was highly localized in scattered sites throughout the country. But, in these isolated areas of solid mineral mining the impact on the local environment was considerable, for example, the accelerated soil erosion caused by tin mining on the Jos Plateau and by coal mining in the scarp lands around Enugu.

Although capital intensive mineral extraction was limited, quarrying of sand, clay, gravel, red earth and building stones was widespread and grew over time to have profound effects on the landscape. Because, in most cases, quarrying requires less financial capital and only the simplest forms of equipment, the mining of these earth materials is engaged in by a large number of unskilled labour throughout the country. The phenomenon of deep rock weathering, which is characteristic of humid tropical regions, has been responsible for the widespread distribution of these materials so that a discussion of the different rock types, their minerals and weathering patterns will give some insight into the scale and environmental impacts of all types of mining in Nigeria.

Rock groups and their significance

There are three major rock groups in Nigeria (Fig. 8.1), namely, (1) the sedimentary rock formations and deposits dating from the Cretaceous to the present; (2) the igneous and metamorphic rocks and old Paleozoic sediments of the Basement Complex; and (3) the volcanic rocks.

The sedimentary rocks

The sedimentary rock formations and deposits of Nigeria date principally from the Lower Cretaceous in the Mesozoic era. They may be grouped into the Lower and Upper Cretaceous sedimentary rocks; the Tertiary marine and continental deposits;

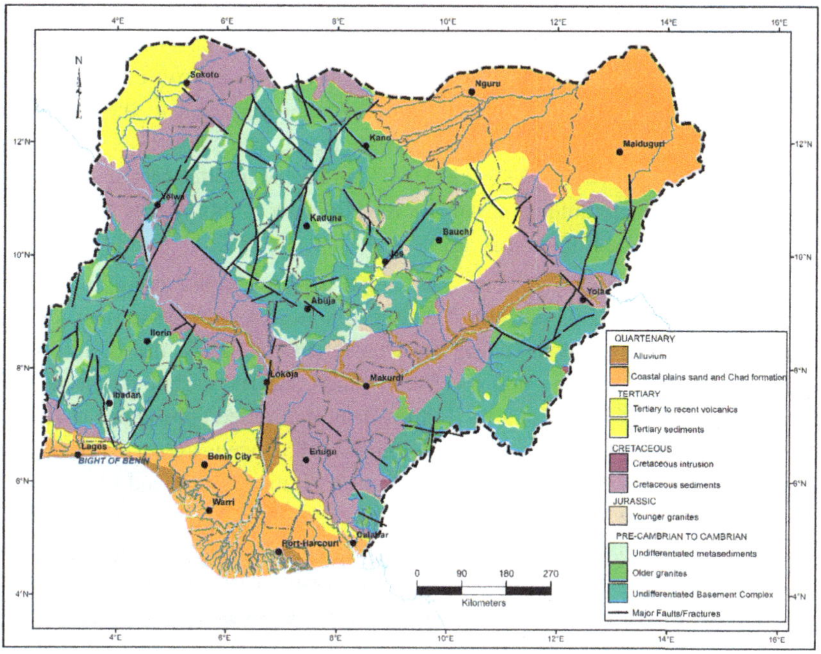

FIGURE 8.1 Major geological formations in Nigeria (Modified from Macmillan, 2010)

and the Recent Quaternary marine and continental deposits. The Lower Cretaceous sedimentary rocks consist of sandstones, shales and thin bands of limestone, which extend inland from the coast up to the Cross river, the Benue and Gongola valleys and beyond to the Chad Basin. Examples of these rocks include the Bima sandstone in Taraba and Gombe States. The Lower Cretaceous sediments were overlain by Upper Cretaceous sediments which extend and are exposed some distance up the Niger trough. The Upper Cretaceous sedimentary rocks consist mostly of sandstones, shales and mudstones and include the well-known Nupe sandstones found in the Bida area of Niger State. Marine transgressions into the northern parts of West Africa from the Cretaceous onward were responsible for the occurrence of the Lower and Upper Cretaceous rocks in Sokoto area.

In the context of this discussion, the Cretaceous rock formations are important for two main reasons. Firstly, the Lower and Upper Cretaceous beds of south-east Nigeria contain the only economic coal deposits of the West Africa region. The scarp lands of south-east Nigeria (Awgu-Enugu-Nsukka) in their lower parts consist of shales, sandstones and coal deposits formed in the Lower Cretaceous. Hence, they are referred to as the Lower Coal Measures. They are overlain by the Upper Coal Measures and the False-bedded Sandstones of the Upper Cretaceous. Secondly, the Lower Cretaceous sediments throughout their exposures in south-east Nigeria, for example, in the Abakaliki and Ishiagu areas and in the Benue and Gongola valleys, contain deposits of lead, galena and zinc as well as salt springs. These mineral deposits and salt springs have long been exploited by the local inhabitants and have considerable potentials for the Nigerian economy now that attention is focused on the solid mineral resources.

The belt of Tertiary marine and continental deposits found along much of the West African coast attains its widest extent in south-west Nigeria, overlying the Cretaceous sediments and including the Coastal Plains sands, the Bende-Ameki clays and clayey sandstones and the Benin sands and red earths. These deposits reach great depths in the delta area as a result of continued subsidence during the period of deposition. The great mass of sediments of Miocene age in the Niger delta area contains natural gas and mineral oil deposits both onshore and offshore. The exploitation of crude mineral oil has played a crucial role in the economic development of the country in the last five decades or so. Furthermore, the Palaeocene limestone formations provide the raw material for the cement manufacturing factories at Nkalagu, Ewekoro and Shagamu. There are occasional deposits of phosphates and extensive beds of lignite which are regarded as potential raw materials for a fertilizer and chemicals industry.

The Chad Basin, like the Inland Delta of the Niger in Mali Republic, is a zone of thick continental clay and sandstone deposits dating back to the Eocene. The Tertiary deposits are referred to as the Kerri Kerri beds, comprising layers of sand and gravel deposits overlain by thick clay beds. These sedimentary layers have intercalations of hard, lateritic crusts indicating a series of fluctuations in the lake level in the past. Overlying the Chad Basin deposits and occurring over wide areas in Northern Nigeria as far as the Zaria district are aeolian deposits from the Sahara Desert to the north.

Tertiary deposits also occur in the Sokoto region overlying the Upper Cretaceous sandstones. The deposits consist of calcareous beds overlying clays, shales, and sandstones. Lateritic ironstones have developed on the surface especially in those areas where the calcareous beds outcrop.

Recent or Quaternary deposits include marine, alluvial and continental sediments. The lagoon system along the coast is fringed by the coastal sand ridges and spits formed by longshore drift. In the river valleys and estuaries alluvial deposits of variable texture may be found including sand, fine silt and mud. Colluvial materials also occur, forming terraces along some of the river flood plains such as the Ogun floodplains. The major alluvial basins are the Niger trough and Anambra valley along the Niger river; the Imo, Cross river basins in the southeast; the Ogun floodplains in the southwest; and Sokoto valley in the northwest. On the Jos Plateau some of the alluvial deposits date from the Tertiary and were buried by later volcanic lava flows. These alluvial deposits contain the alluvial tin ores (cassiterite) being mined on the plateau.

Except on the Jos Plateau and the eastern highlands the sedimentary rocks and deposits have remained relatively undisturbed since they were formed, and have been deeply weathered. Clay, laterite pisolitic gravel and red earth associated with the deep weathering profiles provide materials for construction purposes.

The Basement complex crystalline rocks

The Basement Complex is exposed over nearly fifty per cent of the country; and underlies the volcanic and sedimentary rock formations in the other half. Figure 8.1 delineates three broad classes of Pre-Cambrian to Cambrian Basement Complex rocks, namely, undifferentiated metasediments; Older Granites; and undifferentiated Basement complex. But Oyawoye (1970) described four groups of Basement Complex rocks as follows: (1) Older granites, made up of porphyroblasts, granite gneisses and gneissose granites; (2) migmatitic complex, including banded, migmatitic and augen gneisses and pegmatites; (3) metasediments including schists, quartzites and quartz schists; marble, calcsilicates, metaconglomerates, amphibolite, and metamorphic iron beds; and (4) miscellaneous rock types including charnokites, pyroxenes, syenites, pegmatite and dolerite dykes. The Younger granites, which outcrop primarily in and around the Jos Plateau are of the Jurassic era.

The most prominent and widespread rock types are the granitic gneisses, the migmatites, quartzites and schists. The migmatites are the most common group of rocks forming the ground mass in which all other rocks seem to occur. Quartzites are the most prominent of the metasediments because they usually form very bold ridges as found in Ibadan city and Iseyin area of Oyo State. Schists of various types and grades occur, often in association with the quartzites, and form lowlands between the quartzite ridges. Biotite schists are the most widespread of the schists.

Marble has been located in various parts of the Basement Complex and is being mined in some localities, for example in Minna and Suleja divisions of Niger State. Metasedimentary iron ores consisting of bands and lenses of haematite and crystal aggregates of magnetite occur widely in the Basement Complex. They are associated with quartz and feldspathic gneisses and pegmatites in the Okene region where the first iron and steel mill in Nigeria has long been in building at Ajaokuta.

In other areas, such as Birnin Gwari in Kaduna State, haematite and magnetite grains occur in large quantities in weakly metamorphosed sandstones. Finally, the pegmatite dykes that occur widely usually in the gneisses and in the vicinity of the older granite masses contain numerous valuable minerals, most of which are not of economic importance. The exceptions are cassiterite, columbite and beryl, which are mined on the Jos Plateau.

Thus, the migmatites and metasedimentary rocks are important because of their considerable mineralization. They are the original sources of most of the cassiterite and columbite exported from Nigeria as well as the gold mined in Ilesha area of Osun State and in Sokoto, Kebbi and Zamfara States.

Except in areas of steep rock outcrops, the Basement Complex rocks throughout Nigeria are overlain by a mantle of deeply weathered materials of variable depth. Mean depths ranging from 9 to 13 meters have been observed. The most common type of deep weathering profile is the stratified (duricrust) profile comprising from top to bottom, an indurated zone, a mottled zone and a pallid zone, overlying weathered or unweathered country rock. Knowledge of the process of deep weathering throws light on the formation, precipitation and accumulation of mineral ores in the weathering profile and aids the location of water and various mineral ores and other earth materials (Faniran (1971, 1972, 1974). For example, it is known that iron ore tends to concentrate by capillary accumulation at or near the surface. The most important occurrences of iron ore in Nigeria as in most parts of the world are found on hilltops and slopes. Gravel, for road and other construction purposes, is obtained from the lateritic layer usually of transported and re-cemented weathering profiles. The deep weathering profile is usually more clayey at depth and the pallid layer is a good source of pure clay with little or no coarse grains. Red earth is obtained from the mottled zone and from deep weathering profiles without any layer of hard lateritic crust.

The volcanic rocks

The volcanic rocks of Nigeria date from the Tertiary. The two major areas of extensive lava flow are the Biu Plateau in southern Borno State and the Jos Plateau. Further south there are a number of warm springs and numerous basalt and trachytic cones in the Benue trough and the Gongola basin including the Tangale Peak and Wase rock. On the Jos Plateau the earliest lava flows were weathered, transported and deposited by rivers to give rise to the sedimentary formation known as the Fluvio-volcanic series consisting of decomposed basalts overlying sands, grits and clays. The alluvial sediments buried under these early lava flows are of great economic importance because they contain tin ores and other associated valuable minerals.

Mineral resources of Nigeria

Table 8.2 gives a list of minerals of real or potential economic value, known to occur in Nigeria. Fig. 8.2 shows the distribution of the more important minerals, many of which had not been worked until recent times, even though they could be used in industry. This neglect may have been because the deposits are extremely small and widely dispersed in several locations and/or because of the high cost of extraction, processing and transportation. Improvements in the internal transportation systems and in the financial means of the government in recent years have encouraged state governments, entrepreneurs and the unemployed young men to prospect for and mine some of these minerals including gold, gypsum and uranium-rich ores, which are in high demand by the advanced nuclear countries. Table 8.3 shows the occurrences of solid minerals in each state. Solid mineral mining has also attracted illegal migrant prospectors from neighbouring countries and beyond. The rise in banditry

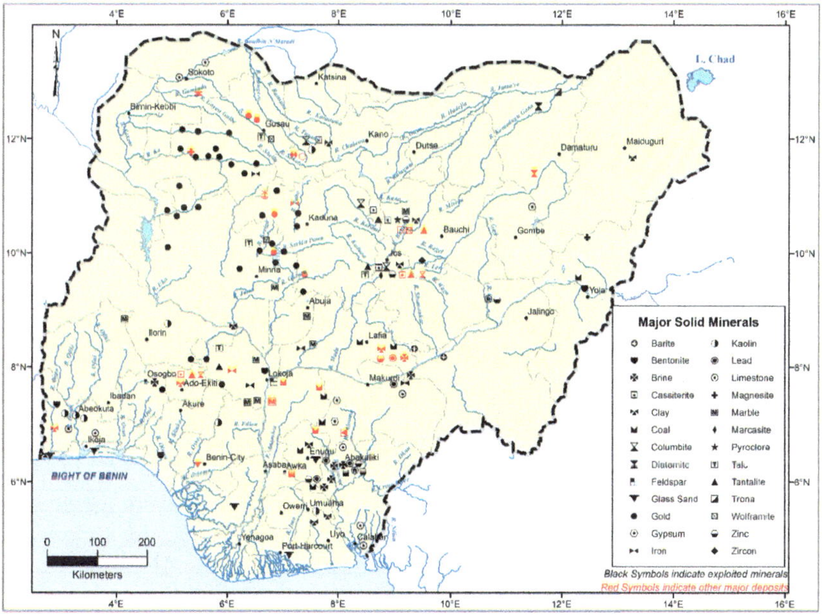

FIGURE 8.2 Known mineral occurrences in Nigeria (Modified from Macmillan, 2010)

TABLE 8.2 Minerals of economic value in Nigeria

Solid Mineral	Reserves	Main Known Locations
Iron and ferro-alloy metals		
Ironstone (oolithic)	209m. tonnes	Agbaja Plateau near Lokoja
Lateritic iron ore	51m. tonnes	Nsude Plateau near Enugu
Columbite and niobium	Unknown	Jos Plateau tinfields
Non-ferrous metals		
Lead and zinc	Unknown (exists in small veins	
Tin	136,000 tonnes	Jos Plateau and environs
Zircon	Unknown	Jos Plateau tinfields
Beryl	Very small quantities	Jos Plateau tinfields
Tantalite	Very small quantities	Jos Plateau tinfields
Precious metals		
Gold and silver	Small alluvial deposits	Widespread but mainly in Sokoto and Ilesha districts
Radioactive minerals		
Pyrochlore (uranium, thorium)	Very small quantities	Jos Plateau tinfields; Liruei Hills in Kano area
Metallurgical and refractory minerals		
Graphite	Very small (in schists)	Zaria Province
Fluorspar	Small quantities in lead, zinc and tin veins	Benue valley, Jos Plateau tinfields
Dolomite	Unknown but considerable	Widespread
Silica	Unknown but considerable	Widespread
Industrial minerals		
Glass sands	Unknown but great	Coastal region; Enugu coalfields
Diatomite	Unknown but great	Chad Basin
Monazite	Very small quantities	Jos Plateau tinfields
Barytes	Unknown but substantial	Benue valley
Mica	Unknown but substantial	Auchi Division, Kogi State
Chemical and fertilizer minerals		
Pyrites and marcasite	Unknown	Lead-zinc ores of Ogoja-Abakaliki
Salt	Unknown but considerable	Brine springs of Benue valley
Phosphate	Not definite but close to 1m. tonnes	Eocene beds of Ogun State
Lignite	Unknown but runs into several hundred million tonnes	Asaba and Onitsha Provinces
Mineral fuels		
Coal	Over 203miilion tonnes	Udi, Kabba and Benue Provinces
Oil and natural gas	Not definite	Niger River Delta region
Ceramic materials		
Kaolin	Small quantities	Basement Complex areas but mainly Osiele near Abeokuta
Clays	Unknown but widespread	Alluvial, sedimentary and Basement Complex areas
Feldspar	Small quantities in pegmatite veins	Basement Complex areas
Structural and building materials		
Hard stone	Abundant	Volcanic rocks and Basement Rock areas
Laterite and gravel	Abundant	Widespread except in coastal areas
Sand	Abundant	Widespread
Limestone	Unknown but widespread	Nkalagu, Ewekoro, Sokoto, Lower Gongola and Kabba areas
Gypsum	Unknown	Sokoto clay-shales
Bitumen	Unknown	Bitumen sand near Lagos, Ilaje-Ese Odo

TABLE 8.3 Solid mineral occurrences in the federating states of Nigeria

S/No.	Mineral	Mineral occurrences (in the Federating States of Nigeria)
1	Tantalite	Cross River, Ekiti, Kogi, Kwara, Nassarawa
2	Kaolin	Akwa Ibom, Anambra, Bauchi, Bayelsa, Ekiti, Imo, Katsina, Kebbi, Kogi, Ogun, Ondo, Plateau, Rivers
3	Mica	Ekiti, Kogi, Kwara, Nassarawa, Oyo
4	Baryte	Benue, Cross River, Nassarawa, Plateau, Taraba, Zamfara
5	Coal	Abia, Adamawa, Anambra, Bauchi, Benue, Cross River, Delta, Ebonyi, Edo, Enugu, Gombe, Imo, Kogi, Nassarawa, Plateau
6	Rutile	Bauchi, Cross River, Kaduna, Plateau
7	Ta l c	Ekiti, Kaduna, Kogi, Niger
8	Bismuth	Kaduna
9	Gypsum	Adamawa, Edo, Gombe, Ogun, Sokoto, Yobe
10	Feldspar	Bauchi, Borno, FCT, Kaduna, Kogi,
11	Gold	FCT, Kaduna, Kano, Katsina, Kebbi, Kogi, Kwara, Niger, Osun, Zamfara
12	Clay	In all the states of the federation
13	Silver	Ebonyi, Kano
14	Ilmenite	Bauchi, Cross River, Kaduna, Plateau
15	Limestone	Benue, Cross River, Ebonyi, Edo, Gombe, Kogi, Ogun, Sokoto
16	Columbite	Bauchi, Cross River Kaduna, Kano, Kwara, Nassarawa, Plateau
17	Cassiterite	Bauchi, Cross River, Kaduna, Kano, Kwara, Nassarawa, Plateau
18	Diatomite	Borno, Yobe
19	Silica sand	Delta, Jigawa, Kano, Lagos, Ondo, Rivers
20	Fluorite	Bauchi, Ebonyi, Plateau, Taraba
21	Bitumen	Edo, Lagos, Ondo, Ogun
22	Lead	Cross River, Ebonyi, FCT, Plateau, Zamfara
23	Zinc	Cross River, Ebonyi, FCT, Plateau, Zamfara
24	Benonite	Borno, Edo, Kogi, Ogun, Ondo
25	Kyanite	Kaduna, Niger
26	Iron ore	Enugu, FCT, Kaduna, Kogi, Nassarawa, Zamfara
27	Lithium	Kaduna, Nassarawa, Niger, Zamfara
28	Magnesite	Adamawa, Zamfara
29	Wolframite	Bauchi, Kaduna, Kano, Kwara, Nassarawa, Niger, Zamfara
30	Phosphate	Ogun, Sokoto
31	Marble	Edo, FCT, Kogi, Kwara, Nassarawa, Oyo
32	Molybdenite	Plateau
33	Manganese	Katsina, Kebbi, Zamfara
34	Gemstones	Bauchi, Kaduna, Kogi, Kwara, Nassarawa, Niger, Ogun, Oyo, Plateau, Taraba

Source: Lar, Uriah. (2018). Geology and Mineral Resources of Nigeria and their Uses.

`https://www.researchgate.net/publication/322963731_GEOLOGY_AND_MINERAL_RESOURCES_OF_NIGERIA_AND_THEIR_USES/citation/download` and other forms of insurgency in northwest Nigeria is partly due to the activities of solid mineral miners. The prospects of exploiting some of the more available and accessible mineral ores has increased with the government emphasis on solid minerals as a means of diversifying the economy. Among the most promising are iron ore, lead and zinc.

According to Lar (2018) Nigeria is endowed with about thirty-four solid minerals identified in 450 locations in the country. Some of these minerals include gold, iron ore, cassiterite, columbite, wolframite, pyrochlore, monazite, marble, coal, limestone, phosphates, clays, barites, lead, zinc, and occur in the different metallogenic provinces. But, less than 5 per cent of the available minerals are currently being mined even with the policy of promoting solid minerals exploitation and in spite of the need for them both within and outside the country. The Federal Government has designated seven of the minerals as "strategic" minerals: they are bitumen, gold, coal, iron ore, limestone, lead/zinc ores, limestone and barites. However, there are still other minerals with good economic prospects, such as, tin, tantalite, niobium, gypsum, gemstones, kaolin etc.

The minerals that are currently being mined on a large scale and are processed for sale both within and outside the country are coal, kaolin, barite, limestone, dolomite, gypsum, feldspar, gold, iron ore, lead-zinc, tin, and niobium and tantalum ores. Still, these solid minerals contribute less than 3 per cent of Nigeria gross domestic product.

Historically, the most important metal mined so far in Nigeria is cassiterite or tin ore (Table 8.4) and it is obtained from alluvial deposits on and around the Jos plateau. The proven reserves in the 1950s were about 137,000 tonnes (Hodder, 1959) so that right from the beginning the industry had been known to have a very limited life. Until the restrictions imposed by the International Tin Agreement (1956) annual exports of tin ore from Nigeria were between 10,160 and 14,224 tonnes. Production rose to a record level of nearly 20,320 tonnes during the Second World War when the Malayan mines were closed to the western world. From the late 1950s yearly production occasionally fell below the 10,160-tonne level as a result of greater competition from other producers, high production costs and the reduction in the number of mining operators (Hodder, 1959; Udo, 1970).

Other minerals mined and exported from the Jos Plateau tinfields include zircon, niobium and columbite, tantalite, and pyrochlore with associated uranium and thorium. Columbite, the most important of these by-products of tin, is found in alluvial deposits and also within certain pegmatites. Moreover, it occurs widely as a primary constituent in the Younger Granites where it is usually associated with varying proportions of tantalite.

TABLE 8.4 Solid Minerals Production in Nigeria, 2014–2017

S/N	Solid Minerals	Tonnage 2017	2016	2015	2014
		Industrial Rocks (in million metric tons MMT)			
1	Limestone	18.5	26.99	21.45	25.65
2	Granite	7.2	7.438	6.638	13.274
3	Laterite	3.7	2.074	2.377	3.802
4	Shale	2.8	1.882	1.075	0.812
5	Clay	0.95	1.646	1.178	1.750
6	Sand	2.4	1.508	2.008	0.776
		Industrial Minerals (in metric tons MT)			
7	Coal	397,000	104,424	121,316	127,466
8	Feldspar	30,000	36,092	13,238	
9	Marble	14,000	22,648		
10	Kaolin	40,000	16,216	25,280	
11	Dolomite	30,000	13,364		
12	Barite	700	1,712	520	550
13	Zircon	4000	1,072	182	1,034
14	Gypsum	40,000	720	34,620	
15	Talc	2000	666		
		Metallic Minerals (in metric tons MT)			
16	Manganese	40,000	70,106	1,796	
17	Lead-Zinc	13,600	5,794	19,657	
18	Tin	12,300	3,625	472	
19	Columbite	1000	1,014	289	960
20	Iron		665		
21	Tantalite	2000	29	31	
22	Wolframite	4	15	52	
23	Gemstones	24	7 T		
24	Gold	0.098	0.025		
25	Copper	300		125	773
	Total Production		41.874 MMT	39.372 MMT	
	Total Value		34.085 Billion Naira (US$=N360.00)	25.560 Billion Naira (US$=N360.00)	

Source: Olade M. A. 2019 Solid mineral deposits and mining in Nigeria: - a sector in transitional change. *Achievers Journal of Scientific Research*, Volume 2, Issue 1, July 2019, p. 1–15. `http://www.achieversjournal.org/`

Gold, containing varying amounts of silver, is widespread in many parts of Nigeria underlain by the Basement Complex. It is associated with quartz veins in schists. But the gold is mined from alluvial deposits derived from these rocks. Many of the occurrences are not sufficiently extensive to support large-scale mining. The chief producing areas are Ife-Ilesha district of Osun State, Minna District, Niger State and the Zamfara valley in Kebbi and Zamfara States.

Coal is a major mineral fuel produced in Nigeria. Nigerian coals are the sub-bituminous or black coal of the Cretaceous age and lignite or brown coal of the Tertiary. The sub-bituminous coals occur in seams, 1-2m thick, in the scarp lands of Udi-Enugu-Nsukka area where they are being mined. They also occur in Kabba area and on the Ankpa Plateau in Kogi State. The estimated total reserves are about 246 million tonnes made up of 74 million tonnes in the scarp lands and 172 million tonnes in the other areas. Production figures at the Enugu coal fields rose steadily until the 1950s when total annual output averaged about 650,240 tonnes. But production has since declined primarily as a result of a fall in the productivity of the mines themselves, but also because of heavy labour and transport costs. Production was also greatly curtailed during the Nigerian civil war and it took a long time to recover; and, even now, it has not attained its pre-civil war level.

Efforts to resuscitate the industry began in the 1980s and coal was being exported to some West African countries. Nigerian coals are noncoking and are suitable only for making gas and tar oils and for the production of liquid fuels by hydrogenation. The establishment of a chemical and fertilizer manufacturing industry based on coal as a raw material has long been under consideration. Hundreds of millions of tonnes of lignites occur in Nigeria but none of these has been mined so far. In Asaba division of Delta State nearly 100 million tonnes of lignite are available in seams varying from 38-58cm in thickness. Its calorific value is not much less than that of the Enugu coals and this has led to the suggestion that it could prove to be a cheaper fuel than coal for large electricity plants. However, the lignites have their greatest potential as raw materials for a chemicals industry.

There are no deposits of high-grade iron ore in Nigeria. But there are many of oxidized ores, which are comparatively thin and of low-medium grade. The deposits were for a long time considered to be uneconomical to work on a large-scale with modern processes. However, changing economic circumstances in the country and political realities in present day international relations, led the Nigerian government to embark on the establishment of an iron and steel mill at Ajaokuta near Okene in Kogi State. The Ajaokuta steel mill has been under construction for about half a century and there is little hope that it will become operational any time soon. There are two major regions of iron ore; the first lying north of Lokoja is the Agbaja plateau, a steep-sided, flat-topped tableland, about 102km^2 in area. It contains deposits of oolithic ironstone averaging from 45 to 50 per cent iron, and estimated reserves of 209 million tonnes. The second area is near Nsude village on the plateau 10km. to the west of Enugu in Enugu State, where an estimated 50 million tonnes of lateritic iron ore are to be found. The lateritic iron ore contains a high percentage of sand and investigations have indicated that they will average 33.2 per cent iron. However, it is possible to raise the figure to 40 per cent by screening off the sand.

Limestone, used in the manufacture of cement, occurs widely in Nigeria but not all the deposits are workable. Large quantities of crystalline limestone have been located at Jakura near Lokoja and the adjoining parts of Edo State further south. Limestone is worked primarily in five areas in the country: Shagamu-Ewekoro axis in Ogun State; Nkalagu in Anambra State; Ukpilla in Edo State; south-west of Sokoto town in Sokoto state; Ashaka in Bauchi State; and Yandev in Benue State.

Finally, mineral oil and natural gas occur extensively in the Niger delta region. Crude oil exports have provided the bulk of the country's foreign exchange earnings since the late 1960s. Production levels per day fluctuate with the fortunes of oil in the world market. Over the years, Nigeria's output has fluctuated between one million and two-and-a-half million barrels of crude oil per day. The bulk of the production is now in offshore locations partly as a result of the operations of insurgent groups at onshore locations. The proven oil reserves are expected to last from twenty to twenty-five years at current rates of exploitation.

Nigeria is said to have vast quantities of natural gas, which has not been touched at all. Indeed, according to experts, Nigeria is more of a natural gas than a crude oil country based on the huge reserves of natural gas. The gas that has been exploited thus far is that associated with crude oil.

The oil and gas sector generally has been a disappointment to most Nigerians primarily because the huge earnings from oil and gas sales have not been used to make a difference in the living standards of the people. Furthermore, the comparative contribution of the sector to employment in the country has been insignificant due to weak linkages with the rest of the economy. Hence the sector accounts for only 1.3% of the total modern sector employment in Nigeria (Odularu, 2007). On top of this dismal performance in contributing to the social and economic well-being of the average Nigerian, there is the devastation of the environment by oil exploration and exploitation activities as discussed in the section.

The ecology of mining

Mining activity involves many operations including the extraction of the mineral, its processing and transportation. Each of these stages or operations may leave its mark on the landscape and considerably influence the ecology of the environment. In order to discuss the various environmental issues arising from mining activities in Nigeria five types of mining operations are considered: tin mining on the Jos plateau; oil drilling in the Delta region; the quarrying of sand, clay, gravel and red earth; stone quarrying; and modern-day gold rush and scramble for solid minerals.

Tin mining on the Jos Plateau

The environmental setting of the Jos Plateau tinfields provides a useful background to any analysis of the impact of tin mining on the landscape. The Jos Plateau lies at a general altitude of over 1,200 metres above sea level. It is Nigeria's major hydrographic centre being the watershed between the Niger, the Benue and the Chad drainage systems. The Plateau surface is dominated by inselberg topography in which steep sided dome-shaped granite masses and flat-topped laterite-capped hills rise abruptly over extensive undulating plains. The plains are heavily dissected by streams and gullies which have cut through thick beds of alluvial and fluvio-volcanic deposits.

The mean annual rainfall is between 1,270 and 1,524mm. Most of the rain falls as heavy torrential downpours during the six months from April to September. In fact, the Jos Plateau experiences some of the worst storms in Nigeria. The torrential rains produce excessive storm runoff in an area where the vegetation is generally low growing open grassland that is also heavily grazed by cattle. The soils are skeletal near the rocky hil1s but get progressively deeper and finer in texture (sandy loams and sandy-clays) towards the streams.

With these environmental conditions it was not surprising that the Jos Plateau suffered extensively from severe sheet and gully erosion. Population pressure on the land, overcultivation and overgrazing were largely responsible for this (Grove, 1952), but tin mining was both directly and indirectly a contributory factor. Indirectly, by taking valuable farmland out of cultivation, tin mining was partly responsible for the widespread population pressure on the land. However, Hodder (1959) felt that this point was probably exaggerated. In the first place, tin mining was most common on land of poor agricultural quality and only occasionally was land diverted from food production to mining. At the time of writing, in the 1950s, Hodder estimated that probably no more than about 90 sq. km. of land had been lost to mining by the farmers. Secondly, government had long enacted laws and regulations to regulate the removal, stacking and replacing of soil disturbed in the process of mining the tin ores. The land was usually reclaimed by replacing the earth removed from the mine pits and by spreading compost on the surface at the rate of twelve to sixteen tonnes per hectare to increase fertility. The reclaimed land eventually reverted to the original owner.

However, Fredriksen *et al.* (2012), showed that over time tin mining on the Jos Plateau not only radically altered the physical landscapes it, also, radically reconfigured forms of social organization on the Plateau. Tin mining took labour out of agriculture and also attracted migrant labour both to tin mining and to agriculture. The tin miners were subjected to poor working conditions and exposed to environmental hazards, which also affected the inhabitants of the Jos Plateau in general. The pressure for farmland by both the indigenous and the migrant population created conflicts between agriculture and mining over access to land.

Tin mining also heightened the destructive impact of population pressure on the environment in another way. By establishing stable social and economic conditions it encouraged local people to abandon the traditional labour-intensive but conservational methods of crop cultivation using terraces and to adopt shifting cultivation, as discussed earlier in Chapter 6.

Tin mining also has had a direct impact on the environment because it opened the land to soil erosion. Tin is mined primarily by the open-cast hand paddocking method although the use of hydro-turbine gravel pumps and heavy draglines became more common over time. The open-cast method is wasteful of land and yields huge dumps of earth which are easily washed away by torrential rains. Deep gullies soon develop on the slagheaps. Strong storm water currents gnaw away at the alluvial deposits at the base while the steep upper walls slump down. In some cases, gully erosion has lowered the groundwater level thereby causing the land surface to dry up sooner than usual at the onset of the dry season.

Gully erosion of the slagheaps and alluvial deposits produces great quantities of sediments which are often deposited at the lower courses of the streams, thus silting up their channels. This has produced one beneficial result, however, because raising the river beds causes the groundwater level also to be raised. Water can often be obtained for part of the dry season by digging into the river beds. Efforts have been made to reduce the rate of soil erosion and sedimentation by constructing storm water ditches that channel the runoff water away from the land surface.

Another aspect of tin mining which is of ecological importance is the creation of ponds (paddocks) and reservoirs either by digging and widening of river channels, by erecting earth banks or by constructing dams. These disrupt the continuity and therefore the normal flow of the streams. The ponds have some uses in that they provide water for livestock and for irrigating small farm plots as well-as for breeding fish, but they are also breeding grounds for mosquitoes and other insects.

A serious negative impact of tin mining and smelting on the environment is the pollution of river water by effluents in the form of suspended solids. Rivers involved in the tin mining process often contain suspended solids of over 5,000 ppm as against 100 ppm in rivers outside the mining areas. While the mining companies have been bound by law to reclaim derelict lands, there has not been any restriction on the dumping of effluents in streams.

A more hazardous environmental impact of tin mining on the Jos Plateau is irradiation. Many homes were found to be highly radioactive and this was traced to the very clean, white sands used in plastering the buildings. As indicated earlier in this chapter, thorium and niobium, by-products of tin mining, are radioactive. Environmental impacts of mining in Nigerria are discussed further later in this Chapter.

Oil drilling in the delta region

Oil prospecting and mining takes place in one of the most difficult and fragile landscapes in Nigeria. The Niger delta is an extensive waterlogged mudflat, covered by swamps and mangroves. Human settlements and arable farmland are located on the sand ridges and islands which are highly susceptible to sheet erosion once cleared of their protective vegetation cover. The islands formed by the numerous distributaries of the delta are gradually being eroded by river floods and storm water runoffs from torrential rains. The islands also suffer from creek erosion by basal sapping and lateral erosion which occur when the level of the sea falls and the creeks erode their banks and cut back gradually.

The first and most noticeable impact of oil exploration in the delta was the modification of the vegetation and the physical landscape. The exploration and drilling activities involved a lot of cutting of traverse lines through forest and swamp and building roads on embankments; transporting heavy drilling rigs and other machines from one location to another; laying pipelines across thick forests, swamps and river channels; and building drilling platforms, storage depots and human dwellings. All these activities resulted in a great deal of forest and swamp clearance, land drainage and consolidation. Tarmac roads with well-built bridges leading to the various remote drilling locations have opened up large areas for cultivation. The extension of agricultural land has also meant further forest clearance and disturbance of the soil. The effect of these changes on the landscape has been an increase in storm runoff causing more frequent and extensive flooding. Some of the roads that have been built in the area are on earth banks raised above the swamps. These raised roads act as dams to flood water in the same way that the

Lagos-Ikorodu road does on the Ogun flood plains (Tahal, 1974). This has exacerbated the flooding phenomenon, which also undermines the stability of the roads, many of which are badly damaged or washed away by the heavy rains during the wet season.

Ecological concerns associated with oil mining operations

The impact of oil mining must be considered in relation to the ecological significance of the Niger River delta environment. According to Imevbore and Zagi (2018), the Niger Delta region harbours the most biologically diverse ecosystems in Nigeria across its main habitats, the freshwater swamps and forests, mangrove swamps, barrier island forests and lowland rainforest. The Niger Delta Environmental Survey (NDES, 2005) documented the following life forms: 70 mammal, 500 bird, 219 fish and 1773 insect species respectively, while the plant communities of the fresh forests harbour up to 250 plant species with the trees constituting the dominant form. Moreover, the delta is home to all of Nigeria's endemic or near-endemic species of mammals including six on IUCN Red List of mammals. Other important animal species include the Niger Delta forest elephant (*Loxodonta Africana cyclotis*), the West African manatee (*Trichechus senegalensis*), the White-throated guenon (*Cercopithecus erythrogaster*), the Sclater's guenon (*Cercopithecus sclateri*), and the pygmy hippopotamus (*Choeropsis liberiensisheslopi*) (UNDP, 2012). It should be easy then to understand why many local and international researchers and organizations have shown so much interest and concern at the impact of oil production operations in the Niger delta region. Unfortunately, biodiversity in the Niger Delta is today, facing tremendous threats (IUCN, 2018). Oil industry threats are in the form of oil spills, gas flaring, access creation and land take. Others not related to the oil industry are hunting, land clearing for agriculture, bush burning, and unsustainable harvest of trees, fish and other biological resources.

The United Nations Development Program (UNDP, 2012) voiced the concern of the international organizations when it noted that biodiversity was a little known and often neglected concept in most government circles in Nigeria. It decried the poor funding of biodiversity by both the federal and state governments. Imevbore (2018) showed that biodiversity conservation initiatives existed for less than 2.5% of the total area of oil and gas concessions. Given the fact that active oil and gas activities in Niger Delta region cover more than 35,000km^2 in terms of oil concessions, it is evident that biodiversity is receiving very little attention from governments and operators alike. This is in spite of current and proposed nature reserves and other conservation areas across the region. Apart from initiatives by a handful of companies, indigenous operators have done little to improve biodiversity conservation. The wildlife studies that are typically carried out as part of pre-project or post-project implementation environmental assessment studies are usually relatively limited in time and geographical coverage. This is a far cry from what is required to provide comprehensive assessment of biodiversity and its seasonal variability. Furthermore, the vast proportion of the delta region outside the oil concession areas has not been touched in these biodiversity and other environmental assessment programmes.

Oil pollution

In general, compared to the past, there has been a decline in the frequency of oil spills and in the quantities of oil lost in the first two decades of the 21st century. However, the fact that there are spikes in oil spills during the tenure of certain administrations and at the transition between one administration and another suggests that the problem of oil spills in Nigeria is due largely to the corruption and indiscipline in governance at both the federal and state levels. Table 8.5 shows the number of oil spills in the Niger delta and their causes in the period 2007–2017 while Figures 8.3a and 8.3b display graphically the numbers and quantities of oil spilled in the period 2010–2017.

Oil pollution of the delta environment had been much discussed but until the

TABLE 8.5 Oil spills and their causes in the Niger delta, 2007–2017

Causes	No. of Incidents	Proportion (%)
Sabotage	7283	73.0
Equipment failure	1392	14.0
Corrosion	728	7.3
Illegal artisanal oil refining	343	3.4
Operational error	232	2.3
Total	9978	

Source: Extracted from Imevbore V.O & M.M. Zagi, 2018: Environmental stewardship in Nigerian oil and gas industry – myth or reality. *Paper presented at Biennial Oil and Gas, Health, Safety and Environment Conference.*

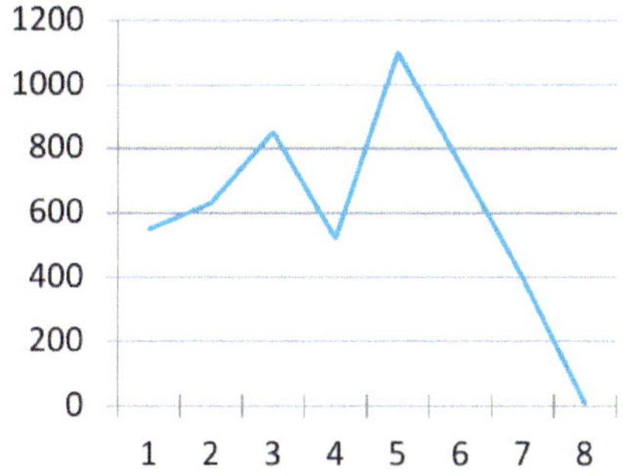

FIGURE 8.3A Number of Oil Spills, 2010–2017

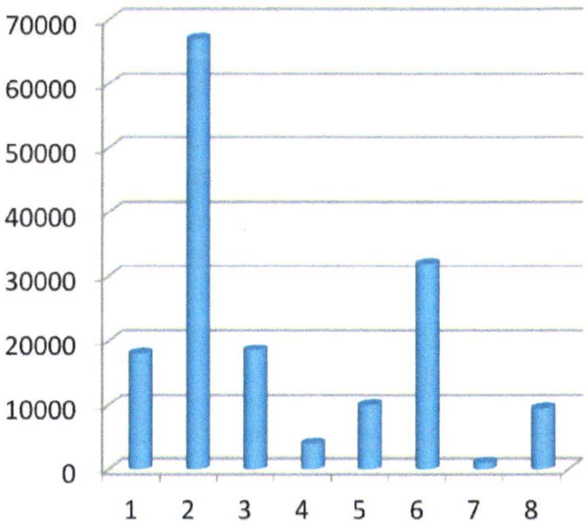

FIGURE 8.3B Quantity of oil spilled (barrels), 2010–2017
Source: Extracted from Imevbore V.O & M.M. Zagi, 2018: Environmental stewardship in Nigerian oil and gas industry – myth or reality. *Paper presented at Biennial Oil and Gas, Health, Safety and Environment Conference*

studies by Odu (1972), the exact effects of such pollution on the environment were unknown. Pollution of rivers, forests, swamps and farmlands occur through oil spillage at the drilling sites and flow stations; at loading jetties, along the pipe lines and from tankers. In July 1978, for instance, an oil spill occurred on one of the pipelines leading to the Bonny terminals. It was estimated that some 500 million barrels of oil were lost and the spill covered a 25 kilometer stretch of pipeline. The thickness of the oil spill was 8cm. on the ground surface and 2.5cm. on the water surface (Daily Times, 1978). Another possible but rare source of pollution, but more devastating than spills in its effects, is blowout, that is, explosion of an oil well. Two early occurrences in the delta region were to the south-east of Bori in Ogoni division in 1970 and at Obagi 21 Oil Well in 1972. The blowouts affected several hectares of arable farmland (Odu, 1977a).

The pollution of rivers and beaches has attracted the most public attention. Spills have been common along the main shipping lanes and in creeks and rivers near drilling areas or straddled by pipelines. The oil spill eventually seeps into the open sea, into fish ponds and onto beaches. Inyang (1973) reported that crude oil spills at the Bonny and other terminals had taken a toll on marine life and that complaints by the local fishermen about reduced catches were widespread along the coast. Studies by several operators in the oil concessions have subsequently shown that the physico-chemical and biological properties of the water, sediments and associated phytoplankton, zooplankton, and macro-benthos are impacted by oil spills. The livelihoods of the inhabitants are affected by oil spill impacts on fish, fisheries and fish spawning grounds. These impacts are due to the substances introduced into the environment by crude oil including heavy metals, oil, grease, saturated hydrocarbons and aromatic hydrocarbons. Habitat loss due to pollution by oil and grease, asphyxiation due to loss of oxygen, acidification of the sediments and water, high dissolved contents and turbidity levels are some of the major impacts suffered by living things in the rivers, creeks and ponds.

Perhaps, more serious than the pollution of water bodies is the pollution of swamps and beaches. Oil leaks can occur from submerged pipelines and flow stations located in swampy areas and may remain undetected for a long time. The oil takes longer to biodegrade because the poor anaerobic conditions do not favour microbial activity. The pollution of the swamps adversely affects aquatic life and village fishermen have complained of a decline in fish catches from these areas. Odu (1977a) also reported that over a 30 km stretch of beach off Furupa on the Atlantic coast, examination of the sand down to a depth of 30 cm revealed successive layers of oil and oil clods of different sizes. The oil could have come from offshore flow stations or from oil tankers being flushed out of the port of Brass. Since then black oil clods have become common occurrences on beaches along the Nigerian coastline from Badagry to the River Cross estuary.

Oil pollution affects agricultural activities in three ways: (1) oil enters both natural and cultivated plants directly through the leaves; (2) it alters soil nutrient composition and status; and (3) it drastically changes the normal behaviour of microorganisms in the soil (Odu, 1972, 1977a and b). When oil gets on to a crop or plant it penetrates the leaves thereby interfering with their function by reducing transpiration and photosynthesis. If they are heavily contaminated the leaves may be shed; where contamination is slight, they turn yellow and wilt thereafter. Usually, plants recover after the removal or degradation of the oi l unless the underlying soil has been heavily contaminated. Slight contamination of the soil of the order of one per cent by weight is actually beneficial to plant growth because the hydrocarbons are attacked by a variety of microorganisms which thereby release nutrients for plant use

Oil in itself is not toxic to plants but it exerts adverse indirect effects by creating certain conditions which alter the nutrient balance in the soil. Nutrients essential to plant growth become unavailable having been exchanged, fixed or swamped by toxic elements. While many micro-organisms in the soil are killed by oil pollution, they still have some value as their residues contribute to the humus content of the soil. After a certain period of adaptation, the population of micro-organisms recovers and grows several times above the normal level. This increase in the population of micro-organisms promotes rapid decomposition of the oil pollutant in the process of which nutrients are released to the soil. Some of the micro-organisms fix atmospheric nitrogen into the soil. So long as the spillage is not too large and the oi l is capable of degradation, it is possible that organic matter and total nitrogen will be increased. The effect on crop yield of these improvements in soil fertility is very dramatic and continues for up to four years thereafter.

Hopes expressed in the 1970s that if urgent actions were taken the Niger delta region still had a good chance to escape the worst effects of oil pollution of agricultural lands have been dashed. Those hopes were based on the fact that unlike most crude oils found in other parts of the world, the Nigerian crude is very light and easily decomposed by soil micro organisms. Moreover, the climatic conditions favour the rapid dispersal of spills; the hot humid weather facilitates microbial activity and the volatilization of hydrocarbons, while the heavy rains help in washing some of it away. Thus, oil spills of the magnitudes experienced up to 1970s did not persist for long on the land or in the soil. The spill from the blowout at Bori was said to have disappeared within two years whereas under temperate climatic conditions it could have persisted for many more years. The cultivation of the land by local farmers was thought to have aided the degradation of the oil spill by improving soil aeration.

Unfortunately, the magnitude of oil pollution in the delta region increased in the 1980s and subsequent years because of insurgency and sabotage of oil wells and pipelines. Even after the insurgency groups laid down their arms, they did so only to turn to more lucrative illegal ventures like oil bunkering and siphoning of crude from pipelines and illegal crude oil refining operations. In recent years, the major oil companies have been cooperating with the Federal Government to clean up the Niger delta environment. For example, in 2018 the Shell Oil Company embarked on the cleanup of Ogoni land, the sector of the delta that had witnessed perhaps the worst levels of crude oil pollution and devastation of terrain, biodiversity and human livelihoods.

Gas flaring and its environmental effects in the Niger Delta

The environmental impact of gas flaring is not talked about as much as the effects of oil pollution. But gas flaring has significant deleterious effects not only on the biota and fauna but also on the human population. Nigeria is regarded as the world leader in the flaring of natural gas associated with crude oil. Nigeria has the most gas flared per tonne of crude oil produced. It is said to be responsible for about 46 per cent of Africa's total gas flared and 19.79 per cent of the global output. According to Emoyan *et al.* (2008) associated natural gas from oil wells in the Niger delta is immediately flared into the environment at a rate that approximates 70million cubic meters per day. This is equivalent to 40 per cent of Africa's natural gas consumption and it is the single largest source of greenhouse gas emissions on earth. Indeed, some of the greenhouse gases, such as methane and carbon dioxide, from the Niger delta contribute to global warming. In addition, gas flares generate heat as high as 1600°C which causes thermal pollution that affects vegetation, animals and inhabitants over 40m. away from the flare site (Alakpodia, 1989). Oluwole *et al.* (1996), showed that the concentration of volatile oxides of carbon, nitrogen and sulphur as well as total particulates exceeded the prevailing Federal Environmental Protection Agency (FEPA 1991) standards. Acid rains have long been associated with the Niger delta region. Rain water shows high levels of lead (0.56mg/l) and low pH values ranging from 5.10 – 6.35 in Warri and environs (Olobaniyi and Efe, 2007). Shallow wells in the region also contain water with low pH values. The combined effects of thermal pollution and gaseous emission can be seen in the human settlements where houses covered with corrugated iron roofing sheets have been corroded. Implements and equipment, metal tanks and vehicles used by workers, artisans and people in the locality also suffer similar corrosion.

The quarrying of earth materials

Earth materials occur throughout the country, and there has been a tendency to take them for granted. The lack of attention, at all governmental levels, to the way in which these resources are exploited and utilized is reflected in the fact that there are no restrictive regulations whatsoever on the quarrying industry similar to those imposed on large-scale mineral mining operations. An operator may choose to dig anywhere he likes as long as he has the consent of the landowner. In many rural areas and, particularly close to motorways, operators hardly bother to seek permission to dig. It is worth noting, also, that large scale users of these materials, such as road and building construction firms and brick and tile manufacturing concerns, have not taken the trouble to initiate any systematic survey of the resources or to evaluate the quality and quantity of the materials. In the Lagos area, for example, most firms rely on local landowners and small-scale mining contractors to tell them the whereabouts of potential quarry sites.

The Lagos coastal region provides a striking example of the impact of quarrying on the landscape. Sand mining is, perhaps, the most important. It is carried out both on land and in the coastal rivers, lagoons and creeks. The ecological implications of river sand mining were discussed in Chapter 5. So far as sand dug from the land is concerned the most important sources are the flat sandy plains running west of Lagos. Although coloured red or brown by iron oxides as compared to the white beach sands or turned black by the high organic matter content in some areas the value of these sands for construction purposes is not much affected. Clay lenses occur within the sands on the same plains and elsewhere along the coast, providing raw materials for industrial concerns at Oregun and Ikeja.

In most parts of the country sand is obtained from valley bottoms, valley heads, *fadama*, floodplains and footslopes with hill wash materials. Quarrying in valley bottoms can gravely disturb the ecology of a whole locality. Wherever sand is dug, the terrain is scarred and rendered useless to farming.

During the wet season sand mining areas in the coastal region are dotted with stagnant pools which collect in the sand and clay pits. Apart from providing breeding sites for mosquitoes and other insects, the pits render many hectares of land waste. Yet these flat sandy plains are recognized as offering very good prospects for the future development of dairy farming. Although quarrying of sand and clay need not be totally eliminated, it should be better controlled, and the reclamation of the exhausted areas should be made compulsory for all operators. It had also been suggested that some of the existing pools could be converted to fish ponds or reservoirs for watering livestock and for rural recreation (Areola and Faniran, 1975).

Gravel and red earth (laterite) quarrying has also rendered many hectares of land derelict and undermined the stability of slopes in the low coastal plateau region of Lagos State from Agege in the west to Epe in the east. The most spectacular quarry was the one at Ojota on the Lagos-Ikorodu road where a whole ridge, several hectares in extent, was completely dug up. The land surface is now some five to ten metres below its original height. Happily, because of the demand for land in Lagos the quarry land has been partitioned, sold and built upon. But many less extensive pits throughout the area, and in other parts of Nigeria, have been abandoned, leaving permanent scars on the land surface. The sides and bottoms of these pits usually harden up following exposure to intense insolation and are thus very difficult for plants to recolonize.

Near large towns some gravel pits have become dumping grounds for scrap metal while others on the outskirts of villages are converted to refuse dumps and are eventually filled in and reclaimed. However, when these refuse dumps are filled by storm runoff, they constitute dangerous health hazards. But, in the sub-humid savanna regions pools collecting in abandoned gravel or clay pits have been very useful in providing water for livestock and for irrigating small gardens. In many parts of the country the water is used for mixing mud and cement for brick-making and other construction purposes.

Inevitably quarrying of earth materials leads to large scale destruction of vegetation. Along with road construction, quarrying is largely responsible for the extensive clearance of forests in recent decades, notably in the Lagos, Benin and Ondo forest regions. As in the case of Ojota, described above, entire interfluves or valley sides may be cleared and quarried. The exposed pits and slag heaps are often subjected to intense erosion resulting from concentrated storm runoff. Sediments removed from such pits and dumps sometimes spill onto highways, block drainage channels and cause gullies to develop by the roadsides.

Rock quarrying

Rock quarrying is less damaging to the landscape and vegetation than the quarrying of earth materials. It is confined to rock outcrops or rocky terrains, which are not useful for agriculture and most of which are bare of vegetation. However, some rocks may have a thin covering of soil which supports woodland and forest type vegetation.

Limestone quarrying for large-scale cement manufacturing is somewhat different from other types of rock quarrying. It involves the clearing of vegetation and digging up of land over extensive areas of former or potential agricultural land.

Perhaps, the greatest environmental hazard of rock quarrying is the vast quantities of dust that pollutes the air in the vicinity of the stone-crushers. The dust often drifts onto the highways obscuring visibility and hurting the eyes of passers-by. The dust problem is also found at limestone quarries and cement factories. In addition to the health hazard to humans, dust particles settle on plant leaves, especially during the dry season, and interfere with the processes of transpiration and photosynthesis.

Marble quarrying is similar in many ways to rock quarrying, both in its operations and in its environmental impact. But marble is usually found on smooth ridges or hillocks which are often well-covered by soil and vegetation. Although, there is some loss of land which could be cultivated or grazed, most marble quarries in Nigeria are very small covering not more than a fraction of a hectare. Marble quarries are to be found around Kwakuti along the Minna-Suleja Road in Niger State; at Jakura in Kogi State; in Auchi district of Edo State; and around Igbeti in Oyo State.

Modern-day gold rush and scramble for solid minerals

It started small and hidden in the era of military government and became full blown in the first two decades of the 21st century. Foreign entrepreneurs from both the advanced and emerging nuclear powers came to Nigeria and other West African countries in search of uranium and other metals rich in radioactive minerals, such as niobium and thorium, pyrochlore, and so on. They went straight to engage the local people in each area of operation. Before long, many unemployed youths, farmers and pastoralists both from within and outside Nigeria joined in this modern-day rush for rare minerals.

In Kebbi, Zamfara and Sokoto states, there is a frenzied, uncontrolled gold rush under way. Gullies, river valleys and terraces, hill tops and slopes and upland plains have been dug up by thousands of prospectors panning for gold. The miners have encroached on farmland, grazing areas and parts of grazing reserves and turned them into derelict land. Although, most operators operate at the artisanal level using the simplest of implements, foreign entrepreneurs have brought in machinery, including giant pumps, and are digging deeper into the ground in search of the rare mineral. The ditches are getting deeper and the slag heaps are mounting up (Plates 28A–D.)

The rush for solid mineral prospecting and mining is thought to have given rise to banditry in the states of Kebbi and Zamfara in north-western Nigeria. Smuggling by gangs of bandits began with the smuggling of solid minerals across borders and grew to cover other items in high demand, including in particular imported rice. The population of bandits has been swelled by the large number of unemployed or unemployable youths in these areas. Wealth acquired from the sale of solid minerals and smuggled goods is used to purchase sophisticated weapons by both bandits and herdsmen militias that have caused so much threat to security in the country.

Environmental impacts of mining and their remediation in Nigeria

Environmental impacts

The environmental impacts of quarrying and solid mineral mining in Nigeria have been described in the foregoing discussions. But, it is necessary to summarise them at this point in order to understand the remediation measures that are being applied or recommended for adoption by various researchers and environmentalists. The major environmental impacts may be outlined as follows:

1. Land degradation involving loss of soil and vegetation cover, creation of extensive pits of various depths on the landscape, leaving huge dumps of slack and dumphills on the land surface, which are then scarred by rill and gully erosion;

2. Sedimentation of rivers, lakes and reservoirs by sediments eroded from mine dumps; these occurred in the tin mines of the Jos Plateau;

3. Mining pits and quarries are filled with water and become stagnant pools that are breeding grounds for insects, some of which are disease vectors or become polluted and exude foul odours;

4. Mine dumps and slack contain heavy metals that sometimes also contaminate the soil, water bodies, growing crops/plants and become health hazards to fish and other aquatic life forms, livestock and wild game that feed on plants, and human beings that consume the contaminated foodcrops and meat products;

5. Dust, smoke, gases emanating from quarries and mines cause air pollution, which may result in health problems not only for the miners but also residents in the vicinity of the quarries and mines;

6. Limestone quarries like those at Ewekoro and Sagamu are notorious for dust pollution that covers the vegetation (including kola nut trees, sugarcane) and

reduces plant transpiration; and reduces visibility to motorists driving by especially during the dry season when there is no rainfall to wash away the dust from the air, vegetation and buildings

7. Air pollution by gaseous emissions that corrode metallic objects including corrugated iron roofing sheets, various types of vehicles and machines; as well as cause acid rain;
8. Wastewater derived from mineral processing also pollute the environment with sediments, heavy metals and obnoxious odours;
9. The use of explosives in quarries and mines, the undermining of hill slopes and the piling up of waste dumps have been known to result in land slides and slumps that may block roads, destroy farmland and forests, and buildings;
10. Earth tremors are common in mining areas such as the coalfields of Enugu
11. In the tin mining areas of Jos Plateau people have used very white sand to mould concrete blocks and plaster their houses only to discover later that the sands were radioactive (Adeyemo and Farai, 2006). Such irradiation had been known to cause health problems, deaths (Aigbedion 2005) and birth defects in pregnant women. Worse still, Jibiri *et al.* (2006) also found accumulation of radioactive elements 40K, 238U and 232Th in local food in the tin mining areas of Jos Plateau. These substances were adsorbed by cultivated food crops from contaminated soils.

Remediation

Remediation is the process of restoring damaged lands to their original states as much as is possible or practicable. The colonial administration enforced it in the tin mining areas of the Jos Plateau ravaged by accelerated soil erosion and sedimentation of rivers. As indicated earlier, the mined lands were usually reclaimed by replacing the earth removed from the mine pits and spreading compost on the surface at the rate of twelve to sixteen tonnes per hectare to increase fertility. The reclaimed lands eventually reverted to the original owners.

It is doubtful if there has been a comparable mine land remediation anywhere in the country since pre-Independence. However, while some quarry sites became waste dumps, mostly illegal, the huge one at Ojota Lagos was developed into a layout for housing development (e.g. Ojota, Lagos). Mpape or Crushed Rock site in a poor neighbourhood on the outskirts of Abuja, is being turned into a tourist site primarily because of the picturesque rock scenery and the pool that has formed inside the abandoned quarry. The area is also said to experience occasional earth tremors.

Odoh *et al.* (2017) described a number of plant based remediation strategies that Nigeria could adopt in redressing the environmental impacts of mineral extraction and processing. If such strategies are ever to be adopted in Nigeria, they are likely to have the best chance of being adopted in the oil producing areas of the Niger delta region, where Shell Oil Company is currently engaged in a big clean-up project in Ogoni land. This is because remediation of the environmental consequences of oil exploration and exploitation activities is built into all planned projects in the region.

The future

It is reasonable to assume that Nigeria's mining operations will increase in the near future. Improvements in the transportation system and the availability of funds and modem mineral refining technology now make it possible to work many sources of valuable minerals at a fair margin of profit. Furthermore, the demand for building materials will continue to rise with the increasing pace of construction works in the country.

The natural environment has not yet been irredeemably damaged by mining activities in many parts of the country. But a catastrophic environmental scenario is building up in the northern savanna lands as a result of the rush for gold and rare metals. At present, everyone seems to be focused on the expanding export market in these solid minerals and on their value in diversifying Nigeria's oil-dominated economy (Table 8.6). The Federal Ministry of Solid Minerals and similar agencies at the state level have their work cut out for them. Thus far, they do not seem to have things under control. The River Basin Development Authorities also should be greatly concerned that the landscape in many a river basin is being destroyed by uncontrolled mining activities, clearing the vegetation cover and generating so much sediments that are likely to wash eventually into reservoirs. The risk of dam bursts and flash floods has increased greatly. All forms of mining should be placed under strict control and should be bound by the same laws and regulations that once operated in the Jos plateau tin fields. In particular regulations pertaining to the reclamation of derelict lands need to be re-introduced and strictly enforced.

There is hope for the oil producing areas of the Niger Delta. In the first place, there are many international organizations and indigenous non-governmental organizations now operating in the region focused on advocacy, funding, and promotion of policies and projects to clean up the oil polluted areas and create more conservation areas for ecosystem and species biodiversity protection. Secondly, there are

TABLE 8.6 Freight on Board (FOB) values of mineral exports (in million US dollars), 2014–2016

S/N	Export Minerals	FOB Value USD		
		2016	2015	2014
Metallic Ores and Concentrates				
1	Lead	12.96	7.72	47.56
2	Iron	10.51	–	–
3	Manganese	8.05	0.08	0.03
4	Zinc	5.04	0.83	20.00
5	Tin	1.31	0.02	0.03
6	Niobium-Tantalum	0.76	0.50	0.72
7	Copper	0.38	0.02	0.47
8	Zircon	0.30	0.41	0.21
9	Gold	-	0.12	0.60
10	Tungsten	0.02		
11	Titanium	0.01		
12	Chromium	0.01		
Non-metallic Minerals				
13	Gypsum	0.12		
14	Crude mica	0.11		0.24
15	Natural sand	0.57		
16	Bentonite	0.01		
17	Kaolin	0.03		0.01
18	Silica sand	0.04		
19	Limestone/Marble	< 0.01		
	Total	\$40.94 million	\$9.73 million	\$26.14 million

Source: The Nigeria Extractive Industries Transparency Initiative (NEITI), 2017, 2018. Quoted by Olade M. A. 2019 Solid mineral deposits and mining in Nigeria: a sector in transitional change *Achievers Journal of Scientific Research* Volume 2, Issue 1, July 2019, p. 1–15. `http://www.achieversjournal.org/`

now so many government and quasi government agencies overseeing oil and gas operations, monitoring and regulating their environmental impacts. They include the Department of Petroleum Resources, the Federal Ministry of Environment (& Federal Environmental Protection Agency), the National Oil Spill Detection and Response Agency, and State Environmental Protection Agencies. Thirdly, the federal government now appears to be more committed to the clean-up of the oil pollution in the delta area. The clean-up operation embarked upon by the Shell oil company is worthy of emulation by all other stakeholders in the Niger delta. The erstwhile insurgency groups have made peace with governments at the federal and state levels. This has created a more peaceful atmosphere for the pursuit of environmental protection programmes. However, although the problem of oil spills has abated somewhat, there are ominous signs in the rise of illegal artisanal oil refining and the fact that sabotage still accounts for over 70 per cent of oil spills. It looks like, having laid down their arms and stopped military action in the delta region, the members of the insurgent groups and others have found a more lucrative and rewarding alternative in illegal oil refining and bunkering.

9

Urbanization and its environmental impacts

Fabiyi and Joseph (2018) enumerated and measured the spatial extent of settlements in Nigeria using the remotely sensed 5-meter resolution SPOT V images acquired in 2005. The analysis highlighted the marked increase in the amount of land that had been gulped up by human settlements in Nigeria. Using settlements with a minimum spatial extent of $2km^2$ they were able to count 5,088 settlements, grouped into five classes (Table 9.1). The largest built-up area, Lagos metropolis, was $110{,}000km^2$ and the smallest was $2.45km^2$. As figure 9.1 shows, the major settlement clusters of cities and towns are in southern Nigeria while the few clusters in northern Nigeria are widely dispersed in a region dominated by towns and minor settlements.

The following general observations may be made about the human settlements: (i) except for Abuja, which is a recent development, most cities predate the colonial period but a few (e.g. Port Harcourt, Aba, Enugu, Jos, Kaduna) originated at strategic points along the transport network (especially ports, railways lines) built by the colonial administration; (ii) since colonial times population has been gravitating toward the south, especially, the coastal areas which have the largest clusters

TABLE 9.1 Classification of settlements in Nigeria based on spatial extent

Class of settlements	Description: Settlements with built-up area between...	Number of cities	Label
I	$10{,}000km^2 - > 100{,}000km^2$	10	Mega cities
III	$100km^2 - 1{,}000km^2$	662	Minor cities
IV	$10km^2 - 100km^2$	2392	Towns
V	$2km^2 - 10km^2$	1912	Minor settlements
Total		5,088	

Source: Fabiyi O. and Joseph O., 2018 Comparative analysis of spatial extent and population sizes of cities in Nigeria: implications for urban space administration. *Journal of Historical Archaeology and Anthropological Sciences, 3(1): 87-95* doi: 10.15406/jhaas.2018.03.00070 http://www.medcraveonline.com

of cities (e.g. Lagos, Warri, Port Harcourt, Aba); and (iii) the phenomenal growth in the number and land area of human settlements has had a huge impact on the Nigerian environment. For example: (a) the mega-cities and major cities have engulfed the surrounding rural areas and encroached on agricultural lands; (b) many cities have grown without or with only the barest of development control resulting in unplanned layouts and development of slums; (c) many towns and cities have spread into wetlands or over hilly terrains without adequate provision for drainage to contain increasing annual floods; (d) most cities are highly congested in relation to the limited municipal facilities; and (e) most cities have problem coping with their sewage and huge wastes generated. These environmental issues are the primary concerns of this chapter. While the cities have offered millions of Nigerians opportunities for employment and means of livelihood, they also have become a huge drain on the country's financial resources partly because of mediocre managerial and administrative skills applied in the management of the cities. A situation where there is one or a few overriding cities implies an imbalance in the economy, which is to the detriment of the cities of lower order. The mega city of Lagos is in a class of its own; it carries a disproportionate volume of Nigeria's trade, commerce and industry and its environmental problems are equally legendary and seemingly intractable.

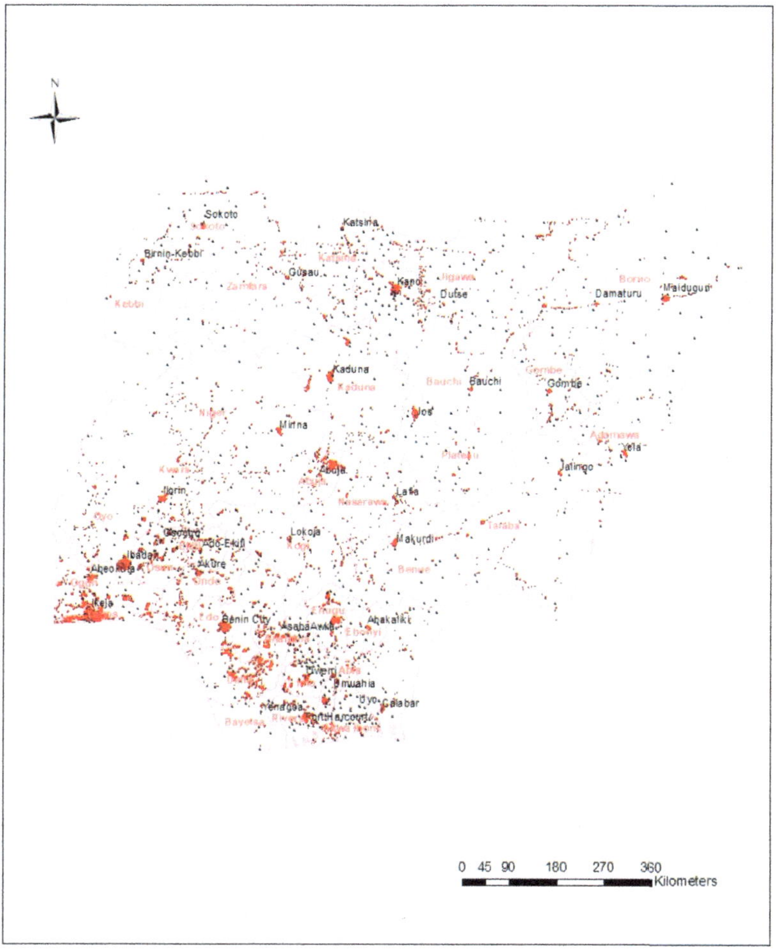

FIGURE 9.1 Distribution of Settlements with at least $2km^2$ built-up area mapped from SPOT V. (Modified from Macmillan, 2010.)

Source: Fabiyi O. and Joseph O., 2018 Comparative analysis of spatial extent and population sizes of cities in Nigeria: implications for urban space administration. *Journal of Historical Archaeology and Anthropological Sciences*, 3(1): 87-95. doi: 10.15406/jhaas.2018.03.00070
http://www.medcraveonline.com

Urbanization

There is more to urbanization than the mere congregation of people engaged in non-rural economic activities in one place. It entails, also, changes in the economic base and in the living habits and standards of the people. It creates conditions in which improvements in production, transportation, housing, waste disposal and the supply of infrastructural facilities must be undertaken, and which require a sophisticated framework within which the process is guided and controlled. The environmental problems arising from the process of urbanization in Nigeria are largely due to the fact that the social, technological and institutional changes necessary for a successful transformation from a rural to an urban lifestyle, or rather from a traditional African urbanism to a modern western or European urbanism, have failed to keep up with the rapid movement of the population into the cities. Nigeria has one of the highest rates of population growth and urbanization not only in Africa but also in the world (Table 9.2) Rapid and uncontrolled urbanization has been a major feature of the Nigerian economy since the establishment of the colonial economy and administration

Nigeria's urban centres, in varying degrees, combine elements of the modern or western city with elements of their former traditional rural base. Mabogunje (1968, p.43) rightly noted that with few exceptions, urban centres in Nigeria are an amalgam of two contrasting levels of urbanization: a traditional, almost mediaeval, pre-industrial urbanization and an advanced, industrial urbanization. The influence of the latter is still much weaker than that of the former. In order to provide the necessary framework within which to examine the ecology of urbanization in Nigeria and the environmental management issues involved, it is important first to discuss the nature of the urban environment as constructed by humans.

The city as an environment

The rise and development of cities is often described in terms of the interplay of socio-economic forces. Theories of urban land use and spatial structure tend to regard the city as an economic mechanism. But, as Guttenberg (1975) noted, a city is certainly more than this. People make their lives in the midst of this mechanism, so the city is also a physical and social environment. Unfortunately, nearly everywhere in the world, concern for the nature and quality of the urban environment has lagged behind the pursuit of economic fulfillment. Guttenberg described the American experience as follows:

> In most of us the aesthetic sense is secondary to the instinct for sur-

TABLE 9.2 Population Growth Rates and Levels of Urbanization in Sub-Saharan Africa and Other World Regions 1950-2015

Country or Region	Total population 2001 (millions)	Population growth rate (%)		Level of urbanization (%)			
		1985-2000	2001-2015	1950	1975	2001	2015
Benin	6.4	2.8	2.6	3.3	21.9	43.8	53.5
Burkina Faso	12.3	2.8	2.8	3.0	6.3	17.4	23.2
Cote d'Ivoire	16.1	2.7	2.0	13.2	32.1	44.0	51.0
Gambia	1.4	3.7	2.2		17.1	26.1	27.8
Ghana	20.0	3.0	2.6	14.5	30.1	45.0	51.1
Guinea	8.2	2.7	2.3	5.5	16.3	34.5	44.2
Guinea Bissau	1.4	2.2	1.9		15.9	33.2	43.5
Liberia	2.9	2.4	3.2	13.0	30.4	44.9	55.5
Mali	12.3	2.3	2.6	8.5	16.2	31.6	40.9
Mauritania	2.7	2.8	2.5	2.3	20.3	60.5	73.9
Niger	11.1	3.2	2.9	4.9	10.6	21.6	29.7
Nigeria	**117.8**	**2.6**	**2.1**	**10.1**	**23.4**	**45.9**	**55.5**
Senegal	9.6	2.6	2.4	30.5	34.2	48.9	57.9
Sierra Leone	4.6	2.9	3.2	9.2	21.4	38.1	47.6
Togo	4.7	2.8	2.5	7.2	16.3	34.5	42.3
West Africa	231.5	2.7	2.5	8.3	20.8	38.0	46.5
Sub-Saharan Africa	669.0	2.5	2.1	n.d.	21.0	34.8	42.8
Africa	784.4	2.5	2.1	15	25	37.9	46.5
Europe	728.9	0.2	-0.1	52	67	74.8	78.6
Asia	3,682.6	1.6	1.1	17	25	36.7	44.7
Latin America	519.1	1.7	1.3	41	61	75.3	79.9
North America	309.6	1.2	0.7	64	74	77.2	80.9
Oceania	30.4	1.4	1.1	62	72	70.2	71.2
World	6,055.0	1.5	1.1	41.8	54	47.0	53.4

Sources: UN-HABITAT, 2005b, UNDP. 2OO2 In Filani M.O. and S. Okafor (eds.) 2006. *Foundations for Urban Development in Africa: The Legacy of Akin Mabogunje*. The Cities Alliance 2006. 1818 H Street NW, Washington, DC 20433, USA. www.citiesalliance.org

> vival. Hungry people pay scant attention to environmental amenities—they are too busy securing their daily bread. But once their hunger is assuaged, the world appears to them in quite a different light no longer as the scene of a perilous or backbreaking struggle for existence, but as a noble setting for human life. So, with the city. People have flooded into American cities from every quarter of the globe in search of economic opportunity, and those who have done well are now eager to enjoy the rewards of their labours. Consequently, the city appears under a double aspect—as a place to make a living and as a place to live well (p. 304).

Nigeria in the 21st century has passed the first stage where people are rushing to the cities in search of economic opportunity. But, while the urban dwellers appreciate and yearn for aesthetically stimulating living environments in the cities, they are nevertheless still enduring harsh conditions of urban poverty, housing congestion, inadequate facilities and poor sanitation. Sadly, Nigeria's major cities have many slums, where living conditions are worse than in the rural areas where most of the slum residents ran away from. Urban dwellers in Nigeria now know too well that financial rewards alone do not guarantee a rich, socially and psychologically satisfying urban life. The urban physical environment deserves closer attention and conscious management so that it does not constitute a limitation to social and economic well-being.

The city, perhaps, represents humankind's most profound transformation of the natural environment. The land surface cover of soil and vegetation is transformed to one of concrete structures and pavements. This greatly modifies the hydrologic cycle because the roof tops and concrete surfaces generate greater runoff from the rains. As a result of the alteration of components of the radiation balance the city develops its own unique climate. The burning of fossil fuels in industrial establishments and the presence of a great many automobiles in the streets introduce gaseous substances into the atmosphere. The urban dwellers generate great quantities of solid wastes and waste water while factories and other work places throw out gaseous, liquid and solid effluents, byproducts of industrial processes.

The net effect of all this is to dramatically alter the natural pathways of energy and matter in the environment, and to create entirely new ones. The aesthetic quality of the urban environment, in any area, depends on the efficiency and stability of the created systems. The spatial structure of each city needs to be harmonized with the physical landscape while at the same time measures and facilities must be put in place to protect the inhabitants against environmental hazards. Adequate and equitable provision of public services and facilities must be planned and environmental amenities such as clean air, clean water bodies and clean open spaces must be actively promoted to create a livable urban environment.

In other words, ideally a city should combine the best of urban and rural life. This is the basic principle on which the Garden City movement in Britain, and the New Town movement in America were based (Guttenberg, 1975). Nigeria also adopted the concept of 'new' or 'satellite' towns in the 1970s, hence the Satellite Town in Lagos, called FESTAC town and the Ajoda New Town near Ibadan were established. The 'satellite' town concept was also factored into the physical development plan of the new federal capital territory, Abuja. Hence, towns such as Gwagwalada, Kubwa, Kuje and Bwari were developed to reduce the influx of workers and migrants into Abuja city itself.

The creation of garden cities or new towns will not solve the environmental problems of the old cities. In fact, a case against the new town concept was that it appeared to regard cities as disposable items that could be abandoned and replaced as soon as they wore out or, in some other respect, failed to conform to the people's desires. The solution seems to lie in a combination of the new town approach and that of urban renewal and redevelopment matched by a dynamic and enlightened management of the urban system in the country as a whole or parts thereof.

Urbanization in Nigeria

The historical perspective on the urbanization process cannot be ignored in any analysis of the problems of Nigerian cities today. Mabogunje (1968) gave a detailed account of the history of urban development in the country. Prior to the colonial era there were two main areas of urbanization, one in the Sudan zone of northern Nigeria and the other in the high grassland plains of Yoruba land. The rise of urbanism in the Nigerian Sudan, as of the West African Sudan as a whole, could be attributed to the extensive open grassland country which, apart from being climatically healthy and agriculturally productive, offered great scope for unobstructed human movement and organization. It was only natural, therefore, that in the mediaeval period interregional and international trade flourished and led to the rise of various kingdoms and empires. The stability provided by these economic and political developments facilitated the rise of urban centres such as Katsina, Kano, Zazzau and Maiduguri. Other towns came later, after the Fulani *Jihad* (holy war) of the nineteenth century. The need for strategic locations on the part of the Fulani elite for the effective control and administration of the newly acquired empire led to the founding of such towns as Sokoto, Gwandu, Gombe, Wurno and Katagum. These urban centres were sustained by trade based on the sale of craft products, textiles, minerals, salt and agricultural crops.

Yoruba towns, by contrast, originated largely as colonies during the intercommunal strives that engulfed the territory in pre-colonial times. They were built by a more aggressive and technologically advanced immigrant group among backward and hostile peoples. The towns were, therefore, planned to dominate the surrounding villages and peoples. Although Yoruba towns evolved basically as administrative centres, they were sustained by trade, based on the sale of minerals, especially salt, crafts and agricultural products. The trade was organized on a system of markets at the local, provincial and inter-kingdom levels.

At the beginning of the British colonial era, therefore, Nigeria had evolved its own systems of cities with elaborate power structures and hierarchical systems of administration. The urban centres were comparatively small, the largest ones having populations between 20,000 and 100,000. Although small the urban populations were relatively heterogeneous in terms of ethnic composition, religion, occupation and social status. In their physical layout, each urban centre evolved around the Oba or Emir's palace, the central market and the principal shrine or mosque in

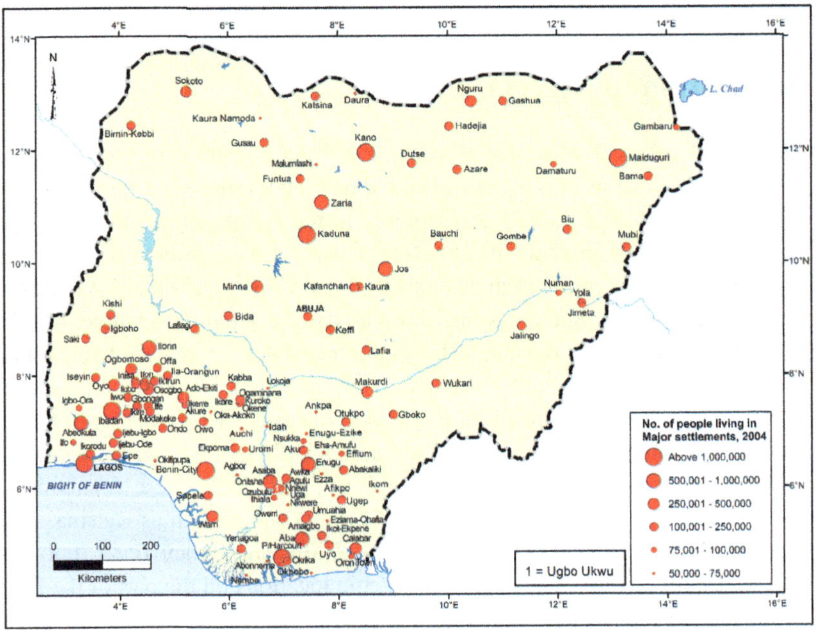

FIGURE 9.2 Urban centres in Nigeria

the town. The town was divided into quarters which accommodated different ethnic groups, chieftains and craftsmen's guilds. Houses were built of mud and had thatched roofs. Each family compound or quarter catered for waste disposal in its portion of the town. Refuse dumps were created at strategic points in the town and close to the market places. These were set on fire at intervals to dispose of the refuse.

The British colonial administration influenced urbanization in Nigeria in three major ways, by: (1) creating new towns, principally at ports, mining camps and along the major transportation routes, notably the railways, to serve the colonial economy and administration (Fig. 9.1&9.2); (2) modernizing the physical structure of the existing towns and introducing western urban amenities; and (3) changing the urban economic base. In most towns, the modern commercial/industrial centres developed outside the traditional town centres, which consequently declined rapidly and became stagnant (Mabogunje, 1968 *op. cit.*) The last two points are worth elaborating upon as they are very crucial to an understanding of present conditions in Nigerian cities.

The modernization of the physical structure of Nigerian towns manifested itself in several ways. One was the replacement of mud walls by baked brick walls which were later gradually displaced by concrete cement blocks. The transition from mud walls to cement blocks was slower in some areas than in others and one finds that up till today buildings in Nigerian towns range from pure mud houses, through mud houses with cement plaster and mud brick houses with plaster, to houses wholly built of cement blocks. In many towns there is a sharp distinction between the old traditional sector and the new modern commercial/industrial sector, not only in the building materials and architectural design of houses but also in the provision of urban utilities. The modern sector is almost exclusively composed of plastered brick or cement block houses which are also more aesthetically designed.

The second major change wrought by the colonial impact on traditional towns was the replacement of less durable and highly combustible thatch or bamboo roofs with corrugated iron roofing sheets, slates or tiles. Unfortunately, quick rusting corrugated iron roofing sheets were the cheapest, most readily available and, therefore, most popular with the people. Thus, within a few years many Nigerian towns with their rusted corrugated iron roofed houses fitted well Mabogunje's dismal description of Ibadan city as having 'the appearance of an extensive, rusty-brown junk-spread'.

Finally, improvements in the economic base of the towns and the emerging affluence of some inhabitants had far-reaching effects on the physical layout of the traditional centres. The new wealthy class included producers and middlemen collectors of agricultural export commodities, traders in imported textiles and other goods, and workers in the service of foreign companies and factories, and employees of the colonial administration. This new wealthy middle class developed different tastes, living standards and expectations; developments which were to weaken family and kinship ties. Thus, affluent members of the family began to acquire portions of the family compounds on which to put their own personal buildings. This led to the disintegration of the family compounds and the rise of slum housing conditions. The space available to accommodate the separate housing units was necessarily limited. Every available space within and around the compound was soon built upon without much consideration for provision of adequate access to the buildings. The outcome can be seen in the traditional section of Ibadan and other cities characterised by unplanned, narrow, irregular streets, and closely packed buildings with comparatively poor social facilities.

The scarcity of space also forced people to build on physically difficult sites such as hillsides and river flood-plains, which was to cause major environmental problems later including accelerated soil erosion and flooding. The disordered nature of the layout of the traditional sections of many towns made the distribution of essential urban utilities such as roads, piped water, electricity and analogue telephones much more difficult. House numbering was an exercise in futility; it served little purpose because it neither facilitated location of places nor the distribution of letters and other items. It has continued to make population censuses and social surveys for planning purposes that much more taxing.

The economic order established by the colonial administration and continued after independence seriously undermined the crafts of traditional Nigerian cities. The modern technologically advanced factories and large multinational commercial enterprises came to hold pride of place in the allocation of land and provision of urban utilities in the cities. The location of these modern commercial/industrial centres influenced each city in its entirety so that expansion took place in the direction of the modern centres, to the neglect of other parts. These new economic institutions also came to exercise an influence on the surrounding rural areas out of proportion to their real sizes and potentials.

This was partly due to the influx of people drawn to the cities by the lure of the job opportunities and the modern developments. For the new urbanites, modern facilities such as piped water supply, electricity, hospitals, schools, improved roads and housing, which were concentrated in the urban centres designated 'townships' during the colonial era, were the 'pull factors' that attracted them to the big urban centres. These utilities attracted, not only more people, but also, more businesses and industrial establishments; a trend that has persisted ever since. Unfortunately, the people that moved into the cities were not only the skilled and well-educated but also the unskilled and the young and inexperienced primary school leavers. The development of slums and a general deterioration of the urban environment have been the inevitable consequences of the overconcentration of people of limited means in the cities.

It is a sad commentary on the slow pace of response of municipal, local and state governments to these problems even in modern times. A study by Dankani and Abubakar (2011) of uncontrolled urban growth in Dorayi area of Kano metropolis lamented the emergence of unplanned urban environment virtually devoid of basic infrastructural facilities and amenities this far into the 21st century. These and other issues raised by urbanization in Nigeria are discussed in what follows.

Impact of urbanization on the rural countryside

The most obvious impact of urban development on the rural population is the encroachment of the cities on the rural landscape. Residential housing and industrial estates, link roads, institutions such as hospitals, schools and colleges made particularly great demands on rural space at the outskirts of cities. For instance, Ibadan city has expanded to engulf settlements that were once distant from it along the major roads leading to Oyo to the north, Iwo to the northeast, Ile-Ife to the east, Abeokuta to the west and Ijebu-ode to the south. Kaduna has more than doubled in area since the 1970s and expansion has been along the major highways leading to Abuja, the Federal capital, Kano, Jos, and Birnin Gwari. Land use/land cover analysis using remotely sensed images showed that between 1986 and 2005 the built-up area of Kano city increased from 13.2 per cent to 19.3 per cent and the 'forested' or vegetated land declined sharply from 33.9 per cent to 13.6 per cent. Only 'rangeland' still appeared to be keeping its stand against encroachment at about 56 percent of the total land cover (Ayila *et al.*, 2014). Lagos expanded beyond the island to engulf such mainland settlements as Mushin, Ikeja, Agege and other settlements all the way to Ota in Ogun State. The metropolis now stretches all the way to Mowe on the Lagos-Ibadan motorway; to Ojoo on the Lagos-Badagry Road and beyond Ajah and Atlantic City on the Lekki-Epe road.

These large urban centres also influence the rural countryside in a more subtle way by dictating the use of land and the exploitation of certain natural resources. The Lagos region provides a good example of this phenomenon (Areola, 1976). In the whole of the Lagos State land is a very scarce resource; the state is among the smallest in the country being only 14,655km^2 in extent. Furthermore, the terrain is poor in large parts of the state placing severe limitations on human settlement and economic activities. The bulk of the population is concentrated in those parts which are relatively free of physical constraints, notably the Ikeja plain and the low dissected plateau region further inland. Migrants from all over the country as well as residents moving out of crowded Lagos island, mainland areas of Yaba, Ebute Meta, Mushin, Agege and other old towns are responsible for the expansion of the metropolis to these areas. In addition, the large commercial and industrial companies have moved from the old industrial estates at Apapa, Iganmu and Ikeja into the same areas to acquire more space to expand their businesses and industrial output. Ota alone has attracted more than 150 commercial/industrial establishments since the last decade of the 20th century. The frenzy generated by the construction of private residences, housing and industrial estates, warehouses, and roads has created a huge demand for building materials. Thus, the pressure of housing, transport and industrial expansion, and the need for building materials, now dictate the use of land and the exploitation of natural resources even in places far beyond the boundaries of Lagos State in the neighbouring Ogun State.

Loss of agricultural land

One of the greatest impacts of urbanization on the rural countryside is the loss of prime agricultural land and agricultural enterprises. Ibadan city used to be the centre of a rich and thriving cocoa producing region. In the 1920-50 period the cocoa plantations had attracted migrant and itinerant farmers from both near and distant places to Ibadan region to work on the cocoa farms. By the 1970s, the cocoa farms close to the city had started to disappear and by the 1980s one could hardly come across any cocoa farms within 10-20km of the city. Today, cocoa farms have all but disappeared from the region except in pockets in remote locations, including parts of Akanran – Onigambari axis to the south and Ijaye-Igbo Oloyin axis to the north. Former agricultural lands around Ibadan have given way to housing estates, educational institutions and, factories although in some places, cassava farms, poultry and piggery farms have replaced the former tree crop farms.

While Lagos State has been working to diversify economic activities particularly by promoting agriculture and fisheries thereby achieving a more broadly-based economy, the sheer aggressiveness and competitive vigour of the urban commercial-cum-industrial sector seem to be defeating that purpose. But agricultural research institutes established by government since colonial ties have continued to survive and thrive in Ibadan area, for example, Cocoa Research Institute of Nigeria (CRIN) at Onigambari, the Forestry Research Institute of Nigeria (FRIN), Jericho, and the Nigerian Horticultural Research Institute (NIHORT), Apata.

In Lagos State, the Ikeja plain is a low dissected but generally flat or gently rolling terrain which is for the most part well-drained. The soils are good and comparatively more productive than in most other parts of the state. The area offers great scope for arable farming and irrigation agriculture. However, because of the favourable topographical and hydrological conditions of the land, the Ikeja plain has fewer physical constraints on urban and industrial expansion. At Agege, Ikeja, Shogunle and Oshodi, the metropolis encroached rapidly on the rural countryside. Land owning families and astute local businessmen are great land speculators and operate in areas far beyond the present urban fringes. The flat sandy area south of Ikeja plain and the coastal riverine and creek areas are other ecological units which have felt the impact of the rapid growth of the metropolis especially along Lekki-Epe and Lagos-Badagry axes. As we saw in Chapters 5 and 8, sand and clay quarrying have laid waste to large tracts of potentially useful land.

All the former provincial towns that became state headquarters during the military administration 1966-1999 are having the same impact on the surrounding areas as in Ibadan and Lagos areas. Research studies have emerged focused on the loss of agricultural land to urban development in some of these state capitals. Dan-Jumbo *et al.* (2018), using Landsat TM and ETM images, carried out land use/land cover change analysis in the Port Harcourt region in Rivers State, comprising 5 major river basins and 39 sub-basins spanning 4820km^2. The analysis which covered the period, 1986–2003, found that 95 per cent of the land use/land cover conversions to urban land occurred at the expense of agricultural land. These conversions to urban land use in a wetland area like Port Harcourt have exacerbated the problem of flooding in the metropolis. The urban area of Greater Port Harcourt expanded from 134.5km^2 in 1986 to 550km^2 in 2003 at a rate of 24km^2 per annum. A similar remote sensing-based study in Makurdi found that 200km^2 of agricultural land representing 19 per cent of the total land mass was lost to urbanization between 1994 and 2014 (Bakoji *et al.*, 2020). In total, the urban built-up area gained 336km^2 representing 32 per cent of the study area. In Ankpa, Kogi State, agricultural land, bare surface and areas of natural vegetation lost ground to urban expansion; and between 1995 and 2016 the area of farmlands declined by 53.1 per cent (Tokula and Ejaro, 2017). The reduction in agricultural land resulted in decline or fluctuations in yields of many of the arable crops in the area. Finally, Lasisi *et al.* (2017) carried out a similar study in 8 peri-urban communities in the two Local Government areas where Osogbo, the Osun State capital is located. Between 1986 and 2014, farmlands/vegetation areas had declined by 8 per cent and built-up areas had increased from 6.6 per cent to 20 per cent in the same period.

Fears have been expressed that these losses of agriculture to urban areas may have grave consequences for food security in the future, especially given the increasing demands for food by the rapidly growing urban populations (Satherthwaite *et al.*, 2010; Bakoji *et al.*, 2020). The conversion of agricultural land to built-up areas also has consequences for greenhouse gas emissions and global warming.

Hydrological consequences of urbanization

A disturbing feature of urbanization in Nigeria is the spread of the cities over the landscape in utter disregard of the physical conditions of the land and the potential environmental hazards that the disturbance of the balance of nature might bring. Thus, in many built-up areas, measures are hardly taken to offset likely environmental hazards and where some attempt has been made the measures taken are grossly inadequate to stem the problem. The most serious environmental hazard facing many Nigerian cities is flooding (Table 9.3). Flooding commonly occurs in two quite dissimilar urban landscapes. First, there is the flat terrain with a high water table and numerous swamps that is characteristic of such coastal cities as Lagos, Port Harcourt and Warri; and secondly, there is the hilly or strongly undulating and dissected terrain typified by Ibadan city.

The perception and the reality are that in the 21st century flooding has become more frequent and devastating as a result of climate change, which has given rise to more erratic and more stormy rains (Idowu *et al.*, 2011). The Benue River valley and its tributaries now experience flash floods almost every rainy season and settlements, farmlands and grazing areas are inundated. Flash floods from sudden rainstorms are also common in the far north, all attributed to climate change and global warming. But, as we shall see in the following discussions on specific cities, the human factor also looms large among the causes of disastrous floods.

TABLE 9.3 Recurrences of flood disasters in Nigeria, 2003-2009

Year	2003	2005	2006	2007	2008	2009
No of flood disasters	47	57	35	88	9	12

Source: National Emergency Management Agency (NEMA), Abuja (from National Bureau of Statistics in Nigeria, 2012 Part III).
Note: Data not available for 2004.

Floods in Lagos, Port Harcourt and Warri

In Lagos and Warri, urban flooding is in the form of flash floods from heavy rainstorms resulting in overflowing rivers and reservoirs, storm runoff overflowing into highways, streets and residential areas and destroying roads, properties and, even, human life in some instances. The problem may be attributed both to the nature of the terrain and to the climatic conditions, especially the heavy torrential rains. But to these must be added such factors as the inadequacy of storm water control and land drainage measures, drainage ditches along highways with inadequate or reduced hydraulic capacity to contain storm runoff, and very poor land use zoning. Ocean surges and high tides rising above their normal peaks also cause more frequent and extensive flooding along the coastal lagoons, creeks and beach fronts (Awosika *et al.,* 1992).

Lagos is located in an area of flat relief which is almost at sea level. This fact coupled with the high water table, makes land drainage particularly difficult. The heavy torrential rains generate great volumes of runoff which do not move freely and quickly off the relatively flat land surface. The capacity of the soil to absorb water is limited by the high groundwater table. Furthermore, many of the swamps around the metropolis, which used to serve as natural reservoirs for excess water from the land surface as well as from the overflowing rivers and creeks, have been filled in and the reclaimed land built upon. In such a physical environment, an elaborate system of storm water and flood control is called for. But, unfortunately many parts of the metropolis, even those areas such as Surulere which were reclaimed from swamps, have very poor drainage systems. The capacity of the drainage ditches and culverts is far too small to contain the runoff from the heavy torrential rains of the wet season. Thus, almost every downpour results in flooding.

Drainage works in the Lagos area have always required huge sums of money to construct and maintain given the nature of the terrain. Funds have not been readily available and this has been partly responsible for their poor quality. An indirect solution to the problem of scarce resources could have been achieved through wiser land use zoning. In metropolitan Lagos, there is the curious situation in which industrial estates are located on some of the best lands while residential areas are given more hazardous zones. Ikeja and Isolo have not been as adversely affected by floods as, say, Surulere, a residential area. In most other parts of the world industrial establishments occupy marginal lands that are lower in quality than those used for residential areas or agriculture. This is not only because of the relatively 'dirty' nature of factories but also because industry can more readily afford the heavy capital expenditure required for storm drainage works and land improvement to provide strong foundation for roads and buildings. Part of the reason for the present situation is, of course, that the industrial estates had to be favourably sited in order to attract foreign industries to the country. This trend should now be reversed and industries be made to contribute more to the development of the lands they occupy.

The 'oil city' of Warri faces the same problems as metropolitan Lagos. Like Lagos, it is located on a flat terrain which is generally less than 8m above sea level while a large proportion of the area is actually less than 2.5m above sea level. There are numerous creeks and swamps. Warri experienced a sudden increase in population during the civil war when circumstances forced the oil companies to shift their operations from the eastern to the western sector of the delta. The sudden increase in population was unexpected and unplanned for and this created serious accommodation problems.

In order to meet the demand for housing, individual local entrepreneurs established housing estates most of which, for want of space, were located in swampy areas or areas liable to flooding. Such estates included Okumagba layout, Essi layout, Mosheshe, Idugbe and Odibo estates (Odemerho, 1976). These local entrepreneurs were not able to finance the elaborate flood control and storm water drainage works needed to free the sites of environmental hazards. In fact, flood risk increased after the urbanization of the terrain. By contrast, the Shell-BP senior staff quarters at Ogunmu, northwest of Warri city are among the best residential complexes in the town. Not only is the land better drained but, the company provided the necessary funds for its proper development. The situation in Warri has worsened over time. According to Aderoju *et al.* (2014) rapid industrial, commercial and economic growth as a result of oil exploration activities led to increasing urbanization over time and by 2007, the impervious built-up area of the metropolis was up to 236.76km^2 representing 43.7 per cent out of the total land area of 499.81km^2. Consequently, there has been an increase in flooding phenomenon over time not only because of the increase in impervious surfaces but also because of indiscriminate dumping of waste in river channels and drains and building construction along the flood pathways. Olanrewaju *et al.* (2017) noted that although annual rainfall in Warri appears to show a declining trend as a result of climate change, the change is not statistically significant; rather the causes of continuing flooding in Warri are still high rainfall, absence of drains and poor urban planning.

In Port Harcourt, house construction by individuals and by indigenous housing estate developers has spread to the wetlands to the east and northeast of the city. Unlike the oil companies, private house owners and indigenous estate developers cannot provide the necessary land drainage and storm water control infrastructure needed to stem the annual floods experienced in these areas.

Floods in Ibadan

Ibadan is a vast city sprawling over the undulating plains and the quartzite hills and ridges of the Ogunpa river basin. The appreciable relative relief, the fairly steep gradients, the numerous drainage lines and the generally light sandy loam to sandy clay soils, give the area a high potential for high runoff and sediment yield. The rapid expansion of the city over the years has greatly modified the natural landscape and vegetation cover. The intensity of soil erosion that can be generated following construction works on steep hillsides was aptly demonstrated by the vast quantity of sediments that clogged the Ogunpa stream and spilled on to the Queen Elizabeth Road from the Premier Hotel hill while the Cultural Centre was being built in the 1970s. In general, there has been an increase in the frequency of floods in the city since the 1980s. The reasons for this increase are twofold: in the first place the amount of runoff following every heavy rain is greater and, secondly, the capacity of the streams, gutters, culverts and bridges to contain the flood flows has decreased over time.

The greater runoff may be attributed to the general loss of vegetation in Ibadan including on the steep hillslopes. According to Areola (2021) the coverage of green spaces in Ibadan metropolis declined from 6,861.1 ha in 1970 to 998.6 ha. in 2015, making a total loss of 5862.5 ha. (85.4%) over the period. But the increased runoff could also be due to the reduced capacity to store water in the soil as a result of increased concreting of the ground surface. Soil infiltration capacity even in the vegetated surfaces has also been impaired by the persistent trampling of the surface by both human and animal traffic.

Existing drainage channels are clearly inadequate to contain flood waters in the city in spite of some improvements made since the flood of August 1980. The Ogunpa river channelization project cleared the river channel of solid wastes and vegetal growth, removed buildings and market stalls from the floodplain and reconstructed some bridges to increase their hydraulic capacity. Unfortunately, the gains of the channelization project have not been sustained: the solid wastes keep coming back; the thick bushes grow profusely on the floodplain and along the channel during the wet season; and buildings and market stalls have been edging unchecked closer and closer to the river channel along some stretches of the river valley.

The Ona River to the west of the Ogunpa River has gained notoriety for its disastrous flooding in the last two to three decades. The Eleiyele reservoir has overflown its banks a number of times and water released from the reservoir at high flood has wreaked havoc downstream destroying both life and property. Except at Eleiyele area close to the dam, nothing much has been done to upgrade the bridges across the river or expand the hydraulic capacity of the river channel as was done with the Ogunpa river channel. Thus, the potential for disastrous flooding is still high if care is not taken.

Waste disposal in urban centres

Waste management is still essentially an urban problem in Nigeria, although Bakare (2019) believes the problem is as serious even in the rural areas. Waste disposal has been one of the most intractable problems of Nigeria's urban centres and one which impinges on the aesthetic quality of the urban environment and human health. Part of the problem seems to be that most urban dwellers simply regard waste disposal as the responsibility of local and municipal government councils. Mabogunje (1968) noted that the general upkeep of a Yoruba town before the British colonial period was undertaken on the basis of each quarter looking after its own sector. Where the administration was efficient, a very clean town resulted. However, even at this time it was noticeable that towns with important trading functions were among the dirtiest in the country. It would appear that as cities grew and the population became more mixed and 'sophisticated', urban dwellers felt less personally responsible for, and became less involved in, keeping the city environment clean.

Another factor which may be partly responsible for the present waste disposal and sanitation problems of Nigerian cities is the vastly increased concentration of people living in them and the enormous quantities of waste generated daily. According to Ike *et al.* (2018) waste generation in Nigeria is of the order of 0.65–0.95 kg/capita/day which gives an average of 42 million tonnes of wastes generated annually. This is about 67.7 per cent of the 62 million tonnes of waste generated in sub-Saharan Africa annually.

Existing arrangements for waste collection and disposal are grossly inadequate so that refuse accumulates at the depots, on the roadsides and in neighbourhoods for days on end before being removed. Only 20 – 30 per cent of the waste generated in Nigeria is actually collected (Bakare, 2019). Ike *et al.* (2018) are of the opinion that the problem lies in the fact that waste management in Nigerian cities is largely monopolized by the agencies of state governments (sub-national governments) which have limited capacity to tackle the problems of solid waste management in their cities. Table 9.4 gives data collected during 2006 national population census on methods of waste disposal by households. Nationally, only about 19 per cent of households enjoyed the services of municipal waste collection personnel and about 28 per cent dumped their waste at illegal dumpsites as compared to 20 per cent that made use of approved dump sites. Altogether 29.5 per cent of households took care of their own wastes by burning, including burning in their compounds. Lagos State

TABLE 9.4 Distribution of Regular Households by Method of Solid Waste Disposal

	State	Total	Method of Solid Waste Disposal (% total no. of households)					
			Collected	Buried by Household	Public Approved Dump Site	Unapproved Dump Site	Burnt by Household	Other
1	Abia	605,987	9.0	11.1	34.4	30.9	12.8	1.7
2	Adamawa	576,348	14.1	8.6	21.2	29.5	24.5	2.2
3	Akwa Ibom	857,436	6.7	21.9	11.9	38.0	19.6	2.0
4	Anambra	882,875	19.0	13.9	19.7	19.0	24.0	4.5
5	Bauchi	847,731	19.2	14.1	19.5	16.7	28.2	2.4
6	Bayelsa	352,025	4.1	4.2	34.8	45.9	6.6	4.5
7	Benue	801,833	9.9	10.3	10.8	38.8	26.7	3.4
8	Borno	787,274	15.4	13.1	24.2	21.3	24.1	1.9
9	Cross River	645,251	5.1	9.9	30.6	42.4	10.7	1.1
10	Delta	890,312	12.7	7.6	16.0	41.6	16.1	6.0
11	Ebonyi	449,709	13.9	8.7	19.4	36.9	17.2	3.8
12	Edo	701,073	18.0	9.9	12.5	32.4	25.5	1.7
13	Ekiti	493,739	6.5	5.4	19.5	40.3	26.7	1.6
14	Enugu	725,767	17.1	10.8	14.4	31.9	21.4	4.4
15	Gombe	419,226	20.0	13.1	20.0	22.0	23.4	1.6
16	Imo	837,195	15.4	14.0	14.4	26.6	25.3	4.2
17	Jigawa	810,310	25.7	13.2	20.0	14.8	22.4	3.9
18	Kaduna	1,115,974	18.2	10.9	24.1	19.2	26.5	1.1
19	Kano	1,603,335	27.0	12.1	21.6	13.9	22.4	3.1
20	Katsina	1,066,316	34.6	12.9	22.2	14.2	13.0	3.1
21	Kebbi	562,827	34.7	11.9	17.8	19.6	12.5	3.4
22	Kogi	641,556	7.1	8.0	19.4	45.8	18.7	1.1
23	Kwara	468,780	7.6	5.0	25.3	43.3	16.4	2.3
24	Lagos	2,195,842	53.4	2.8	22.1	12.6	8.0	1.1
25	Nassarawa	342,711	11.5	10.0	23.9	34.9	18.6	1.1
26	Niger	729,964	10.0	6.9	27.4	38.8	15.4	1.4
27	Ogun	880,970	17.5	4.8	16.7	29.0	30.9	1.1
28	Ondo	763,020	9.4	5.6	14.0	43.9	25.5	1.6
29	Osun	730,313	7.1	5.9	16.8	38.4	30.1	1.8
30	Oyo	1,248,105	10.2	3.8	13.6	43.2	26.9	2.4
31	Plateau	604,491	10.7	9.1	14.8	30.8	33.4	1.3
32	Rivers	1,123,998	9.9	9.0	39.8	30.5	8.8	2.0
33	Sokoto	688,710	34.2	13.3	18.5	19.0	11.3	3.6
34	Taraba	431,385	11.8	9.6	16.3	37.1	21.9	3.1
35	Yobe	418,999	18.7	11.9	23.1	22.8	20.8	2.8
36	Zamfara	592,106	34.3	13.2	19.4	17.0	11.9	4.0
37	FCT Abuja	303,592	24.5	3.5	17.3	42.5	11.7	0.4
	Nigeria	**28,197,085**	**19.3**	**9.6**	**20.4**	**28.2**	**19.9**	**2.5**

Source: Nigeria Census Board: Priority Tables Volume 1 Update, 2006 Census was the only state where over half of the households (53%) had their waste collected by municipal waste collection services. The worst states in terms of waste collection services were Bayelsa (4%), Cross River (5.1%), Ekiti (6.5%), Akwa Ibom (6.7%), Kogi (7.1%), Osun (7.1%) and Kwara (7.6%).

The quality of the waste materials is also very low with 52 per cent comprised of organic wastes that decay, attract flies and give off very bad odours and greenish to brownish leachates. This low quality of the wastes is thus responsible for much of the negative impact which unattended waste accumulations at depots, roadsides and neighbourhoods have on the surrounding areas. In Ibadan City, for example, the bulk of the waste materials in the 1970s consisted of leaves, garbage and the residue of various raw food crops such as yam and cassava peels and maize cobs (Table 9.5a). The situation has not changed much in fifty years except for the increase in volume of plastic wastes. The preponderance of leaves among the solid wastes indicates that the problem is partly a socio-cultural one. The inhabitants of Ibadan and Yoruba towns in general used a lot of leaves in wrapping food sold to the public, but nowadays plastic (cellophane) bags and disposable containers have been substituted for the leaves.

The rubbish dumps are often wet even during the dry season and decay soon sets in. Unfortunately, most refuse collection depots are located in conspicuous positions along the roads for ease of access. The dumps are usually infested with flies which can contaminate food in homes, shops and other public places. It is not surprising that dysentery, diarrhoea and cholera epidemics are more prevalent in crowded cities than in the smaller towns and villages.

The quality of waste materials tends to improve with a rise in the quality of the residential or commercial area as a comparison of Tables 9.5a and 9.5b would show.

Higher quality waste material is generated from high-income residential areas,

TABLE 9.5A Composition of solid wastes by weight in old Ibadan town (%)

Component	Sample 1	Sample 2	Sample 3	Average
Leaves	85.3	70.6	86.0	80.7
Paper	4.4	4.0	1.7	3.4
Food waste	5.2	11.8	10.0	9.0
Tins	1.3	4.3	1.3	2.3
Glass	1.6	0.0	0.0	0.5
Rags	2.0	9.3	1.0	4.1

TABLE 9.5B Composition of solid wastes by weight in residential reservations in Ibadan (%)

Component	Sample 1	Sample 2	Sample 3	Average	Toronto[a]
Leaves	7.2	9.6	12.1	9.6	6.5
Paper	9.8	12.6	7.7	10.0	39.5
Food waste	74.7	70.5	65.8	70.3	32.4
Tins	3.3	4.0	10.4	5.9	5.5
Glass	1.7	2.8	3.0	2.5	8.0
Rags	3.3	0.5	1.0	1.6	1.5
Miscellaneous	0.0	0.0	0.0	0.0	6.6

Source: Maclaren International *et al.* (1971)
[a] *Note*: Toronto for comparison.

institutions and modern commercial districts. However, this is not necessarily a welcome thing as wastes in these neigbourhoods have a higher proportion of plastics than in the poorer neighbourhoods. The polyethylene bags have a way of flying off when blown by winds; they litter and constitute a nuisance in neighbouring compounds and surroundings.

Scrap metal, discarded tyres, beer cans, iron rods and wire mesh are prominent waste materials in certain locations, particularly open spaces such as river banks and undeveloped plots of land in housing estates. The proliferation of illegal waste dumps is evidence of the poor waste collection and disposal arrangements in most cities.

Moreover, most of the gutters or drainage channels in Nigeria's major urban centres reflect the poor waste disposal system. Gutters are clogged with all kinds of materials including leaves, paper, plastic bags and bottles, human waste, cans and so on. Water from bathrooms and urinals filters into the gutters so that the liquid mass flowing through them is heavily polluted. Many gutters are too narrow and shallow for the amount of water and wastes they have to carry. Thus, during every downpour of rain storm water spills out of the gutters littering the streets with waste materials. Even major thoroughfares are covered in litter after every downpour. In some cities, including Ibadan, Kano and Kaduna, the refuse collection vans dump their loads in open pits, usually in some remote part of the city. In order to reduce obvious risks to human health the dumps are occasionally set on fire or covered with earth. Some are walled round to hide them from public view as was done, for example, at the Ring Road refuse dump in Ibadan.

Abuja, Nigeria's new capital city, has managed to avoid some of the environmental problems associated with older major cities in Africa. But, according to Imam *et al.* (2008), the same problems that had bedevilled waste collection and disposal in the older cities were beginning to manifest by the first decade of this century. Table 9.4 shows clearly that as at 2006, 42.5 per cent of households in Abuja dumped their waste at unapproved dump sites and only 24.5 per cent had their waste collected by the municipal waste disposal vans. Rather than take advantage of its newness and fresh start, to develop a modern waste disposal system with waste sorting at source and a well-developed system of waste recycling, this has not been the case. There are technical and operational constraints related to insufficient number of waste collection and disposal vehicles, and use of inappropriate waste collection vehicles. In most cities in Nigeria an assortment of vehicles is used for waste collection and the waste collection personnel employed by most private contractors do not wear any protective gloves or mouth and nose covers. The vehicles themselves loaded with raw waste under tattered jute or tarpaulin covers are eyesores to behold as they ply the city roads.

Dump sites are poorly maintained throughout the country, especially during the rainy season when they are wet and soggy. Waste trucks are wary of getting stuck in the muck and so tend to dump their loads close to the entrances. Occasionally, municipal waste management authorities bring in tractors to level the waste dumps and cover them with earth materials. Dry conditions in the dry season encourage the authorities to set dump sites on fire to reduce the accumulated waste load. Scavengers also frequent the dump sites to recover recyclable materials for personal use or for sale. So do many rodents, birds, and domestic animals, such as, stray dogs and goats, frequent the waste dumps in search of food. Dump sites are also used for open defecation by the common people. The general conditions in the dump sites make them health hazards, especially as they are infested by flies, which can carry germs that can infect visitors to the sites or people living in the vicinity of the dump sites.

Waste dumps are usually sited far from the city because of the health hazards associated with them. But, then, urban sprawl catches up with the waste dumpsites. Hence, waste dumps are forever being closed and relocated. Sometimes the relocation takes far too long to effect, primarily because it is not easy to find suitable locations for waste dumps. Even in a developing country like Nigeria where most governments rule by fiat, local people do resist the siting of waste dumps in their neighbourhoods. Communities refuse to give out land for dump sites. Hence, abandoned quarries are favorite locations for siting waste dumps in many parts of the country. Sometimes, it is the waste collection companies or factory owners that start off the process by secretly and illegally dumping their waste in abandoned quarry pits. Almost every city has these unofficial waste dumps all around.

Slum development and the despoliation of open spaces

Two modes of slum development have already been discussed first in connection with the breakup of traditional family compounds and, secondly, in relation to the mass movement of the rural population to the urban centres. The immigrants concentrate in the low-quality residential areas of each city which are invariably the least planned and the least provided with urban utilities. The congestion of these areas by immigrants only compounds the problem by putting undue pressure on the limited facilities available. The quality of the living environment deteriorates very rapidly. The slum areas generate huge quantities of low-quality waste materials. Streams and open spaces in such areas rapidly lose their scenic beauty and aesthetic value. The slum areas are usually the foci for the spread of epidemic diseases in the urban centres as evidenced by past cholera epidemics in Ibadan, Kano and other cities in the north.

Apart from the slum areas described above, certain kinds of land use tend to degrade the quality of the urban environment. The most notorious are traditional open markets, motor parks, tinkers' workshops, and roadside mechanics' yards. The sites usually generate as much litter as the slums, and since they occupy many of the open spaces in the cities, it is not surprising that these areas have added very little to the scenic beauty and aesthetic quality of our urban environments. In Ibadan, some effort was made to enhance aesthetic quality in the urban environment by relocating automobile mechanics to designated "mechanics villages" where they would not constitute a blight on the city landscape. The mechanics villages were fenced. But the experiment has broken down in some places and mechanics have been forced to move out of the villages to relocate elsewhere. But it is gratifying that those who have moved out relocated to fenced sites reminiscent of the mechanics villages. Those that could not find fenced open spaces have moved to secluded locations away from the major thoroughfares.

Open spaces are poorly preserved in Nigerian cities. In fact, there has been a tendency to destroy every patch of green vegetation without any thought for amenity or aesthetics. On Lagos Island, the wide green field that was once the race-course in front of the Parliament buildings is now occupied by the Lagos Parade Grounds and Shopping Centre, a huge concrete structure. In Ibadan, many of the erstwhile wooded ridges, which gave such a unique beauty to the urban landscape, have been cleared and built upon. The state government made up for this by developing the Agodi Gardens along the Ogunpa River close to the Secretariat as a recreational park. The beautiful landscape and recreational facilities attract many visitors to the park especially at festive seasons and public holidays.

It is important at this point to draw attention to a welcome development in some cities where sections of the river valleys are being saved the indignity of waste accumulation and degradation by horticulturists raising seedlings of ornamental trees, hedge shrubs and fruit trees for sale to homeowners. The plant nurseries are located beside rivers in order to have a ready source of water for watering the seedlings. In Chapter 6, we also referred to the use of valley bottomlands in fadama agriculture. More developments such as these would help promote better use of open spaces and enhance the aesthetic quality of cities

The disregard for open spaces in many cities and towns can be attributed partly to the present low level of economic development in a country which places a high premium on the economic and physical growth of the city than on enhancing its aesthetic value. However, as the economy develops and the aspirations of the people are increasingly realized, there is likely to be a re-awakening of the aesthetic sense and an interest in green belts and open spaces within and around the nation's major cities. This is already happening in some of the large, prosperous cities including Metropolitan Lagos, Abuja, and Port Harcourt. Cities in the northern savanna zone have characteristically paid greater attention to tree planting along streets to ameliorate the harsh local climatic conditions. Hence, they are 'greener' than cities in most other parts of the country.

The future

The foregoing discussions have shown that urbanization of the natural landscape can be looked upon as a complex resource process which has important repercussions on the people and on the urban environment. The ecological consequences of urbanization are not necessarily negative, but in Nigeria the impact has been rather devastating because adequate measures have not been taken to offset emerging environmental hazards. Nigeria's major cities are therefore faced with serious human-induced problems, the solution to which would depend on the pattern of future development. What Mabogunje (1968) noted long ago is still very true even at the beginning of the third decade of the 21st century, that the single most important problem of Nigerian cities is urban management which has remained largely in the hands of the unenlightened and the inexperienced. Few cities had independent governing councils in the past but were treated as part of districts (the domains of specific traditional rulers) whose administration was frequently dominated by officials based in the rural areas. The situation improved with the new local government structure adopted in the country in the mid-1970s. Some of the large urban centres were constituted municipal government areas, and both indigenes and non-indigenes have been able to participate in the administration of these cities.

The areas in which urgent action is needed to minimize the adverse environmental effects of urbanization include land use zoning, taking note of the constraints of the land, on the one hand, and the likely economic returns from the different types of development, on the other. Other needed actions include provision of adequate waste disposal and drainage systems; the establishment and maintenance of open spaces; the formulation and enforcement of regulations on personal hygiene among urban residents and environmental cleanliness of workplaces among tradesmen, artisans and shop/stall holders in the traditional markets. Open defaecation must be eradicated by providing public toilets at strategic locations (e.g. near markets, motor parks) and improving the public water supply and sanitation facilities.

A participatory approach that incorporates the views of local residents at the grassroots level might be a more effective way of dealing with some of the urban environmental problems. With particular reference to flooding phenomenon in Warri metropolis, Odemerho (2014) emphasized the need to take the views and accumulated flood knowledge of the local residents into consideration in developing coping mechanisms to flooding. He recommended a "bottom-up" adaptation strategy in flood risk management that begins with the views and coping strategies proffered by the people at the grassroots level.

10

Perspectives on resource management in Nigeria

As we conclude the discussions in this book, it is important to review the major issues that have arisen in the analysis of the ecology of natural resources in Nigeria and see what the future portends for the management of the natural environment and natural resources in a rapidly changing Nigerian social, economic and political landscape. We have described the major natural resources in Nigeria, their intrinsic characteristics and environmental settings. We have looked at the resource processes associated with each major natural resource. The picture that has emerged is that of a degrading environment due to intense pressure of population on the land and available resources. In the eyes of development planners, the traditional indigenous resource processes that had evolved probably over several centuries have been slow to modernize and embrace new ways of doing things in a deliberate and systematic manner. The revolutionary approach to change is abhorred and, perhaps, rightly so because of the dire consequences on social order and political stability in the country. Thus, the people involved in the resource processes have been coaxed along in the hope that, somehow, they would adopt more and more rational approaches that would ensure a more enlightened management and conservation of the disappearing natural landscapes and dwindling natural resources. But the circumstance now is such that Nigeria needs to give top priority to environmental and resource management. The era of laissez faire approach to development created more problems than it solved.

Need for environmental management in Nigeria

For most of the colonial era and the period of nearly 30 years after political independence, environmental and natural resource management in Nigeria was done primarily on an *ad hoc* basis as successive governments were forced to respond to one environmental crisis after another. Such responses included the creation of River Basin Development Authorities and the Operation Feed the Nation campaign of the 1970s partly in response to the Sudano-Sahelian droughts of 1968-72 and the scarcity of food occasioned by the drought and the civil war. As Fair Observer (2020), noted it was not until the Koko incident of 1988, in which toxic wastes were unscrupulously dumped on the country's shores by an Italian businessman, that environmental protection became an issue of national interest. The event led to the establishment of the Federal Environmental Protection Agency (FEPA) as it made government to be alive to its responsibility in ensuring environmental protection and sustainability in Nigeria.

The Federal Environment Protection Agency was overwhelmed by its mandate and barely succeeded in achieving anything of lasting value in terms of promoting environmental accountability and sustainability on the part of land resource users and managers. Not long after its establishment, the devastation of the oil producing regions of the Niger delta and the resultant violent reactions by elements of the affected communities drew the attention, not only of Nigerian federal and state governments but also of the international community. The situation in the Niger delta and the clamour of the local peoples for action to stem the environmental disaster caused by oil exploration and exploration activities led to the creation of the Niger Delta Environmental Survey (NDES) and, subsequently, the Niger Delta Development Commission (NDDC). Even so, failures and lapses in environmental management have continued to constrain sustainable development in the Niger delta region (Ejimudo and Uwador, 2014). Opinion in the region is more or less unanimous that the Nigerian state and oil conglomerates have a poor sustainable development mentality and action plan to achieve same. The need for environmental and natural resource management and sustainable development is stronger today than ever before. The great concern of international organizations about the sustainability of nature conservation initiatives in the Niger delta region was discussed in

Chapter 8 in connection with the devastation caused by oil pollution in the region.

In the late 1980s and early 1990s, there were other initiatives in biodiversity conservation to stem the pressure of population growth and land use on the few remaining areas of high forest in the country. The highly limited population of game in Nigerian forests including the game reserves was under relentless pressure from hunters and rural dwellers who trap and kill game for consumption. In order to demonstrate its commitment to natural resources conservation the Federal Military Government in 1968 had joined other countries to sign the African Conservation Convention for the Conservation of Nature and Natural Resources (ACCCNNR), Addis Ababa. Nigeria followed with the signing of at least 3 other related agreements, namely, (1) the Convention for Protection of World Cultural and Natural Heritage, Paris, 1972; (2) the Convention on International Trade in Endangered Species of Wild Fauna and Flora, Washington, 1973; and (3) the Convention on Migratory Species of Wild Animals, Bonn, 1979. The last treaty led to the promulgation of the Endangered Species Decree of 1985 for the protection and preservation of wildlife in the country. Various departments of government joined in efforts to protect wildlife in the country; the efforts were fragmented and uncoordinated. However, in 1980, the Treaty on World Conservation Strategy came into being, which, in 1985, propelled the Nigerian government to develop a National Conservation Strategy. The Natural Resources Conservation Council of Nigeria was established by Decree No. 50 of 1989 based on the recommendations in the National Conservation Strategy. The Decree called for a Conservation Action Plan to arrest the rapid degradation and depletion of renewable natural resources. The Natural Resources Conservation Council did produce the Conservation Action Plan and also promoted the community-based approach to the management of conservation areas and their resources.

The implementation of the multifaceted treaties, protocols and action plans faced financial and technical challenges in addition to problems of overlapping mandates and politicization of environmental and resource management issues. International organizations moved in to provide much needed financial and technical aid that yielded valuable base line data and models for the implementation of projects on the sustainable management of the environment and natural resources. Unfortunately, the Natural Resources Conservation Council did not live long before it was merged into something else.

The Federal Government signed many treaties and protocols and entered into agreements on many sustainable environmental and resource projects, such as the NEWMAP discussed in Chapter 6 and the many studies and projects sponsored by FAO, The World Bank, the World Wildlife Fund and IUCN. Eventually, the

Federal Government established the National Environmental Standards and Regulations Enforcement Agency (NESREA), to begin to give much-needed attention to issues of environmental neglect in the country and to guide and promote coordination of actions and activities in environment and natural resource management and conservation. A similar initiative in water resource quality assessment, the Nigerian Standard for Drinking Water Quality (**NSDWQ),** was discussed in Chapter 5.

In the human environment, climate change and environmental pollution and associated health hazards galvanized efforts at environmental management to achieve sustainable cities (Adebowale, 2019). The highly increased interest in urban green spaces and urban forestry is driven by the urgent need to promote sustainable cities in the era of increasing emissions from motor vehicles, factories and waste dumps, not to mention the degradation of urban environments by mechanics workshops, motor parks and traditional market places (Adekola and Ogundipe, 2017). Adekola and Ogundipe (2017) used five criteria to determine the environment-friendly cities among Nigeria's 36 state capital cities and Abuja, the Federal Capital. These criteria were (1) effective and efficient waste management board; (2) non-moribund Ministry of Environment; (3) embracing effective horticulture and greening of the environment; (4) effective and efficient transport agency/authority; and (5) a central and well-managed motor park. Only Lagos metropolis met all five criteria. But 4 other capital cities, Abuja, Minna, Ilorin and Makurdi also scored enough to be classified as being environment-friendly. Urban beautification initiatives in Ibadan city that included the forced movement of motor mechanics workshops to assigned mechanics villages in certain locations, were discussed in Chapter 9.

The United Nations Environment Program and United Nations Human Settlements Programme (UN-Habitat), provided both financial and technical aid in setting up and implementing the Sustainable Cities Nigeria Programme 1994-2006 in Ibadan, Oyo State; Enugu, Enugu State; Kano, Kano State; and Karu, Nassarawa State (Yari, 2008). Assessment of the programme after ten years of operation gave mixed results: while Ibadan had achieved significant results and Kano and Karu made appreciable progress, Enugu was still struggling. The introduction of a geographical information system (GIS) in Kano was most welcome to the government departments. Overall, however, the four cities in the Sustainable Cities Programme were able to achieve some results mainly at the community level, and also developed a positive attitude to urban environmental management among the people and the officials.

From the few projects reviewed, it is clear that Nigeria still has a long way to go in getting its acts together in the area of environmental and natural resources management. The impatience of both local and international advocacy and research organizations and financial and technical aid agencies with the slow pace of change is palpable and understandable but now is not the time to give up. There is hope and the advocacy, research and development aid institutions are a large part of this. Massive amounts of technical data on the environment, natural resources and resource processes have been and are being generated. There is urgent need to collate and correlate data and information on many elements of the environment and resource processes from different parts of the country for development planning purposes. The process has begun; the National Bureau of Statistics for one has modernized its database management system. The government departments and research institutes also are building up databases in the areas of their several mandates. The nagging and seemingly intractable problem has been about how to carry the results of research and demonstration projects to the field, to the users and managers of the natural resources and resource processes down to the grassroots level. The newcomers into the rural economy, for example, in the fields of crop and animal husbandry, silviculture, horticulture, ranching and fisheries are likely to play a pivotal role in promoting new ways of doing things. But these young, literate and enterprising innovators need to be properly engaged by governments at all levels and facilitated by the Nigerian and international financial and technical aid agencies.

Pre-requisites to achieving sustainable environmental management

Over the years, the population of Nigeria has grown very rapidly and is now projected to be over 203 million (2020). This rapid growth has put undue pressure on the land and its resources, and this has been compounded by inadequate production and management techniques. The sustained yield of Nigeria's natural resources and the ability of the resource processes to support the growing population will depend on improvements being made in the following areas:

1. Knowledge and understanding of the natural resources;
2. Reorganization of the land tenure system and land administration;
3. Involvement of the people at the grassroots level in decision-making on all aspects of resource use and development; and
4. New approaches to fostering innovation in various resource processes.

In the remaining sections of this Chapter, we shall discuss each of these issues at length. None of the measures is sufficient in itself and the effectiveness of each depends upon the success of the others, but taken together they do offer a possible way towards overcoming many problems associated with environmental and natural resources management in Nigeria.

Scientific knowledge of Nigeria's natural resources

To a casual observer, it may seem that nothing has changed in Nigeria that things have remained virtually the same whichever area of life one looks at. Yet, positive changes have taken place, which a discerning observer would see as portending well for the future if only the level of governance and administration could be upgraded somewhat. For example, in the area of natural resources, many studies have been carried out in different parts of the country especially with the wider geographical spread of universities and research institutes and the sustained motivation by international organizations offering both financial and technical assistance in natural resource studies. The reason these efforts have not resulted in more enlightened resource management and utilization in the country is due to limited success in carrying the results of research to operators and managers of the natural resource processes throughout the country (FAO, 2018).

In the past, there were serious constraints of paucity of real technical data on which to plan the management and exploitation of Nigeria's vast natural resources. Development planning in Nigeria as in other Third World was described then as 'planning without facts' because of the paucity of technical data, not only on natural resources but also on vital aspects of the social and economic life in those countries. But, these limitations or constraints of data have been pushed back progressively over the past six decades since political independence from Britain in 1960. Modern, sophisticated field and laboratory procedures based on the use of state-of-the-art equipment and analytical processes have increasingly been applied in the study of the Nigerian environment and resources.

Nigeria joined other countries and regions of the world in developing her own space programme, which has yielded dividends in terms of resource mapping and monitoring, information and communications technology and the like. Airborne and orbital space borne imaging systems are being applied in monitoring, analyzing and mapping the changing character of Nigeria's landscapes, natural resources and spatial organization of human society and structures using the techniques of digital image analysis and geographical information systems. Unmanned aerial vehicles (UAVs), air, ground and ocean based geophysical techniques are being used in diverse fields including precision agriculture, wildlife management, flood mapping, and groundwater, mineral ore, oil and gas exploration. Individual businesses and organizations have successfully carried the outcomes of these developments into day-to-day activities in resource exploitation, management and utilization. The oil companies and international agencies with interest in nature and biodiversity conservation have applied the latest resource monitoring and analytical techniques and equipment, particularly, to the Niger delta environment.

All stakeholders in the Nigeria project, including the federal, state and local governments need to learn from these organizations about how to carry research results and modern approaches to the field in order to achieve real, measurable changes in environmental and natural resource management in the country. Fortunately, Nigeria is learning to collate data on different aspects of its economy and resource base such that the data can be made available to users for various purposes. For example, the former Federal Office of Statistics has metamorphosed into the National Bureau of Statistics having computerized its operations and developed a database management system that allows users ready access to valuable data through the internet and the World Wide Web. The river basin development authorities (RBDAs) have developed geographical information systems to collect, manipulate, analyse, model and map data in order to plan and manage their operations in river basin and rural development. Many other departments at both the federal and state levels are also moving in the same direction, especially as the national assembly and federal government have been promoting e-government.

In spite of these welcome developments, there is still a long way to go to make access to data and information as free and as smooth as possible. Even more important is the task of carrying the outcomes of research to impact positively on environmental and natural resource management and utilization in the country. Intense agricultural extension services and agricultural promotion campaigns are needed to get the people and state and local government to adopt innovation from the research institutes and international organizations.

Moreover, much of the data collection had been carried out at the reconnaissance level so that for each resource the density of data, the number of parameters measured, the mode and units of measurement may not be very suitable for detailed planning purposes. Furthermore, much of the available data have not been fully analyzed or synthesized. But the situation varies from one natural resource element to another since some have received greater attention and have been more amenable to study and evaluate than others.

Climate

In comparison with other elements of the natural environment there is a considerable body of data on the climatic parameters in Nigeria. The length and quality of the records vary from place to place but weather stations with fairly reliable and up-to-date records are fairly well distributed throughout the country. Such weather recording stations are located at airports, meteorological stations, universities, and agricultural research institutes like the ones in Samaru, Umudike and Moor Plantation Ibadan. Others are sited at the district stations of the ministries of agriculture and natural resources; sea and river ports; municipal waterworks; and river basin development authority project sites. Most secondary schools have weather stations but very few of them keep up-to-date records; usually there are breaks in record keeping during the school holidays.

The weather stations in Nigeria had been grouped into four categories by Ayoade (1971) as follows:

1. Synoptic stations sited at the Meteorological Office stations, airports and universities, and supervised by full-time professional observers who maintain continuous weather watch and take hourly instrumental measurements or readings. Temperature, humidity, pressure, rainfall, duration of sunshine, cloud cover, wind speed and direction, evaporation, radiation and soil temperature are observed and recorded. These standard and top-grade weather stations account for less than five per cent of the total number of weather stations in Nigeria.

2. Agricultural stations, manned by part-time observers who make twice daily instrumental observations of temperature, rainfall, humidity, wind speed and direction. Some stations also measure radiation, evaporation and soil temperature. Some of these parameters are autographically recorded.

3. Climatological stations, manned by part-time observers who make once or twice daily instrumental observations of temperature, humidity and rainfall.

4. Rainfall stations, manned by part-time observers who take readings of rainfall only. These account for more than 85 per cent of the weather stations in Nigeria.

The gauge density of Nigeria, at 1 per 863km^2 of territory, is said to compare favourably with the minimum density of 1 to 600 - 900 km^2, suggested by the World Meteorological Office but is far less than that of West European countries which have densities of one rain gauge to between 50 and 85km^2 of land.

Climatic records are more easily obtained than most other kinds. Apart from the fact that they can be collected at the recording stations themselves the Meteorological Office in Lagos is the repository for all climatic records in the country and is a ready source of data. Some recording stations such as the Department of Geography, University of Ibadan, publish their daily and monthly records of climatic observations but at irregular intervals. Unfortunately, of all the climatic parameters, only the data on rainfall and, to some extent, temperature, have been subjected to detailed statistical analysis. This partly reflects the availability of data. Data on evaporation, radiation and sunshine hours exist for only a few stations and values for most parts of the country are no more than extrapolations from these few stations. However, a number of research studies exist based on mathematical manipulations of rainfall and temperature data or on instrumentation to measure certain climatic parameters. Such studies include those of evapotranspiration and the water balance (Ayoade, 1976) and of radiation flux over different land use surfaces (Oguntoyinbo, 1974).

But the world has moved beyond the use of archival climatic statistical data to produce series upon series of isoline maps of climatic elements. The availability of orbiting geostationary meteorological (e.g. Moderate Resolution Imaging Spectroradiometer (MODIS)) and other satellites continuously capturing and transmitting data to receiving stations on the earth's surface has made it possible to analyse and model atmospheric processes and predict future trends in weather, climate and water resources, among other things. From these sophisticated space imaging and recording systems, atmospheric data can now be combined with vegetation data for modelling and prediction purposes in agriculture, forestry, nature reserve management, fire and wildlife monitoring, and climate change studies. Nigeria is behind the countries of Eastern and Southern Africa in developing and applying these capabilities in environmental and natural resource management. Yet, climatic change has the potential to negatively impact almost every facet of life and livelihood in the country ((Idowu *et al.*, 2011; Ebele and Emodi, 2016; Ikpe *et al.*, 2017).

Climate change

In spite of protestations to the contrary in certain countries and governmental circles, climate change is a fact of life in contemporary 21st century environment. Its effects are felt in every area of human endeavour. In Nigeria, the recurrent droughts in the Sahel/Sudan savanna zones since 1968 till the present; the shifting sand dunes encroaching on and gulping up settlements and farmlands in these regions; and the near total disappearance of the Lake Chad from the surface of the earth are some of the more glaring signs of climate change in Nigeria. But there are many more evidences in the country. Reduced annual rainfall totals, late onset of the rains in the wet season, reduced number of rain days, shortened length of rainy season and high temporal and spatial variability of rainfall are causing farmers to adapt to the reality of climate change. Ikpe *et al.* (2017) found that changing weather and climatic conditions in the Sudan zones have favoured millet cultivation at the expense of guinea corn (sorghum); and affect "all stages of agricultural production from land clearance and preparation through crop growth and management to harvesting, storage, transportation and marketing of agricultural products." The frantic and vicious scramble for grazing land by the nomadic herdsmen and their militia backers can be attributed partly to the adverse effects of climate change in the northern savanna lands. Flash floods have become more frequent and widespread in the country and the losses in life and property have escalated over the years. The risk of dam bursts has increased as a result of flash floods and the build-up of sediments against river dams.

There have many studies on weather and climatic indices of climate change and on the local people's adaptations to mitigate the impacts of climate change. But Nigeria needs to go further to model climate change and its possible long-term effects not only in cropping systems, but also in forestry, wildlife management, water resources and other sectors of the economy. Nigeria's satellite system promises to facilitate the continuous monitoring of the impacts of climate change on a long-term basis. Improved service delivery by the Nigerian Meteorological Agency and availability of data from weather satellites should facilitate climate change modelling and prediction

Surface water

River basin studies were first embarked upon during the colonial period in connection with river transport particularly on the Niger, the Benue and the 'oil rivers' of the delta. Investigations were geared towards measuring the depth of water and tracing the positions of shifting shoals along the rivers and their estuaries. Later, further studies were carried out for road and railway bridge construction, and up till today some river gauging stations are located at bridges. The regularity with which readings were made at these points varied and some of the earlier gauging stations have not been maintained. Furthermore, in some stations only records of peak flows were kept rather than a continuous all-year reading of the water level. The gauging stations were concentrated on the major river basins including, among others, the Ogun, Osun, Sokoto, Kaduna, Gurara, and Cross river basins.

In general, because of the sub-humid climatic conditions and because of the crucial role of water in the life of the savannazones, greater strides had been made in hydrometeorological studies in the northern than in the more humid southern parts of Nigeria. Later the policy and practice shifted towards integrated and comprehensive development studies of entire river basins. Examples of well documented river systems are the Kaduna, the Sokoto-Rima and the Hadejia. Measurements of river discharge, the water level and sediment yield were also kept. Water samples were taken for laboratory analysis.

In other parts of the country spot hydrological studies of small or medium sized river basins were usually carried out in connection with domestic water supply schemes. These studies were usually carried out without much preparation and over very short periods of time. In most cases once the dams had been built and the reservoirs filled, little or no further hydrological data were recorded. Reservoirs located in the more remote areas (like Owo reservoir on the Osse River in Ondo State), had no regular attendants at all. It is only on the bigger schemes such as Asejire or Kainji that ecological and hydrological conditions have continued to be monitored.

The creation of river basin development authorities (RBDAs) that covered all the river systems in the country has brought some much-needed improvement in the measurement and documentation of the climatic elements and water resources in many more river basins (figs. 5.8 & 5.10). It is difficult to know how much priority the RBDAs have given to this aspect of their mandate. What is clear is that the RBDAs have helped in spreading systematic scientific studies of the landscape and its features over much of the country's territory. They are helping in building up continuous time series data on large parts of the country. But there is the need for the Federal Department of Water Resources to collate the data and develop a database management system and geographical information system to make data and information open to the general public.

Rocks and minerals

The search for minerals led to the first geological explorations in Nigeria. For example, as early as 1905 a 'mineral survey' of southern Nigeria was carried out by the colonial government (Dunston, 1910), followed by that of northern Nigeria in 1907 (Dunston, 1911). The mineral-rich and the potentially mineral-rich areas were the earliest to be surveyed in any detail among them the Jos Plateau tinfields (Falconer, 1921, 1926); the Enugu coalfields (Bain *et al.*, 1927), Calabar district (Raeburn, 1927), and the tinfields of Zaria and Kano Provinces (Raeburn *et al.*, 1927). Individual prospectors also searched for minerals and there were studies of other areas too including the Younger Granites of the Jos area (Jacobson, 1947) and the sedimentary belt of southern Nigeria (Parkinson, 1907). The geology of both the western and eastern railways were described and reported in the bulletins of the Geological Survey of Nigeria (Wilson and Bain 1922, 1925; Wilson, 1925). The reports gave a broad picture of the succession of geological formations and physiographic units from the coast to the northernmost parts of the country.

At first geological information was obtained from observations of rock outcrops, or of weathering profiles along gullies and river channels. Also, a good deal was inferred from observations of land forms and structural trends. Later the method of drilling bore holes was adopted. This had been particularly useful in studying the stratigraphy of the sedimentary rocks and of the deep weathering profile on the Basement Complex rocks. Such boreholes also yielded data and information on the depth of the water table and the sources of underground water. Detailed descriptions of these boreholes were kept in the files of the Geological Survey of Nigeria (now known as Nigeria Geological Survey Agency). Many borehole records were obtained from the wells dug by commercial and industrial firms in their premises to ensure adequate water supply. Such wells are common, for example, in the industrial areas of metropolitan Lagos.

In more recent times, the oil companies have been responsible for much of the geological survey of the coastal and deltaic areas. They were among the first to employ air photo-interpretation on a large scale in geological exploration and mapping in Nigeria. Other commercial firms such as Hunting Surveys, carried out a lot of photo-geological studies of large pockets of the country most notably in the Chad basin. Nowadays, modern techniques and equipment are increasingly being employed. Thus, apart from air photogeological interpretation, digital analysis of more sophisticated remotely sensed satellite images, electrical resistivity and seismic surveys and gravity measurements are being carried out especially in the Basement Complex country. Also, from time to time the Nigerian Geological Survey Agency received technical assistance from overseas. In this regard airborne geodetic survey techniques and radar imaging systems were introduced not only in mineral prospecting but also in geological and hydro-geological explorations.

The Nigerian Geological Survey Agency is now producing digital geological maps. It produces seven types of maps to meet the needs of the Nigerian economy, namely, geological sheet maps at 1:100,000 and 1:250,000; mineral resources maps; geological and mineral resources maps of each state in Nigeria; geochemical maps showing the chemical concentrations of metals of industrial value and precious metals (gold, silver, copper, aluminium, cerium); gravity and aeromagnetic maps. But it has also begun to produce geological maps at 1:50,000 to match the topographic map series. Its website indicates that between 2005 and 2013 178 topographic sheets on a scale of 1:50,000 were mapped. Perhaps, the most significant development is the fact that the Nigerian Geological Survey Agency is building up digital databases including a mineral occurrences database and a core log database that contains subsurface data on the basement complex rocks and the major sedimentary basins in Nigeria. It also has four specialized centres that promise to yield or collate very valuable data for environmental monitoring purposes. They are (1)

National Geohazards Monitoring Centre; (2) National Geoscience Research Laboratory; (3) National Centre for Marine Geosciences; and (4) National Geosciences Information Centre.

The Nigerian Geological Survey Agency has one unique feature. It controls and coordinates all geological/geophysical survey work in the country: the same techniques and classification schemes are used on all survey maps. This is unlike soil surveying, for example, which is carried out by different state-based organizations, each with its own techniques and system of classification.

Soils

Possibly no other natural resource in Nigeria has received as much attention as the soil, and yet our knowledge of soils is still comparatively limited and most soil maps are of the small-scale reconnaissance type. The paucity of detailed information is reflected in the scarcity of soil maps at the local level. But there is high hope that things will change for the better. This optimism is hinged on certain welcome recent developments. First, we have mentioned above the role of the RBDAs in improving knowledge of soils and the environment in general over most of the country. Secondly, the young and enterprising Nigeria Institute of Soil Science (NISS) is working to partner with the Federal Ministry of Agriculture and Rural Development (FMARD) to carry out a semi detailed soil survey of Nigeria as well as develop and maintain a functional soil information system in collaboration with the Federal Department of Lands and Climate Change Services. Thirdly, and by far the most exciting is the on-going GlobalSoilMap.net Project, being funded by the Melinda and Bill Gate Foundation and aimed at global digital mapping of soils, from which Nigeria stands to benefit a great deal. According to Odeh (2012) the project objectives are (1) to produce spatial predictions of key soil properties at continuous depth intervals at a spatial resolution of 100m for the entire world; (2) derive management recommendations to the stakeholders; and (3) link the GlobalSoilMap.net information to the end-user through cyber infrastructure. In preparation for the Nigeria sector of the project Odeh (2012) has looked at past efforts at soil survey and mapping in the country and given a very good summary of what has been achieved. This is because existing records, data and maps (termed "legacy soil data or maps") would serve as major sources of data for the global soil mapping project. Odeh (2012 *op. cit.*) listed the diverse institutions and organizations that had been involved in soil survey and mapping in the country as follows:

1. Departments within a number of Federal Ministries- variably include Agriculture, Environment and Natural Resources - examples are the FDALR (Fed-

eral Department of Agricultural Land Resources) and National Program for Food Security—both now within the Federal Ministry of Agriculture;

2. Overseas Development Aid Organizations such as the British-owned Land Resource Division of the Overseas Development Administration and USDA;

3. Former Regional Government Ministries of Agriculture and affiliated Institutions (e.g. Institutes of Agricultural Research at Samaru, near Zaria; Moor Plantation, Ibadan; and Umudike, near Umuahia);

4. River Basin Development Authorities of which there are currently 12;

5. Government-funded Research Institutions within or outside Universities;

6. Individuals within Faculties of Agricultures in various Universities and Colleges of Agriculture;

7. Private sector—large farm projects; reports indicate that only a limited number in this category have some form of soil profiles data

The activities of some of these institutions have been described in Chapter 6. Table 10.1 gives a summary of the extent of soil mapping in Nigeria through decades of effort by the various groups itemized above (see also fig. 10.2). The prospects are high that soil survey and mapping in the country will be approached in a more systematic and results oriented manner. The need is particularly great and urgent to increase the coverage of detailed soil surveys and mapping in the country to aid development planning. The collation and computerization of all existing soil data would inexorably lead to the development of a soil information system for the whole country in addition to whatever is available at the global level. This would be good for soil correlation and dissemination of information from one part of the country to another. The issue of a systematic soil classification scheme for the country should not stand in the way of the imminent progress in soil survey and mapping in the country. The United Nations Food and Agriculture Organization (FAO) has contributed immensely to the understanding of soils and their geographical distribution in Nigeria. The FAO World Reference Base for Soil Resources (WRB) soil classification scheme has been easier to understand and apply to Nigerian soils than the United States Soil Taxonomy. It should suffice as the basis for future soil survey and mapping activity in Nigeria. The legends of existing soil maps are easy to translate and correlate with the WRB scheme.

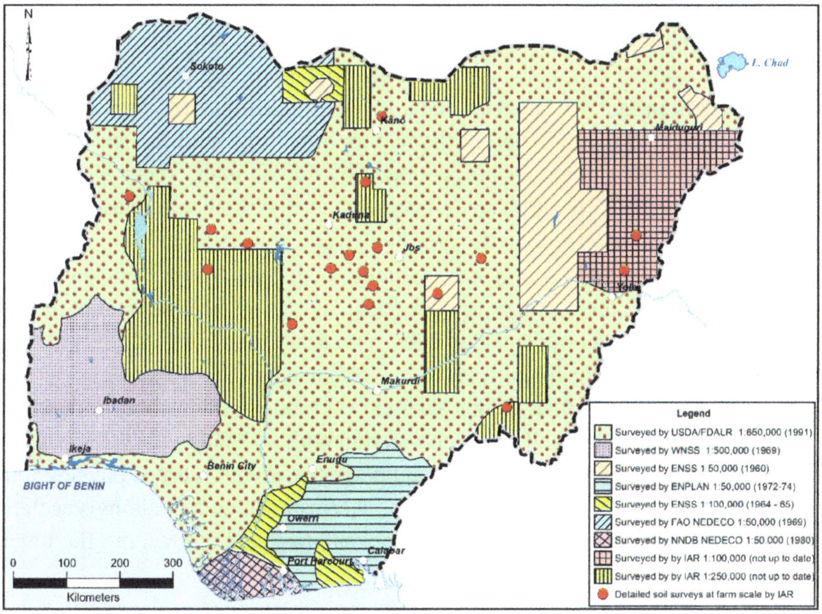

FIGURE 10.1 Extent of legacy soil mapping in Nigeria (after Odeh, 2012)

TABLE 10.1 Legacy of soil surveys in Nigeria: Proportion of soil surveys

Type (scale) of survey	% coverage
Reconnaissance	33
Very detailed	29
Exploratory	16
Semi-detailed	15
Detailed	6
Compilation of exploratory	1

Source: Extracted from Inakwu O.A. Odeh: The Legacy of Soil Mapping in Nigeria: from analogue to digital soil mapping. *Prepared for DES Seminar Series June 29 2012, Department of Environmental Science, Faculty of Agriculture and Environment, University of Sydney*

Vegetation

The broad pattern of vegetation distribution in Nigeria is rather simple following closely the major climatic belts. But this broad pattern conceals a great deal of internal variation. In fact, the great diversity of floristic associations which has made classification very difficult is one of the factors militating against vegetation mapping in Nigeria. The situation is compounded by human interference with the natural environment and vegetation so that anthropic vegetation communities are now more varied and diversified than the original natural communities.

However, the paucity of detailed vegetation maps should not be taken to mean that there had been no vegetation or ecological studies in the country. On the contrary since the colonial era there had been numerous small-scale studies of vegetation relating to specific forest reserves, geological formations or physiographic landscape types (Ross, 1954; Richards, 1957). These studies varied in depth and some, especially those carried out in connection with soil and land system surveys, were no more than a listing of the dominant plant species and a broad physiognomic classification of the vegetation. Some areas were shown as containing vegetation complexes because of the difficulty of differentiating and delineating the diverse vegetation types (Tuley and Alford, 1975).

Studies carried out by forestry divisions mostly involved evaluating the major economic or merchantable tree species and the soil and climatic conditions under which they grew. More detailed empirical studies were carried out by researchers in the universities and research foundations. The ecological impact of annual bush burning on forest and savanna was widely studied (Hopkins, 1975, Charter and Keay, 1960). The most thoroughly studied of the vegetation types had been the savanna both because of its grazing value and because it attracted attention by its rapid expansion at the expense of the forest (Adejuwon, 1968, 1971). Scientists at the Shika Agricultural Research Station, Ahmadu Bello University were in the forefront in the evaluation studies of the grazing value of Nigeria's savanna communities (de Leeuw, 1976).

The availability of remotely sensed images since 1972 has made it easier to study and map vegetation and land use in Nigeria. The Federal Department of Forestry also commissioned the Nigeria Radar (NIRAD) project that resulted in the complete coverage of the country with radar images. The radar images combined with aerial photographs made it possible to classify and map vegetation and land use at the national level. The usefulness of the vegetation and land use maps was limited by the lack of a systematic scheme of vegetation and land use classification. However, the radar images were also useful in geological, physiographic and hydrographic surveys and mapping.

In the twenty-first century, the country owes much to the financial and technical assistance of international organizations and research teams for the survey and evaluation of the biodiversity and resource potentials of Nigeria's ecological zones. Some of these studies were mentioned in Chapter 4. Most of the studies have been to determine (1) the productivity of the forests in terms of timber yield; (2) the ecological status of the ecoregions in terms of ecosystem and species biodiversity; and (3) the conservation status of animal species found in these locations. The international organizations have been very active in advocacy campaigns and in promoting the creation of nature reserves, wildlife sanctuaries and similar conservation areas.

Future data collection and management strategy

If Nigeria is to establish a firm foundation for sustainable development based on solid databases on the environment and natural resources, it would be very necessary to resuscitate or revitalize the relevant government departments at the national level which have the mandate to oversee the different natural resources. They include the Federal Ministry of Agriculture and Rural Development and its agencies such as the Federal Department of Forestry, Federal Livestock Department, Department of Fisheries and the like. Important units established and operating at the regional level must also be involved including the Institutes of Agricultural Research at Samaru, Umudike, and Moor Plantation, Ibadan, which are largely responsible for the survey and mapping of soils. The Nigerian Geological Survey Agency, the Federal Ministry of Water Resources and its agencies such as the RBDAs are also involved. They must all be developed to start playing the type of roles played, for example, by the United States Geological Service, the Forest Service and the USDA Soil Conservation Service in America.

Each of the resource units must become a repository of all data and information on the areas of their mandate. It is important that each unit establish a database management system for each resource element under its purview linked to a geographical information system as a decision support system in problem solving and development planning. In order to facilitate the development of such database management and information systems on each resource element, each unit must develop a dynamic scheme for the classification and mapping of each element of the environment and natural resources

There can be no significant progress in natural environment and resource management in Nigeria where there are no agreed rational schemes for classifying and mapping the different elements of the resource base. Without such a classification scheme it would be difficult to correlate and combine data from different sources and different parts of the country for effective planning either at the national, state or local government levels. In order to succeed these government and quasi-government establishments must be open to collaboration among themselves and with professional and academic associations in relevant fields. Unlike the situation in the past, these government departments and their officials must be more receptive to the idea of working with each other and with relevant professional bodies and experts outside government. They must be less obstructive to new initiatives by academics, professionals and professional associations, all in the name of protecting jurisdiction and mandates.

In order to derive maximum benefits from the new arrangements the resource units must depend less on *ad hoc* resource studies and consultants for the survey and mapping of the natural environment and resources. There are competent national and international institutions with solid foundations and significant footprints in the field of environmental studies that can provide much needed technical assistance. Finally, these units do not need to look far in search of geospatial technologies to apply in the survey and mapping of natural resources. As Areola (2012) noted Nigeria has not fully utilized the capabilities of existing readily available geospatial technologies, including aerial photography, satellite imagery (e.g. LANDSAT, SPOT), Radar, digital image processing and analysis and geographical information systems capabilities. Like the waste dumps of discarded desktop computer terminals and monitors found in some cities, Nigeria is like a junkyard of inadequately mastered and utilized geospatial technologies. Individual professionals and practitioners have benefited but the country at large has benefited much less. Many practitioners, both qualified and not so qualified, are doing a lot of work for clients both in the public and private sectors but the outputs are of variable quality. Quality control is very poor. The lack of departments comparable to the USGS and the US Census Bureau or the Ordnance Surveys in Britain, to provide standard databases, base maps and other products is largely responsible for such poor-quality control among practitioners in geospatial information systems. The newly revitalized government departments and agencies such as for example the Nigerian Geological Survey Agency, the Federal Department of Agricultural Land Resources, the Federal Department of Water Resources, should play that role once they are firmly established

Areola (2012 *op. cit.*) also noted that new technologies cannot help any country where access to the data and derived products is highly restricted either directly through denial of access or indirectly through the charging of exorbitant prices to gain access. The countries that have benefited immensely from geospatial technologies are those which have facilitated access to the products and which also actively promote their use by the general public, apart from educational, public and private sector institutions. Free flow of data and information is necessary for Nigeria to derive maximum benefits from the new technologies being applied in resource development and development planning.

Land tenure systems and land reform

A land use decree (later changed to Land Use Act) was promulgated in 1978 by the Federal Military government, which vested all land in the federal and state governments to hold in trust for the people. The intention of government was to put a halt to land speculation; regularize and ensure equitable allocation of land for both rural and urban needs; and facilitate the acquisition of land for government projects. While the provisions took hold in major urban centres, the traditional land tenure systems have continued to hold sway throughout the country. Even, in the large urban centres, traditional land tenure operates at the primary level of land acquisition by individuals. An individual who has purchased land then applies to government for a certificate of occupancy ('C of O') to satisfy the provision of the Land Use Act! Therefore, not much has changed in terms of traditional land transactions and their numerous shortcomings.

The essential features of the land tenure systems in Nigeria were described in detail long ago by Adegboye (1966) and Oluwasanmi (1966) among others. Basically there are two major land tenure systems in the country: (1) *customary tenure* practised in the southern states which vests ownership of the land in the family, clan or community whose individual members have the right to the use of the land held by them, for any purpose, for as long as they effectively occupy it; and (2) the land tenure system in the northern states based on the *Land Tenure Law of the Northern Region, 1962*, which gives an individual only usufructuary right or 'right of occupancy' but not the right of ownership, which is vested in the minister or commissioner for land affairs (Udo , 1977).

It has been said that the system in the north is better and more organized than that in the south which has caused so much confusion in both rural and urban areas. Even so, for lack of adequate supervision, the northern system has been abused in many areas notably near the large urban centres where many people sell or transfer the land allocated to them without the commissioner's consent.

Of greater relevance to this discussion is the impact which the land tenure systems have had on the use of land and natural resources. A system (especially as operated in the south) which only gives the individual right of use but not proprietary rights to the land, dampens initiative and discourages the individual from investing in the long-term development of the land. Also, it creates a type of mentality which makes the individual want to extract as much as possible from the land during the period he occupies it without due regard to the danger that might be done. It is the sharecropper's mentality described in Chapter 6 in respect of rubber tapping.

Another aspect of the land tenure systems is that the landowners lay claim to everything in and on the land. These resources can be exploited and disposed of at the will of the landowners; there is no mechanism for controlling what they do with the land and resources. This has resulted in indiscriminate exploitation and wastefulness in the cultivation of farmlands and exploitation of earth materials, forests and wild game.

Customary land tenure entitles a member of the family, clan or community to as much land as he needs to make him self-sufficient, at least in food production. In practice, this has meant that the individual gets a share of every piece of land owned by the group. Thus, it is not uncommon for an individual to have from two to four farms located in different places. This dispersal of farm holdings does not make for efficient husbandry. The holdings hardly constitute viable economic production units. This pattern of land holding is responsible for the perpetuation of the subsistence rural economy that stultifies efforts at specialization in the systems of production.

Land reform properly carried out should improve the planning and control of rural land, and ensure the best use of natural resources. Such controls, effectively used, could also facilitate the introduction of new production techniques.

The Land Use Act, 1978 and Land Reform Policy, 1999

When the land use act was promulgated in 1978 as Land Use Decree by the military administration, there was optimism that it would impose some order in rural and urban land use. Adalemo (1978) listed the following possible benefits that could accrue from this decree if it were properly implemented: (1) an end to land disputes, land speculation and arbitrary inflation of land values; (2) orderly development of urban areas; (3) access to agricultural land for those actually engaged in agricultural production with the possibility of consolidating fragmented farm plots and the encouragement of large-scale agricultural enterprises; (4) prevention of developments incompatible with the natural attributes of the land and maintenance of the intrinsic quality of the total environment.

Although there was no doubt about the urgent need for land reform in Nigeria, lack of political will and executive capacity on the part of government at all levels and the lukewarm attitude and lack of support of the people, has meant that the decree to all intents and purposes ended up on the scrap heap, along with other still born attempts at tenancy reform in various parts of the Third World. Indeed, in the first few years of its operation, several interest groups had called either for modifications in the provisions of the decree or for its outright repeal. Every attempt to modify the provisions of the Act since the return to civilian rule in 1999 has failed as members of the national assembly have not been able to reach an agreement on terms that they could sell to the general public and the traditional rulers in particular. As King (1977) noted one of the great ironies of land reform was that it was most needed in those developing countries which were least able to carry it through. This was not only because of a lack of funds and management skills but also because the dynamic social and political movements necessary to induce and sustain the reforms were either lacking or very weak.

The civilian administration that succeeded the military regime in 1999 had sidestepped the Land Use Act by putting up a Land Reform Policy that addressed some of the people's concerns. The Land Use Act had more or less dispossessed families and communities of their land and ignored the authority of traditional rulers in land matters, a sore point, particularly in the southern parts of the country. The failure of the Land Use Act to recognize and accommodate the traditional and customary basis of land ownership has been the major cause of its ineffectiveness in promoting land reform in the country. To the common man the Land Use Act "appropriated" the people's lands and "nationalized" them thus depriving them of their most treasured possession.

Thus, in spite of the land use act, land transactions at the community and individual family levels have proceeded unchecked. This is how newcomers to the rural economy have been accommodated. Landowners in the rural areas are selling land to individuals, companies and industrial concerns for various purposes. Indeed, a novel introduction to the rural economy in the 21st century is the emergence of agricultural business entrepreneurs. They acquire extensive areas of land for modern mechanized farming and then advertise for individuals to invest by paying for farm plots. The investor gets the proceeds from the sale of the crops grown in each plot after deducting production costs. The crops vary from vegetables to fruits and grain crops that yield dividend annually. There are similar arrangements for those interested in animal husbandry including poultry, piggery, fishery, snails, and ruminants.

Apart from the agribusiness entrepreneurs described above, many other people, especially urban dwellers, have gone into the rural hinterlands to lease or purchase land from the local inhabitants to grow crops on a commercial scale. Cassava and maize are the most favoured crops grown. There is still another group of entrepreneurs who go about buying crops in the field and harvesting them to process for sale in the urban markets. These entrepreneurs usually focus on such crops as cassava which is processed into garri; yam, processed into yam flour; and maize, sold either green or shelled and dried. Some people have established plantations of oil palms, cashew, citrus, mango, and exotic trees such as teak (*Tectona grandis*) and gmelina (*Gmelina arborea*).

The Land Reform Policy, 1999, to all intents and purposes, recognized and accommodated these developments in land ownership and transactions. It recognized "the importance of customary and ancestral lands and the uniqueness of people's primordial cultures and belief systems in land ownership" (Nkwunonwo, 2020). But, significantly, it sought to promote and confer individual property rights rather than the erstwhile communal tenure, to the present generation of users and tillers of the land. Furthermore, it took steps to facilitate the inventory and documentation of every parcel of land in every nook and cranny of the country as a first and necessary step towards poverty reduction and eradication at the grassroots level (Adeniyi, 2013). Such development would facilitate land administration and make it easier, for example, for individuals to use their land holdings as collateral for bank loans and similar transactions.

Thus, the Federal Government set up the Presidential Technical Committee on Land Reforms to: (i) collaborate and provide technical assistance to States and Local Governments to undertake land cadastral nationwide; (ii) determine individuals' "possessory" rights using best practices and most appropriate technology to determine the process of identification of locations and registration of title holdings; (iii) ensure that land cadastral boundaries and title holdings are demarcated in such a way that community, hamlet, village, village area, town, etc. will be recognizable; (iv) Encourage and assist States and Local Governments to establish an arbitration mechanism for land ownership conflict resolution; (v) make recommendations for the establishment of a National Depository for Land Title Holdings and Records in all States of the Federation and the Federal Capital Territory; (vi) make recommendations for the establishment of a mechanism for land valuation in both urban and rural areas in all parts of the Federation; and; (vii) make any other recommendations that will ensure effective, simplified, sustained and successful land administration in Nigeria. In furtherance of its mandate, the committee began to install Continuously Operating Referencing Stations (CORS), which are key cadastral infrastructure, in all states of the federation where such are not available starting with Ondo, Kano, Katsina and Imo States. In order to put in place a more transparent, accountable and timely land administration process, a feasibility study for the adoption and customization of FAO's Solutions for Open Land Administration System (SOLA) was carried out by the Committee with the aim to computerize land registration and other land administration activities and make them affordable and sustainable (Adeniyi, 2013).

It remains to be seen whether the foundation of land registration and land administration being laid by this technical committee would be sustained as governments change from time to time. Lack of continuity has been the bane of many a government initiative in Nigeria.

Involving the rural population in environmental and natural resource development programmes

As we observed in the first chapter of this book, it is most important to consider the issue of poverty and its consequences in any intervention measures that may be proffered to improve land and natural resource management in Nigeria. The problems facing environmental and natural resources management in Nigeria require that policy makers and development agencies focus more on what is happening at the grassroots level. Indeed, this was the essence of the MDGs and the successor SDGs. Although, we have noted above the increasing interest and participation of educated and business-oriented entrepreneurs in agricultural enterprise in the rural areas, they are still too few to replace the millions of traditional rural farmers and pastoralists, who will certainly still be producing a large proportion of the food and raw materials required by the urban-industrial sector for the foreseeable future.

The operators at the grassroots level are moved less by appeals for the preservation of the productive capacity and the aesthetic quality of the environment than by campaigns that focus on the provision of the basic necessities of life and survival in an increasingly hostile urban-dominated social and economic system. Indeed, it would make great sense to weave strategies for achieving sustainable natural environments and natural resources into poverty alleviation programmes at the grassroots level by adopting community-based participatory approaches to sustainable rural development. The concept of Integrated Natural Resources Management (INRM) emerged at the beginning of the 21st century, as a way of improving agricultural productivity in a sustainable manner in order to maintain the integrity of the environment while at the same time alleviating poverty. According to Hagmann *et al.* (2002), the Consultative Group for International Agricultural Research (CGIAR-INRM-Group)' defined INRM as "the responsible and broad-based management of the land, water, forest and biological resources base (including genes) needed to sustain agricultural productivity and avert degradation of potential productivity." The concept of INRM came into being because of the failure of an earlier approach, the farm systems research (FSR) approach, which did not make the local people co-participants in project development and implementation processes.

INRM emphasizes community-based participatory approach in development planning and implementation.

The philosophy of natural resource development programmes is that the local people will not only be involved in project formulation, planning and implementation but they will also be the first beneficiaries of the proceeds of such development projects. The concept and philosophy have been accepted and recommended for application not only in agriculture but also in forestry and wildlife management. In wildlife management, local hunters have been incorporated into the programme of protecting wild game against poaching or illegal game harvesting. Local hunters protect their areas of operation against intrusion by hunters and poachers from outside and they follow restrictions on the types and ages of animals they can hunt in forest reserves. They only pay a small proportion of their proceeds as fees to the managers of the reserves.

Dealing with the millions of operators in farming may require more planning and work than dealing with hunters in a relatively small number of forest or game reserves. Farming requires more regimentation than hunting. In fact, the crucial problem to overcome may not be that of finding the right alternative cropping and farm management approaches but that of getting them accepted by the rural inhabitants and convincing them that they stood to benefit by adopting them. A feature of many of the new cropping practices which has often made them unpopular is the very rigid schedule that has to be adhered to, for example, in applying irrigation water and chemical fertilizers or in spraying herbicides and insecticides on the farms. Failure to keep up with the proper scheduling of farm operations usually has the effect of reducing the impact which some of the new techniques have on crop yields. However, if the outcomes are good and profitable to the people, that would be an incentive to them to persevere.

But, some of the modern measures can only be truly effective if structural changes in rural land use are made. For instance, the small production units that are characteristic of most resource processes in rural Nigeria would make mechanization rather expensive and less efficient than if the production units were much larger. Indeed, high costs relative to expected returns may discourage farmers from adopting new techniques or from following guidelines exactly. It is common knowledge that many cocoa farmers do not spray their crops the recommended number of times during the year due to the cost of fungicides. Many farmers also dilute the herbicides and pesticides with water to make them cover more ground. The consequence of these cost cutting gimmicks is that crops continue to be lost due to weeds, pests and diseases.

In conclusion then, what emerges from a study of Nigeria's natural resources is a picture of a country with immense indigenous wealth already considerably dissipated by ignorance and mismanagement and still under threat from those who would further reduce the resource base for the sake of short-term benefits. The break-up of the country into smaller administrative units (36 states) was meant to create more centres of decision making and responsibility for the better control and development of the nation's natural resources. The new local government system brought into being in the 1970s should have augured well for the development of local resources if the local governments (774 in number) had been provided with the necessary funds, expertise, statutory powers and the freedom to function. A well-equipped local government is best placed to monitor the local environment and plan land use and resource development based on the needs and preferences of the community. Unfortunately, political corruption and ineptitude on the part of successive governments at federal, state and local levels have stultified progress in almost every sector of the economy. The natural environment and resources are in greater danger of abuse and misuse today than at any time in the country's history. There is little appreciation among the political elite of the dire need to put the welfare of future generations into consideration in the management of the environment and resources. In their scramble for wealth and position, they have connived among themselves and with foreigners to virtually destroy the economy and the natural environment and resources that sustain their greed and shameless pursuit of wealth.

Although, many have complained for good reason that Nigeria has been too dependent on foreign money, institutions and organizations for much of her development planning and programmes, it is in the interest of the people of this country at the moment that such external interventions continue. Much of the progressive developments in agriculture, forestry, water resources, nature conservation and such like have come about as a result of financial and technical assistance from regional and international financial and research institutions, especially United Nations agencies. For example, much of the rational inputs in the rehabilitation and sustainable management of the oil producing areas of the Niger river delta in the last two decades or so has been guided by international laws, conventions, agreements, protocols and guidelines to which Nigeria has been signatory. The sustained drive to promote ecosystem and species biodiversity in the country by creating more conservation areas has been led by international organizations especially the World Wildlife Fund (WWF) and the International Union for the Conservation of Nature and Natural Areas (IUCN). Although, Nigeria never succeeded at meeting fully the Millennium Developments Goals (MDGs), 2000-2015 and the succeeding Sustainable Development Goals (SDGs), the mere presence of those development goals galvanized action at both the federal and state government levels that focused on environmental sustainability and the welfare of the people at the grass-roots level. The country needs to take ownership of these internationally sourced or driven, people-oriented and environment friendly programmes, and adopt their modes of operation, financial discipline and application of modern equipment and techniques. Nigeria needs to build and sustain digital databases and geographical information systems as decision support systems in development planning on each natural resource element and in each sector of the economy.

Abbreviations and Periodicals

ABBREVIATIONS AND ACRONYMS

ACCCNNR	African Conservation Convention for the Conservation of Nature and Natural Resources
ACP	African, Caribbean and Pacific countries
AQUASTAT	FAO's Water Information System
C of O	Certificate of Occupancy
CA	Conservation agriculture
CBD	Convention on Biological Diversity
CBN	Central Bank of Nigeria
CGIAR	Consultative Group for International Agricultural Research
CILSS	Comité permanent Inter-Etats de Lutte contre la Sècheresse dans le Sahel (Permanent Interstate Committee for Drought Control in the Sahel)
CORS	Continuously Operating Referencing Stations
COVID-19	Corona Virus Disease, 2019
CRIN	Cocoa Research Institute of Nigeria
CT	Conservation Tillage
CVF	Climate Vulnerable Forum
EC	European Commission
EC-FAO	European Commission – United Nations Food and Agriculture Organization
ERTS	Earth Resources Technology Satellite
FAO	United Nations Food and Agriculture Organization
FCT	Federal Capital Territory
FDLAR	Federal Department of Agricultural Land Resources
FEPA	Federal Environmental Protection Agency
FESTAC	Festival of Arts and Culture

FMARD	Federal Ministry of Agriculture and Rural Development
FMC	Forest Management Committee
FMWR	Federal Ministry of Water Resources
FOB	Freight on Board
FRIN	Forestry Research Institute of Nigeria
FSR	Farm systems research
GBD	Global Burden of Disease
GCC	Grassroots Consultative Committee
GCP	Google Cloud Platform
GDP	Gross Domestic Product
GEF	Global Environment Fund
GRASS	Gully Rapid Action and Slope Stabilization
IHME	Institute for Health Metrics and Evaluation
IITA	International Institute for Tropical Agriculture
ILCA	International Livestock Centre for Africa
INRM	Integrated Natural Resources Management
ISO	International Organization for Standardization
ITD	Inter-Tropical Discontinuity
IUCN	International Union for the Conservation of Nature and Natural Resources
KEISA	Kenya Improved Stove Association
LANDSAT	Land Satellite
Landsat ETM	Landsat Extended Thematic Mapper
Landsat TM	Landsat Thematic Mapper
MDGs	Millennium Development Goals
NAFDAC	Nigeria Agency for Food and Drug Administration Control
NASA	National Aeronautics and Space Administration
NBSAP	National Biodiversity Strategy and Action Plan
NCDC	Nigeria Centre for Disease Control and Prevention
NCF	National Conservation Foundation
NDDC	Niger Delta Development Commission
NDES	Niger Delta Environmental Survey
NEEDS	Nigerian Economic Empowerment Development Strategies
NEITI	The Nigeria Extractive Industries Transparency Initiative
NEMA	National Emergency Management Agency
NESREA	National Environmental Standards and Regulation Enforcement Agency
NEWMAP	Nigeria Erosion and Watershed Management Project

NFDP	National Fadama Development Project
NGOs	Non-Governmental Organizations
NIHORT	Nigeria Institute of Horticulture
NIR	Near Infra-Red
NIRAD	Nigeria Radar project
NIROWI	Nigeria-Romania Wood Industry
NISS	Nigeria Institute of Soil Science
NPK	Nitrogen-Phosphorus-Potassium fertilizer
NRCC	Natural Resources Conservation Council
NSDWQ	National Standards for Drinking Water Quality
NSDWQ	Nigeria Standard for Drinking Water Quality
NSO	Nigeria Standards Organization
NTFR(s)	Non-Timber Forest Resource(s)
NWFP(s)	Non-Wood Forest Product(s)
O&G	Oil and Gas sector
PAPs	Project Affected Persons
PSP	Private Sector Participation
PWI	Presidential Water Initiative
RADWQ	Rapid Assessment of Drinking Water Quality
RBDAs	River Basin Development Authorities
RETAP	Renewable Energy Technology Assistance Programme
RIM	The Resource Inventory and Management Limited
RUGA	Rural Grazing Area
SARS	Severe Acute Respiratory Syndrome
SDGs	Sustainable Development Goals
SFDs	State Forestry Departments
SOLA	FAO's Solutions for Open Land Administration system
SPOT	Satellite pour l'Observation de la Terre
TT	Traditional Tillage
UAVs	Unmanned Airborne Vehicles
UNDP	United Nations Development Program
UNEP	United Nations Environment Program
UNFCCC	United Nations Framework Convention on Climate Change
UN-HABITAT	United Nations Agency for Human Settlements
UNICEF	United Nations Infant and Child Education Fund
USAID	United States Agency for International Development
USDA	United States Department of Agriculture
USGS	United States Geological Survey

WHO	World Health Organization
WRB	World Reference Base for Soil Resources
WWF	World Wildlife Fund
WWH	World-Wide HYPE

Bibliography

LIST OF REFERENCES BY CHAPTER

CHAPTER 1

Abbas I. I. (2009) An overview of land cover changes in Nigeria 1975–2005. *Journal of Geography and Regional Planning*, Vol 2(4), pp. 062–065 April 2009. `http://www.academicjournals.org/JGRP ISSN 2070-1845`

Ahmed A. (ed.) (2019) *Climate Change and the Dryland Resources of Nigeria.* LAP LAMBERT Academic Publishing. ISBN: 978-613-9-44966-8 Mauritius: International Book Market Service Ltd.

Ajayi S. S. (1991) *Natural Resources Conservation Action Plan.* The Presidency: National Resources Conservation Council. 220p.

Areola O. (2001) Deforestation in Nigeria: the pressures at the grassroots level. **In** Dhirendra K. Vajpeyi (ed): *Deforestation, Environment, and Sustainable Development: A Comparative Analysis.* ISBN 0-275-96989-4, Westport CT: Praeger Publishers (261 pages), pp. 173–198.

Chokor B. A. (2018) Cultural ethics and social mediation of environmental action and use of space in Nigeria. *Environmental Ethics*, Vol. 40, pp. 325–342 Winter 2018. DOI:10.5840/enviroethics201840432.
`https://www.researchgate.net/publication/335861763`

Chokor B. A. (2019) Trends in environmental valuation: the challenge to Nigerian geography and sustainable development. **In** Ikporukpo C. I, Ajayi D. D. and T. Osayomi (eds.); *Geography and Sustainable National Development.* Ibadan: Connel Publications ISBN: 978-978-55890-0-9. Chapter 2, pp.12–25.

Clapham Jnr. W. B. (1973) *Natural Ecosystems.* New York: Macmillan 248p.

Comité permanent Inter-Etats de Lutte contre la Sècheresse dans le Sahel (CILSS) 2016. *Landscape of West Africa–A window on a changing world.* Ouagadougou, Burkina Faso, CILSS, 219p. `https://doi.org/10.5066/F7N014QZ`

U. S. Geological Survey EROS, 47914 252nd St, Garreston, SD 57030, United States.

Daly, Herman E. (1990) Sustainable development: From concept and theory to operational principles. *Population and Development Review Volume 16, Supplement: Resources, Environment, and Population: Present Knowledge, Future Options* pp. 25–43. Population Council.
`http://www.jstor.org/stable/2808061` accessed June 25 2020.

Firey W. J. (1960) *Man, Mind and Land.* Glencoe, Illinois: Free Press.

Fish R. (2016) *Human Ecology in the 21st Century* Think Kent, International Thinkers, Global Impact Mar 18, 2016 STUDY BSc Human Ecology at the University of Kent. `http://bit.ly/2dp6iat`
ISBN 0023225009, 9780023225000.

Goodland R. and Daly H. (1996) Environmental sustainability: universal and non-negotiable. *Ecological Applications*, 6(4) 1002–1007. ©The Ecological Society of America.

Megbolugbe I. (2013) *A Program Development Proposal for the Creation of a Scalable Healthcare System for Ondo State.* Technical Proposal by GIVA International, October 10 2013. 52p.

Miller G. T., Jnr. (1975) *Living in the Environment: concepts, problems and alternatives.* Belmont, California.

National Agricultural Extension and Research Liaison Service (NAERLS) (1992) Trees and Shrubs species for fodder in Nigeria. *Extension Bulletin*, 57(3):1–27.

Odum E. P. (1959) *Fundamentals of Ecology.* Philadelphia: Saunders.

Owen O. S. (1975) *Natural Resource Conservation: an ecological approach* 2nd Edition. New York: Macmillan.

Phillips J. (1959) *Agriculture and Ecology in Africa.* London Faber.

Simmons I. G. (1974) *Ecology of Natural Resources.* London: Edward Arnold.

World Wildlife Fund (2016) *Western Africa: Southern Nigeria: Ecoregions.*
`http://www.worldwildlife.org/ecoregions/at0122.`

CHAPTER 2

Adeyoju, S. K. (1975) *Forestry and the Nigerian Economy.* Ibadan: Ibadan University Press.

Aiyeloja A. A. and Ajewole O. I. (2006) Non-timber forest products marketing in Nigeria: a case study of Osun State. *Educational Research and Reviews*, Vol. 1 (2), pp. 52–58. `http://www.academicjournals.org/ERR`

Ajayi S. S. (1991) *Natural Resources Conservation Action Plan.* The Presidency: National Resources Conservation Council. 220p.

Alao J. S., Ali B. and E. Okrikata (2016) Types of non-timber forest products in Yobe State, Nigeria. *Gashua Journal of Irrigation and Desertification Studies*, Vol. 2 No. 1, pp. 183–192. `www.gjidsfugashua.org.ng`

Areola Olusegun and Wanyama Henry (2008) *Mid-Term Evaluation Report on the Market Transformation for Highly Efficient Biomass Stoves for Institutions and Medium-Scale Enterprises in Kenya*. United Nations Development Program GEF Project. Nairobi, Kenya.

Arnold J. E. Michael (2004) Foreword **In** Alexiades M. N. and Shanley P. (eds.) 2005 *Forest products, Livelihoods and Conservation: case studies of non-timber forest product systems*. Volume 3: Latin America. Centre for International Forestry Research (CIFOR). SMK Grafika Desa Putera, Indonesia ISBN 979-3361-24-7. `http://www.cifor.cigiar.org`

Aweto A. O. (1978) *Secondary Succession and Soil Regeneration in a Part of the Forest Zone of South-western Nigeria*. Unpublished PhD. Thesis, University of Ibadan.

Borokini T. I., Babalola F. D., Amusa T. O., Ivande S. T., Wala Z. J., Jegede O. O. and J. O. Ihuma (2012) Community-based forest resources management in Nigeria: case study of Ngel Nyaki Forest Reserve, Mambilla Plateau, Taraba State, Nigeria. *Journal of Tropical Forestry and Environment*, Volume 2 Number 01, pp. 69–76. ISSN 2235-9370 Print / ISSN 2235-9362 Online © University of Jayewardenepura.

Enabor E. E. (1973) The prospects of forest industries development in Nigeria. *Nigerian Geographical Journal*, Vol. 16, pp.51–65.

Etowa B., Ojogu O. E. and S. S. Odunlami (2015) Leveraging rural livelihoods with forest conservation in Nigeria: the role of non-timber forest products. *Review of Agricultural and Applied Economics*, the Successor of the Acta Oeconomia et Informatica. ISSN 1336-9261, XVIII (Number 1): 33–42.

Faniran A. and Areola O. (1976) The concept of resources and resource utilization among local communities in Western Nigeria. *African Environment*, Volume 2(3) pp. 39–51.

FAO (1966) *Agricultural Development in Nigeria, 1965–80* Rome.

FAO, 1999. *The Non-Wood Forest Products of Nigeria* EC-FAO PARTNERSHIP PROGRAMME (1998–2000), Tropical forestry Budget line B7-6201/97-15/VIII/ FOR PROJECT GCP/INT/679/EC (prepared by Osemeobo G. J. and Ujor G.)

FAO (2005) Global Forest Resources Assessment 2005 (FRA 2005) Results. `http://www.fao.org/forestry/fra2005`

Federal Office of Statistics: *Review of External Trade, Nigeria* 1965, 1970, 1975, 1985 *Forest Machine Magazine*, May 2020 "Sustainable Logging". `https://forestmachinemagazine.com/sustainable-logging/` [accessed October 19, 2020]

Ingram V. J. (2014) *Win-wins in Forest Product Value Chains? How governance im-*

pacts the sustainability of livelihoods based on non-timber forest products from Cameroon. `http://www.researchgate.net/publication/260751691` (accessed October 1 4 2020) ISSN 2224-3186

Keay R. W.J. (1952) *An Outline of Nigerian Vegetation* Lagos: Federal Ministry of Information.

Ladipo D. (2010) The state of Nigeria's forests *IITA–R4D Review* No. 4, pp.24–25.

Ladipo D. O. and Boland D. (1994) Trade in *Irvingia* kernels. Paper presented at International Workshop on Nigeria in West and Central Africa. International Centre for Research in Agroforestry (ICRAF).

Larinde S. L. (2010). Secondary wood processing and the Nigeria sawmill industry: issues, challenges and opportunities. **In** Adeyoju S. K. and Bada S. O. (eds.): *Readings in Sustainable Tropical Forest Management.* Ibadan: Zenith Book House Publishers. Chapter 14, pp.277–291.

Longman K. A. and Jenik J. (1974) *Tropical Forest and Its Environment.* London/New York: Longman.

Forestry Department: Nigeria Forestry Statistics (1983).

Nigerian Trade Statistics WITS. (*wits.worldbank.org*)

Nye, P. H. and Greenland, D. S. (1965) *The Soil under Shifting Cultivation.* Technical Communication No. 51, Commonwealth Agricultural Bureau, England.

Ogunwusi A. A. (2012) The forest products industry in Nigeria. *African Research Review: An international multidisciplinary Journal*, Ethiopia, Volume 6 (4), Serial No. 27, October 2012, pages 191–205. ISSN 1994-9057 (Print) ISSN 2070-0083 (Online). DOI: `http://dx.doi.org/10.4314/afrrev.v6i4.13`

Ohagwu C. J. and B. O. Ugwuishiwu (2011) Status of wood processing and storage in Nigeria. *Nigerian Journal of Technology* Volume 30 No. 2 pp. 94–104.

Omoluabi A. C. and Abang S. O. (1994) Marketing margins in non-timber forest products trade in Cross River State of Nigeria. *Cross River State Forestry Strategic Plan. Working Paper 12. Prepared for the Cross River State Forestry Project* (ODA Assisted). Forestry Department Headquarters, Calabar.

Redhead J. F. (1971) The timber resources of Nigeria *Nigerian Journal of Forestry*, Volume 1 (1), pp. 102–117.

Richards P. W. (1952) *The Tropical Rain Forest.* Cambridge: Cambridge University Press.

Schulz, J. P. (1960) *Ecological Studies on Rain Forest in Northern Suriname.* Amsterdam: North Holland.

Shasanya *et al.* (2015): Shasanya O. S, Agbe F. J., Okanlawon J. O., Amoo, A. O., Odunela, T. O, Faniyi, I. B, Ademuwagun, A. A, Abiodun, F. O, Ekunola, J. N., Oke, O. O, Okumodi, B. O, and Awoleye, M. O. 2015. Some useful Nigerian timbers, their destroying agents and measures for their prevention. *Journal of Natural Sciences Research*, 5(15): 15–20. www.iiste.org

Sunderland T. C.H., Besong S. and J. O. Ayeni (2003) *Distribution, Utilisation and Sustainability of Non-Timber Forest Products from Takamanda Forest Reserve, Cameroon*. SI/MAB Series (8): 155–172

CHAPTER 3

Ajayi S. S. (1991) *Natural Resources Conservation Action Plan*. The Presidency: National Resources Conservation Council. 220pp.

Aliu Y. O. (1975) Dry season Fulani transhumance and cattle trypanosomiasis: the nature of the relationship. *Nigerian Journal of Animal Production*, 2(2), pp.204–211.

Aregawi T., Melaku S. and L. Nigatu (2008). Management and utilization of browse species as livestock feed in semi-arid district of north Ethiopia. *Livestock Research for Rural Development*, Vol.20 (6), 1–9.

Areola O. (1977) Problems of agricultural development and innovation in the derived savanna zone: a case study of Ibarapa Division. **In** Adejuyigbe, O. & F. M. Helleiner (eds.): *Environmental and Spatial Factors in Rural Development in Nigeria*. Nigerian Geographical Association. Section II, 60–66.

Babalobi O. O., Ogundipe G. A.T. and G. O. Esuruoso (2000) Livestock Population Dynamics and Problems in a Grazing Reserve in Northern Nigeria *Proceedings of the 9th International Symposium on Veterinary Epidemiology and Economics 2000*. www.sciquest.org.nz

Bourn D. (1983) *Tsetse Control, Agricultural Expansion and Environmental Change in Nigeria*. D. Phil. Thesis June 1983 University of Oxford.

Bourn D., Wint W., Blench R. and E. Woolley (2018) Nigerian Livestock Resources Survey. Animal Genetic Resources (AnGR). https://angr.org/2018/03/28/nigerian-livestock-resources-survey/

Brinckman W. L. and de Leeuw P. N. (1975) The nutritive value of browse and its importance in traditional pastoralism. In Ajayi S. S. *et al.* (eds.) Wildlife Management in Savanna Woodland. Pp. 190–205.

Darling F. F. (1960) *Wildlife in an African Territory*. Oxford: Oxford University Press.

de Leeuw P. N. (1975), Species preferences of domestic ruminants grazing Nigerian savannas. **In** Ajayi S. S. *et al.* (eds.): *Wildlife Management in Savanna Woodland*, pp. 142–152.

de Leeuw P. N. (1976) Fodder resources and livestock development in north-east Nigeria. *Savanna* Vol. 5(1), pp. 61–74.

FAO (2018) *Africa Sustainable Livestock 2050: Transforming Livestock Sector: Nigeria What do long term projections say?* Brief prepared by Wisser D., Odunze E., Boussini H., Mickecz O. and U. Pica-Ciamara October 2018.

Harris, Peter S. ((2000)) *Grassland Resource Assessment for Pastoral Systems*. Food and Agriculture Organization of the United Nations. ISBN: 9251045372

Idowu O. (2018) How Climate Change Influences Herdsmen in Nigeria. Climate Scorecard © (2019). `https://www.climatescorecard.org/who-we-are/` [accessed October 20 2020]

International Livestock Centre for Africa (ILCA) (1979) *Grazing reserves and development blocks: A case study from Nigeria development requirements*. Rome: FAO.

Ismail Iro, (2015) Grazing Reserve Development: A Panacea to the Intractable Strife Between Farmers and Herders Founder, `gamji.com`. [USA] webmaster@gamji.com

Mortimore M. J. (1967) Land use and population pressure in Kano close-settled zone, Northern Nigeria. *Advancement of Science*, Vol. 23, pp. 677–686.

Nye, P. H. and Greenland, D. S. (1965) *The Soil under Shifting Cultivation*. Technical Communication No. 51, Commonwealth Agricultural Bureau, England.

Ogboru T. and O. Adejonwo-Osho (2018) Towards effective cattle grazing and rearing legal framework: an imperative for environmental protection. *Journal of Sustainable Development Law and Policy (The)*, Volume 9 No. 1 Special Issue pp. 58–79.

Ogunbosoye Dupe O., G. O. Tona and F. K. Otukoya (2015) Evaluation of the nutritive value of selected browse plant species in the southern guinea savannah of Nigeria for feeding to ruminant animals. *British Journal of Applied Science & Technology*, 7(4): 386–395 2015, Article no. BJAST.2015.156. ISSN: 2231-0843. SCIENCEDOMAIN international. `www.sciencedomain.org`

Owen D. F. (1973) *Man's Environmental Predicament: an introduction to human ecology in Tropical Africa*. Oxford: Oxford University Press.

Oyenuga V. A. (1967) *Agriculture in Nigeria: an introduction.* Rome: FAO.

Tolera A., Khazaal K. A. and E. R. Orskov (1997) Nutritive value of some browse species. *Animal Feed Science and Technology*, 67, 181–195.

van Raay J. G.T. (1975) *Rural Planning in a Savannah Region: the case of the Fulani pastoralist in the North Central State of Nigeria*. Rotterdam, the Netherlands. Rotterdam University Press.

Williamson G. and Payne W. J.A. (1965) *An Introduction to Animal Husbandry in the Tropics.* London: Longman.

CHAPTER 4

Adejuwon J. O. and Ekanade O. (1988) Soil changes consequent upon the replacement of Tropical Rain Forest by plantations of *Gmelina arborea, Tectona grandis* and *Terminalia superba. Journal of World Forest Resource Management* 3, 47–59.

Adekola B. O. and Adetoro A. O. (2014) Threats to biodiversity conservation in Cross River National Park, Nigeria. *International Journal of Conservation Science*, Volume 5, Issue 4, October-December 2014, pp. 547-552. ISSN: 2067-533X `www.ijcs.uaic.ro`

Afolayan T. A. and Pelinck E. (1975) Wildlife resource utilization in Nigeria. Paper presented at the International Symposium on Environmental Resource Management, University of Ife (1973). (Unpublished).

Afolayan T. A. (1975) Wildlife management in Nigeria *Obeche*, 11 pp. 12–20.

Akinsorotan O. A., Ogunjemite B. G. and T. A. Afolayan (2011) Assessment of the large mammals of Arakhuan range, Okomu National Park, Nigeria. *Ethiopian Journal of Environmental Studies and Management*, Volume 4 No. 3, pp. 25–37.

Aladejana K. (1971) Plantation establishment practices in the high forest region of Nigeria. *Nigerian Journal of Forestry*, 1(1) pp. 21–26.

Ayeni J. S.O. (1972) Chemical analysis of some soil samples from the natural mineral licks of Yankari Game Reserve, North-Eastern State, Nigeria. *Nigerian Journal of Forestry*, Vol. 2(1), pp.16–21.

Ayeni, J. O.S. (1975) Further aspects of big game utilization of natural mineral licks. **In** Ajayi S. S. (1975) *Wildlife Management in Savanna Woodland*: Proceedings of the Ibadan/Garoua International Symposium on Wildlife Management, 22–26 September 1975 pp. 105–121.

Ball, J. B. and Daniyan, C. O. (1977), *Summary of Plantation Areas in the Southern States of Nigeria*. August 1977. Ibadan: Federal Department of Forestry.

Beak Consultants (1998). *Forest Resources Study of Nigeria, Draft main Report—volume 1 (Overview)*. Unpublished report prepared by Beak Consultants Limited of Canada for FORMECU, Federal Department of Forestry, Abuja, Nigeria 69pp.

Bergl, R. A., Dunn, A., Haruna, S., Mshelbwala, J, and G. Nyanganji (2011) *Aerial Survey of Elephants and Other Large Mammals at Yankari Game Reserve, Bauchi State, Nigeria* Wildlife Conservation Society, North Carolina Zoo, MIKE.

Charter J. R. (1970) The economic value of wildlife in Nigeria. Paper presented at the First Conference of the Forestry Association of Nigeria, Ibadan.

Child G. S. (1973) Wildlife utilization and its relevance in West Africa. *Proceedings of the Symposium at the Seventh Biennial Conference of West African Science*

Association

Eludoyin O. S., Ugbana T. and S. O. Otali (2019) Extent of forest depletion from 1986–2015 as a consequence of land use in Andoni Local Government Area of Rivers State, Nigeria. *Journal of Applied Science and Environmental Management*, Vol.23(6), pp. 1065–1073. Print 1119-8362 Electronic ISSN 1119-8362 `https://dx.doi.org/10.4314/jasem.v23i6.11`

Faniran A. and Areola O. (1974) Landform examples from Nigeria—a gully *Nigerian Geographical Journal*, 17, 57–60.

FAO (1970) *Wildlife Management in Nigeria* . Department of Forestry, University of Ibadan.

FAO (2017) *6 ways indigenous peoples are helping the world achieve #ZeroHunger.* `http://www.fao.org/zhc/detail-events/en/c/1028010/`

Federal Department of Forestry: *Nigerian Forestry Statistics 1983, 1984. First National Biodiversity Report July 2001.* `https://www.cbd.int/doc/world/ng/ng-nr-01-en.pdf`

Henshaw J. and Ayeni J. (1971) Some aspects of big game utilization of mineral licks in Yankari Game Reserve. *East African Wildlife Journal*, Vol. 9, pp. 73–82.

Iyamabo D. E. and Ojo G. O.A. (1971) Plantation establishment techniques in the savanna areas of Nigeria. *Nigerian Journal of Forestry*, 1(1) pp. 16–20.

Kala C. P. (2017) Conservation of nature and natural resources through spirituality. *Applied Ecology and Environmental Sciences*, Volume 5(2), 24–34. DOI: 10.12691/aees-5-2-1

Lameed A. G. and Adetola J. (2012) Species-diversity utilization of salt lick sites at Borgu sector of Kainji Lake National Park, Nigeria **In** Gbolagade Akeem Lameed (2012) *Biodiversity Enrichment in a Diverse World*, Chapter 2. IntechOpen, DOI: 10.5772/51089

Lasan A. H. (1975) Habitat management for wildlife conservation with respect to fire, salt lick and water regime. **In** Ajayi S. S. (ed.): *Wildlife Management in Savanna Woodland: Proceedings of the Ibadan/Garoua International Symposium on Wildlife Management*, 22–26 September 1975 pp. 142–152.

Magama Y. A., Babagana M., Usman A. U., A. Gujja A., Adamu A. and A. E. Karachi (2018) Assessment of wildlife species mostly involved in human-wildlife conflict around Yankari Game Reserve, Bauchi State, Nigeria *International Journal of Contemporary Research and Review.*

Msuya T. S. and Kideghesho J. R. (2009) The role of traditional management practices in enhancing sustainable use and conservation of medicinal plants in West Usambara Mountains, Tanzania. *Tropical Conservation Science*, Vol. 2(1), 88–105. ISSN 194-0829 `http://www.tropicalconservation.org`

New Vision Reporter (2014) The unwritten law: traditional ways of conserving environment. NewVision Printing and Publishing Co. Ltd. (UG). `https://www.newvision.co.ug/` [accessed October 27, 2020]

Nigeria National Parks, *Annual Calendar 1998*. 13pp. ISSN 0976-4852 September, 2018, Volume 09 Issue 09. `https://doi.org/10.15520/ijcrr/2018/9/09/605`

Nigerian Conservation Foundation (NCF) (2020) *Biodiversity Conservation Action Plan—Edo*. `https://www.ncfnigeria.org/projects/biodiversity-action-plan-edo` [accessed October 28 2020].

Oates J. *et al.*, (2007) Regional Action Plan for the Conservation of the Cross River Gorilla *Gorilla gorilla diehli*. IUCN/SSC Primate Specialist Group and Conservation International, Arlington, VA, USA. ISBN: 978-1-934151-02-0

Ogunsanmi, A. O. (1997) Serum Chemistry and Biochemical in dices in some Small Ruminant and captive reared Grey Duiker (*Sylivicara grimmia*) infected with *Trypanosoma congolense* PhD. Thesis, University of Ibadan, Ibadan.

Olaleye O. A. and Ameh, C. E (1999): *The EC-FAO Partnership Programme (1998–2002)—Project GCP/INT/679/EC Data Collection and Analysis for Sustainable Forest Management in ACP Countries—Linking National and International Efforts*. `http://www.fao.org/3/ab578e/ab578e00.htm`

Osunsina I. O.O., Ogialekhe P., and O. A. Adebayo (2018) Bird species diversity and abundance in Borgu sector of Kainji Lake National Park, Nigeria *Nigerian Journal of Wildlife Management* (NJWM), 2(2), 2018, 10–10.

Redhead J. F. (1971) The timber resources of Nigeria *Nigerian Journal of Forestry* Volume 1(1), pp. 102–117.

Rim-Rukeh A., Irerhievwie G. and I. E. Agbozu (2013) Traditional beliefs and conservation of natural resources: evidence from selected communities in Delta State, Nigeria. *International Journal of Biodiversity and Conservation*, Volume 5(7), pp. 426–432. DOI: 10.5897/IJBC2013.0576 ISSN 2141-243X. `http://www.academicjournals.org/IJBC`

Udo R. K. (1970) *Geographical Regions of Nigeria*. London: Heinemann Educational Books

UNDP/GEF (2009) Niger Delta Biodiversity Project *United Nations Development Programme Country: Nigeria Project Document* PRODUC 2047 Niger Delta Conservation FINAL 180712.

Wikipedia : *Natural Areas in Nigeria*. `https://en.wikipedia.org/wiki/Natural areas in Nigeria` [accessed October 28 2020]

Wikipedia: Cross River National Park.

https://en.wikipedia.org/wiki/Cross River National Park [accessed October 29 2020]

World Bank, June (1992). *Federal Republic of Nigeria: Forestry Sector Review.* Confidential Report No. 10744-UNI.

World Wildlife Fund (2014), *Living Planet Report, 2014.* http://www.eoearth.org/view/article/154857/

CHAPTER 5

Adams W. M. (1987) Approaches to water resource development, Sokoto Valley, Nigeria: the problem of sustainability. **In** Anderson David and Grove Richard (eds.): *Conservation in Africa: people, policies and practice.* Cambridge: Cambridge University Press. Chapter 15, pp. 307–325.

Adelalu G. T. (2012) Climate variability and River Benue discharge in Jimeta, Yola area, Nigeria. **In** *Hydrology for Disaster Management. Special Publication of the Nigerian Association of Hydrological Sciences*, pp. 375–382. https://www.researchgate.net/publication/280131578

Akintola F. O. and Nnyamah N. C. (1979) land use and surface water pollution in Ibadan. **In** Oguntoyinbo J. S., Filani M. O. and O. Areola (eds.). *Resources and Development in Africa*, III, pp. 338–342.

Ayoade J. O (1974) A statistical analysis of rainfall over Nigeria *Journal of Tropical Geography*, Vol. 39, pp. 11–23.

Ayoade J. O. (1975) Water resources and their development in Nigeria. *Hydrological Sciences Bulletin*, Vol. 20, pp. 581–591.

Bennet A. J. (1967) A Survey of River Water Quality in Northern Nigeria with particular reference to Irrigation Agriculture *Samaru Miscellaneous Paper No. 17.* Institute for Agricultural Research, Ahmadu Bello University, Zaria.

Berit Arheimer, Rafael Pimentel, Kristina Isberg, Louise Crochemore, Jafet C. M. Andersson, Abdulghani Hasan, and Luis Pineda (2020) Global catchment modelling using World-Wide HYPE (WWH), open data, and stepwise parameter estimation. *Hydrol. Earth Syst. Sci.*, 24, 535–559 (2020). https://doi.org/10.5194/hess-24-535-2020

Carter J. D. and Barber W. (1956) The rise in the water table in parts of Potiskum Division, Bornu Province. *Records of the Geological Survey of Nigeria*, pp. 5–13.

Charles Kumolu https://www.vanguardngr.com/2013/04/river-basins-how-unending-policy-reversals-abet-inefficiency/

du Preez J. W. and Barber W. (1965) The distribution and chemical quality of groundwater in Northern Nigeria. *Bulletin of the Geological Survey of Nigeria*, 36.

Faniran A., Sada P. O. and Areola O. (1976) River sand mining, its organization and the building industry in the Lagos area. *Quarterly Journal of Administration*, Vol. X (4) pp. 423–435.

FAO (2016) AQUASTAT FAO's Water Information System Nigeria: Water Resources. `http://www.fao.org/nr/water/aquastat/countries regions/Profile segments/NGA-WR eng.stm`

Federal Ministry of Water Resources (FMWR) (2013) *Dams and reservoirs operations* Federal Ministry of Water Resources. [Accessed April 2020]

Frank Eric Tasting Foka, Clarence S. Yah, Marie Ebob Agbortabot Bissong ((2018)) Physico-chemical properties and microbiological quality of borehole water in four crowded areas of Benin City, Nigeria, during rainfalls. *Shiraz E-Medical Journal*: 19 (11); e68911 `https://dx.doi.org/10.5812/semj.68911`

Ngene B. U. (2009) Nigerian Rain Gauge Station Optimization and National Development: The importance of Head Count. *IOP Conference Series Materials Science and Engineering* 413(1):012022. September (2018). `doi:10.1088/1757-899X/413/1/012022`

Nnaji J. C & S. Omotugba (2014) Physico-chemical quality of drinking water in New Bussa, Niger State, Nigeria *International Research Journal of Pure & Applied Chemistry*, 4(4), 437–446. `www.sciencedomain.org`

Odunuga O. A. (1977) *Urban Pollution with particular reference to River Pollution in Ibadan.* Diploma in Town Planning Dissertation, The Polytechnic, Ibadan.

Oribabhor B. J. and Udo M. T., 20 Impact of mangrove forest destruction on the fisheries resources of the Niger delta, Nigeria. Nature & Faune Volume 25 Issue 2, pp 72–76 ISSN-L:2026-5611 **In** Bojang F. and Ndeso-Atanga A., (eds.) *Economic and social significance of forests for Africa's sustainable development.* FAO Regional Office for Africa. `http://www.fao.org/africa/publications/natuure-and-faune-magazine/`

Simmons I. G. (1974) *Ecology of Natural Resources.* London: Edward Arnold.

Tahal Consultants (Nig.) Ltd. (1974) *Ikeja Irrigation and Drainage Project Reconnaissance Survey.* Lagos: Lagos State Government.

The World Bank ((2000)) *Water Supply & Sanitation Interim Strategy Note.*

The World Bank (2000) *Water Supply & Sanitation Interim Strategy.*

Uduma-Olugu N and Onukwube H. N. (2012) Exploring the coastal tourism potentials of Lagos. *Journal of Sustainable Development*, Vol. 5 No. 7, pp.156–165. `http://dx.doi.org/10.5539/jsd.v5n7p156`

Ujih O. U., Dahiru M. M., Yelwa S. A., Shiru M. S., Abubakar M. and I. N. Ishaya (2014) Promoting sustainable use of rivers for tourism and recreation in Nigeria. *Journal of Research in National Development*, 12 (2), 278–281. ISSN 1596-

8308. www.transcampus.org/journals; www.ajol.info/journals/jorind

Wali S. U., Umar K. J., Abubakar S. D., Ifabiyi I. P., Dankani I. M., Shera I. M. and S. G. Yauri (2019) Hydrochemical characterization of shallow and deep groundwater in Basement Complex areas of southern Kebbi State, Sokoto Basin, Nigeria. *Applied Water Science*, 9:169. https://doi.org/10.1007/s13201-019-1042-5

WHO (2010) Rapid assessment of drinking-water quality in the Federal Republic of Nigeria: country report of the pilot project implementation in 2004–2005. Prepared by M Ince, D Bashir, OOO Oni, EO Awe, V. Ogbechie, K. Korve, MA Adeyinka, AA Olufolabo, F. Ofordu and M. Kehinde WHO-UNICEF-RADWQ-Nigeria-Report.pdf World Health Organization and UNICEF. ISBN 978 92 4 150060 9

Wikipedia (2020) *Tourism in Nigeria.* (accessed October 14 2020)

CHAPTER 6

Adeyemi O. and Adelabu S. (2020) implications of rainfall variability in the light of climate change on crop production in the Sudano-Sahelian belt of Nigeria. *Conference: WaterEnergyNexus 2020 (3rd Edition) At: Virtual Springer Nature* https://www.researchgate.net/publication/345134680 Implications of rainfall variabilities in the light of climate change on crop production in the Sudano-Sahelian belt of Nigeria

Agboola A. A. and Fayemi A. A. (1971) Effect of interplanted legumes and fertilizer treatment on the major soil nutrients. *Proceedings of the International Symposium on Fertility Evaluation*, New Delhi, 1, pp. 529–540.

Agboola A. A. (1970) Preliminary investigation on the effect of continuous cropping of maize on grain yield on total nitrogen, available phosphorus and exchangeable potassium on three Nigerian soils. *Nigerian Journal of Science* 4, pp.89–99.

Agboola, S. A. ((1979)) *An Agricultural Atlas of Nigeria* Oxford: Oxford University Press.

Areola, O. (1978) Soil and vegetal resources. **In** Oguntoyinbo J. S., Areola O. and M. O. Filani (eds): *A Geography of Nigerian Development.* Ibadan: Heinemann Educational Books Plc. Chapter 5, 72- 91.

Areola, O. (1980) Some issues and problems in studying savanna fallows. *African Environment*, No.3 Vol. IV, 1, 51–62.

Areola, O. (1984) The characteristics and fertility status of the soils of the old cocoa farms of Ibadan region, *Nigeria. Malaysian Journal of Tropical Geography*, 10,

1–11.

Areola, O. (1985) Extractable copper content of soils under peasant coca farms in Ibadan region, Nigeria. *Turrialba*, 35(3), 229–232.

Aweto, A. O. (1987) Physical and nutrient status of soils under rubber (*Hevea brasiliensis*) of different ages in southwestern Nigeria. *Agricultural Systems Vol,*. 23, pp. 63–72.

Ayodele O. P. and O. A. Aluko (2017) Weed management strategies for conservation agriculture and environmental sustainability in Nigeria, *IOSR Journal of Agriculture and Veterinary Science*. (IOSR-JAVS) e-ISSN: 2319-2380, p-ISSN: 2319-2372. Volume 10, Issue 8 Ver. August 2017, pp. 01–08.

Buchanan K. M. and Pugh J. C. (1955) *Land and People in Nigeria*. London: University of London Press.

Dahiru S. (2020) *Nigeria Erosion and Watershed Management Project* (NEWMAP). Federal Ministry of Environment.
https://newmap.gov.ng/index.php/from-the-national-coordinator/

Ekanade O. (1988) The nutrient status of soils under peasant cocoa farms of varying ages in southwestern Nigeria. *Biological Agricultural and Horticulture*, Vol. 5, pp.155–167.

FAO (1966) *Agricultural Development in Nigeria, 1965–80* Rome *FAO Statistical Yearbook, 2014*. http://www.fao.org/3/a-i3620e.pdf *Part 3: Feeding the World*, pp. 61–114.

FAO (2014). *FAO Statistical Yearbook 2014*. Part 1: The Setting, pp.1–30.
http://www.fao.org/3/a-i3620e.pdf

FAO (2018) FAO in Nigeria: Nigeria at a glance. http://www.fao.org/nigeria/fao-in-nigeria/nigeria-at-a-glance/en/

FAO. (2015) Country Fact Sheet on Food and Agriculture Policy Trends. *FAO Country Report 2015*.

Farooq M. & K. H.M. Siddique (2015) eds. *Conservation Agriculture*. 1 edition Springer 665p. ISBN-10 3319350552 ISBN-13 978-3319350554.

Federal Ministry of Agriculture and Rural Development, (2008) *National Food Security Programme*. Federal Republic of Nigeria.

Floyd, B. (1964) Terrace agriculture in Eastern Nigeria: the case of Maku. *Nigerian Geographical Journal*, Vol. 7, pp. 107.

Gleave M. B. (1966) Hill settlements and their abandonment in tropical Africa particularly Nigeria. *Transactions and Papers of the Institute of British Geographers*, Vol. 40 pp. 39–40.

Gleave, M. B. (1965) The changing pattern of settlement in the uplands of Northern Nigeria. *Nigerian Geographical Journal*, Vol. 8pp. 127–141.

Idris Mahmoud (2020) Why Katsina farmers are reviving yam cultivation. *Daily Trust Newspaper*, October 18, 2020, at 3:06:40 AM.

Iortyom E. T., Abawua J. I. and T. Shabu, (2018) The economic impact of Fadama III Project on beneficiaries in Benue State: a study of Makurdi Local Government Area. Paper presented at the 59th Annual Conference of the Association of Nigerian Geographers, Department of Geography, University of Ibadan, 4–9 November 2018, pp. 1–17.
`https://www.researchgate.net/publications/331036393`

Jones M. J. and Wild A. (1975) *Soils of the West African savanna: the maintenance and improvement of their fertility*. Technical Communication No. 55, Commonwealth Agricultural Bureau, England.

Klinkenberg K. and Higgins G. M. (1968) An outline of northern Nigerian Soils. *Nigerian Journal of Science*, 2(2), pp. 91–115.

Lal, R. (1973) *Soil Erosion and Shifting Cultivation*. Draft Report on FAO/SIDA/ARCN Regional Seminar on Shifting Cultivation and Soil Conservation in Africa, Ibadan, 2–21 July 1973, pp. 55–64.

Manshard W. (1974) *Tropical Agriculture: A geographical introduction and appraisal* London: Longman.

Mortimore M. J. (1967) Land use and population pressure in Kano close-settled zone, Northern Nigeria. *Advancement of Science*, Vol. 23, pp. 677–686.

National Bureau of Statistics: *2009 Annual Collaborative Survey of Socio-Economic Activities in Nigeria Main Statistical Report Volume 1*. August 2010.

Nye, P. H. and Greenland, D. S. (1965) *The Soil under Shifting Cultivation*. Technical Communication No. 51, Commonwealth Agricultural Bureau, England.

Oguntoyinbo J. S. (1971) Irrigation and land reclamation projects in Nigeria. *Nigeria Agricultural Journal*, Vol. 7, pp. 53–69.

Reicosky, D. C. (2003) Conservation agriculture: global environmental benefits of soil carbon management. In Garcia-Torres L., Benites, J., Martinez-Vilela A., Holgado-Cabrera A. (eds) *Conservation Agriculture. Springer, Dordrecht Online*. ISBN 978-94-017-1143-2
`https://doi.org/10.1007/978-94-017-1143-2 1`

Richards P. (1975) Alternative strategies for the African environment. **In** Richards P.(ed.) *African Environment: problems and perspectives*, IAI Special Report No. 1 pp. 102–117.

Smyth A. J. and Montgomery R. F. (1962) *Soils and Land Use of Central Western Nigeria*. Ibadan: Government Printer.

The World Bank (2015) Fadama Project Turns Nigerian Farmers into Agro-preneurs.
`https://www.worldbank.org/en/news/feature/2015/04/02/fadama-`

takes-nigerian-farmers-to-higher-level [accessed October 27 2020]

The World Bank (2019) NationalFadamaDevelopmentProject(NFDP) III https://projects.worldbank.org/en/projects-operations/project-detail/P096572

Vanguard Newspaper (2020). Poor returns force farmers to abandon food crops in North West–Survey. November 1 2020 3:52 PMIN NEWS https://www.vanguardngr.com/2020/11/poor-returns-force-farmers-to-abandon-food-crops-in-north-west-survey/

CHAPTER 7

Adejuwon J. O. (1978) Pests and diseases, **In** Oguntoyinbo J. S., Areola O. and M. O. Filani (eds): *A Geography of Nigerian Development.* Ibadan: Heinemann Educational Books Plc. Chapter 6, 92–112.

Adeyemi S. A.O., Donnelly J. and J. A. Odetoyinbo (1966) Studies on chemical control of the stem borers of maize. *Nigeria Agricultural Journal*, 3, pp. 61–66.

Alabi R. A. Esobhawan A. O and M. B. Aruna (2006) econometric determination of contribution of family poultry to women's income in Niger delta, Nigeria. *Journal of Central European Agriculture*, Volume 7 No. 4 pp. 753–760.

Amusa N. A., Adegbite A. A., Muhammed S. and R. A. Baiyewu, (2003) Yam diseases and its management in Nigeria. *African Journal of Biotechnology*, 2, no. 12, 497–502. https://doi.org/10.5897/AJB2003.000-1099

Areola O. (1977) Problems of agricultural development and innovation in the derived savanna zone: a case study of Ibarapa Division. In Adejuyigbe, O. & F. M. Helleiner (eds): *Environmental and Spatial Factors in Rural Development in Nigeria.* Nigerian Geographical Association. Section II, 60–66.

Auta A., Banwat S. B., Dayom D. W., Shalkur D. and M. O. Avu (2012) Occurrence and Treatment of Common Health Problems in a Nigerian Community. *Journal of Young Pharmacists*, Vol 4 / No 1 pp. 49–53. https://www.ncbi.nlm.nih.gov/pmc/articles/PMC3326783/pdf/JYPharm-4--49.pdf

Bandiya H. M., Majeed Q., Ibrahim N. D. and H. A. Shindi (2012) Density and dispersion of mounds of *Macrotermes bellicosus* [Isoptera: Termitidae] in some Local Government Areas of Sokoto State, semi-arid zone of Nigeria. *Research in Zoology*, 2(1), 1–4.

Bandiya H. M., Majeed Q., Ibrahim N. D., Yahaya M. M. and M. A. Yahaya (2013) Intra-colonial population of *Macrotermes bellicosus* (Smeathman) [Isoptera; Termitidae] in Sokoto, semi-arid zone of Northwestern Nigeria. *Nigeria Journal of Basic and Applied Science* (March 2013), 22(10: 55–59.
http://www.ajol.info/index.pho/njbas/index
http://dx.doi.org?10.4314/njbas.v21i1.8

Barker D., Oguntoyinbo J. S. and P. Richards (1977) *The Utility of the Nigerian Peasant Farmers; Knowledge in the Monitoring of Agricultural Resources,* General Report MARC Report No. 4, Chelsea College, University of London.

Booker R. H. (1963) the effect of sowing date and spacing on the rosette disease of groundnuts in Nigeria with observation on the vector *Aphis craccivora. Annals of Applied Biology* Vol. 52 pp. 125–131.

Booker R. H. (1965) Pests of cow peas and their control in Northern Nigeria. *Bulletin of Entomological Research*, 55(4), pp. 663–672.

Boyetchko S. M. and G. Peng (2004) Challenges and strategies for development of mycoherbicides. In A. D. K. (ed) *Fungal Biotechnology in Agricultural, Food and Environmental Applications.* Marcel Dekker, New York

Brooke Borel in the *Scientific American* Newsletter June 8, 2018.

Brown A. (1955) Diseases as an element of the Nigerian environment. **In** Buchanan K. M. and Pugh J. C. (1955) *Land and People in Nigeria* London: University of London Press. pp. 41–57.

Chhokar R. S., Ram H. and V. Kumar (2014) Integrated weed management. In Shukla R. S., Mishra P. C., Chatrath R., Gupta R. K., Tomar S. S. and I. Sharma (eds) *Recent Trends on Production Strategies of Wheat in India* (pp. 155–171). India: Jawaharlal Nehru Krishi Vishwa Vidyalaya (JNKVV), Jabalpur and ICAR-Indian Institute of Wheat and Barley Research.

Das T. K. (2008) *Weeds and their control.* `http://nsdl.nicair.res.in/jspui/bitstream/123456789/568/1/Weeds%20and%20their20control%20methods%20-%20Formatted.pdf` [Accessed May 2020]

de Gier J., Cecchi G., Paonea M., Dede P. and W Zhao (2020) The *continental Atlas of tsetse and African animal trypanosomosis in Nigeria.* Food and Agriculture Organization of the United Nations (FAO), Viale delle Terme di Caracalla, 00153, Rome, Italy.

Dede P. M., Zaria L. T. and F. A.G. Lawani (2007) current status and the new strategy in combating tsetse and trypanosomiasis in Nigeria based on the PATTEC initiative. In 58th International Scientific Council for Trypanosomiasis Research and Control ISCTRC, Luanda, Angola.

Desquesnes M. (2018) Animal Trypanosomosis (including tsetse-transmitted, but excluding surra and dourine), OIE, Terrestrial Manual 2018.

Ehi-Eromosele C., Nwinyi O., and O. Ajani (2013) Integrated pest management. **In** Soloneski S. and M. Larramendy (eds.) *Weed and Pest Control: conventional and New Challenges* (pp. 105–115). Croatia: Intech.

Ezcurra E., Rapoport E. H. and C. R. Marino (1978) The geographical distribution of insect pests. *Journal of Biogeography*, 5, pp. 149–157.

FAO (2020) Nigeria: FAO in Emergencies (helping to build a world without hunger). `http://www.fao.org/emergencies/countries/detail/en/c/213439`

Federal Ministry of Health, Abuja (National Bureau of Statistics, *Social Statistics in Nigeria 2012–Part III (health, employment, public safety, population & vital registration)*.

Global Burden of Disease Compare (2018) Institute for Health Metrics and Evaluation's (IHME's) Global Burden of Disease. `http://www.healthdata.org/nigeria`

Golding F. D. (1946) *The Insect Pests of Nigerian Crops and Stock.* Lagos: Agricultural Department of Nigeria, Special Bulletin No. 4.

Hall J. B., Kumar R. and Enti A. A., (1972) The obnoxious weed *Eupatorium odoratum* (Compositae) in Ghana. *Ghana Journal of Agricultural Science*, Vol. 5, pp. 75–78.

Hobbs, P. R., Sayre, K., & Gupta, R. ((2008)). The role of conservation agriculture in sustainable agriculture Philosophical *Transactions of the Royal Society* B: Biological Sciences, 363(1491), 543–555. (Quoted by Ayodele & Aluko (2018)).

Ishag H. M. (1971) Weed control in irrigated groundnuts (*Arachis hypogea* L.) *Journal of Agricultural Science*, Vol. 77 pp. 27–42.

Jeje L. K. (1977) A simple comparison of soil nutrient status on cultivated hillsides in parts of Ekiti Division. **In** Adejuyigbe O. and Helleiner F. M. (eds.) *Environmental and Spatial Factors in Rural Development in Nigeria* Nigerian Geographical Association. Section II, pp. 72–79.

Kingsley P. (2015) Inscrutable medicines and marginal markets: tackling substandard veterinary drugs in Nigeria Pastoralism 5, 2 `https://doi.org/10.1186/s13570-014-002-6`. [Accessed July2 2020]

Lana A. O. and Adeniji M. O. (1976) *Diseases of food crops in Nigeria, their causes, spread and methods of control*. Synopsis of Lectures delivered at *Operation Feed the Nation Training Programme* (OFNTP), Department of Agricultural Biology, University of Ibadan (Unpublished).

Moloo S. K. (1993) A comparison of susceptibility of two allopatric populations *of Glossina pallidipes* for stocks of *Trypanosoma congolense*. *Medical and Veterinary Entomology*, Volume 7 Issue 4. `https://doi.org/10.1111/j.1365-2915.1993.tb00707.x`

Ndubuaku T. C.N. and E. U. Asogwa (2006) Strategies for the control of pests and diseases for sustainable cocoa production in Nigeria. *African Scientist*, 7:209–216.

Obigbesan G. O. (1976) *Root and tuber crops.* Lectures delivered at Operation Feed the Nation Training Programme (OFNTP), Department of Agricultural Biology,

University of Ibadan (Unpublished).

Odukwe D. O. (1965) *Eupatorium odoratum* (Siam weed). *Bulletin of the Nigerian Forest Department*, April 1964.

Oerke E. C., Dehne H. W., Schonbeck F. and Weber A. ((1994)). *Crop production and crop protection: Estimated losses in major food and cash crops*. Elsevier p. 808.

Okaisabor E. K. (1965) Preliminary studies on the epidemiology of *Phytophthora palmivora* I: Outbreak of black pod disease on cocoa. *Nigerian Agricultural Journal*, Vol. 2)2), pp. 67–70.

Onyekwelu K., Ejezie F., Eze A., Ikekpeazu J., Isaac C., Ibegbu D., and P. Ogbunude (2017) Molecular identification of trypanosomes in tsetse flies trapped from Onicha Ugbo in Delta State, Nigeria. *Biomedical Research* 28 5463–5467.

Onyekwelu K., Ejezie F. E., Eze A. A., Ikekpeazu J. F., Isaac C., Ezeh R. C. and G. C. Edeh (2017) Prevalence of trypanosome infection in tsetse flies from Oji River and Emene axis of Enugu State, Nigeria: a preliminary report. *Tropical Parasitology* 7:98–102.

Onyiah, J. A. (1995) *Tsetse Distribution in 30 States and the Federal Capital Territory*. N.I.T.R. Publication Series.

Rachie K. O. and Roberts L. M. (1974) Grain legumes of the lowland tropics. *Advances in Agronomy*, Vol. 26 pp. 1–132.

Sence Agric (2020) in `http://www.agriculturenigeria.com/production/pest-diseases/pesticides-part-1`

Sheldrick R. D. (1968) The control of Siam weed, *Eupatorium odoratatum* (Lian.) *Journal of the Nigerian Institute for Oil Palm Research*, Vol. 17, pp. 7–19.

Sutherland J. R.G. (1952) Survey of stem borers of graminaceous crops. *Annual Report on the Department of Agriculture*, Part II 1952–53. Lagos: Federal Government Printer.

Swarbrick J. T. & B. L. Mercado (1987) *Weed science and weed control in Southeast Asia*. FAO Plant Production Paper 81. FAO Rome.

Taylor T. A. and Ezedinma F. O.C. (1964) Preliminary investigation on field pests of cowpea and methods of control. *Nigerian Agricultural Journal* Vol. 1, pp. 8–11.

Tobih F. O., Adegbite A. A. and C. C. Ononuju. (2011) Evaluation of some plant materials as organic mulch for the control of root-knot nematode (Meloidogyne spp.) and its impact on growth and yield of *Celosia argentea L. International Journal of Agri Science* Vol. 1(4), 195–202 September, 2011. ISBN 2228-6322.

Tomori O. (2019). Why Nigeria is battling to control disease outbreaks like Lassafever. The Conversation. February 21 2019 4.17pm SAST. `https:`

//theconversation.com/why-nigeria-is-battling-to-control-disease-outbreaks-like-lassa-fever-111771 [Accessed July 3 2020]

Toye S. A. (1969) A severe damage to a banana plantation by *Zonocerus variegatus* (L). *Nigerian Entomological Magazine*, 1 pp. 125–128.

Toye S. A. (1974) Feeding and locomotive activities of *Zonocerus variegatus* (L) Orthoptera; Acridoidea. RZA, Vol. 88(1), pp. 205–212

Weber J. S., Ngomtcho S. C.H., Shaida S. S., Chechet G. D., Gbem T. T., Nok J. A., Mamman M., Achukwi D. M. and S. Kelm, (2019) Genetic diversity of trypanosome species in tsetse flies (*Glossina spp.*) in Nigeria. *Parasites & Vectors*, 12:481 pp. 1–12. https://doi.org/10.1186/s13071-019-3718-y

Wei Zhang, Edward Kato, Felix Bianchi, Prapti Bhandary, Gerrit Gort, Wopke van der Werf (2018) Farmers' perceptions of crop pest severity in Nigeria are associated with landscape, agronomic and socio-economic factors *Agriculture, Ecosystems and Environment*, 259 (2018) 159–167.

Yahaya A. T., Baba K. M., Kyiogwom U. B., and A. Singh (2008) Effects of weeds on profitability of millet (*Pennisetum glaucum*) and groundnut (*Arachis hypogaea*) crops in mixed cropping systems. *Nigerian Journal of Basic and Applied Sciences*, vol. 16 no. 2 December 2008 pp. 213–220.

Youdeowei A. and M. W. Service (1993) Pest and Vector Management in the Tropics. Lagos: Longman Nigeria.

CHAPTER 8

Ademola J. A. and Farai I. P. (2006) Gamma activity and radiation dose in concrete building blocks used for construction in Jos, Nigeria. *Radiation Protection Dosimetry*, Volume 121, Issue 4, pp. 395–398.
https://doi.org/10.1093/rpd/ncl052

Aigbedion, I. N. (2005) Environmental pollution in the Niger delta, Nigeria. *Interdisciplinary Journal of Enugu Nigeria*, 3(4), 205–210.

Alakpodia I. J. (1989) *The Effects of Gas Flaring on the Micro-climate and adjacent Vegetation in Isoko Area of Bendel State*. Unpublished MSc. Thesis, University of Ibadan.

Areola, O. & A. Faniran (1975) Ecology and economics in the Lagos state. *Nigerian Journal of Social and Economic Studies*, 17(1), 63–76.

Bridge G. and T. Fredriksen (2012) Order out of Chaos: resources, hazards and the production of a tin-mining economy in northern Nigeria in the early twentieth century *Environment and History*, Vol. 18, No. 3 (August 2012), pp. 367–394.
https://www.jstor.org/stable/23250941

Daily Times Nig. Ltd. (1978) *Oil Spillage*. Tuesday, 25 July. p .3.

Emoyan O. O., Akpoborie I. A., and E. E. Akporhonor (2008) The oil and gas industry and the Niger Delta: implications for the environment. *Journal of Applied Science and Environmental Management*, September 2008 Volume 12(3), 29–37. JASEM ISN 1119-8362 www.bioline.org.br/ja

Faniran A. (1975) Rural water supply in Nigeria's Basement Complex: a study in alternatives. *Proceedings of the Second World Congress of the International Water Resources Association*, New Delhi, III, pp. 89–100.

Faniran A. (1972) River basins as planning units. **In** Barbour K. M. (ed.) *Planning for Nigeria: a geographical approach*. Ibadan: Ibadan University Press.

Faniran A. (1974) The extent, profile and significance of deep weathering in Nigeria. *Journal of Tropical Geography*, Vol. 38, pp. 19–30.

Federal Environmental Protection Agency (1991) *National Policy on the Environment*. The Presidency, Abuja 43p.

Frederiksen, T, Bridge, G & Fredriksen, T (2012), "Order out of chaos": Resources, hazards and the production of a tin-mining economy in northern Nigeria in the early twentieth century, *Environment and History*, vol. 18, no. 3, pp. 367–394. https://doi.org/10.3197/096734012X13400389809337

Grove A. T. (1952) Land use and soil conservation on the Jos Plateau. *Bulletin of the Geological Survey of Nigeria*, 22.

Hodder, B. W. (1959) Tin mining on the Jos Plateau of Nigeria. *Economic Geography*, 35, pp.109–122.

Imevbore V. (2018) Spatial Analysis of Biodiversity Initiatives in the Niger Delta Region Quoted in Imevbore V. O & M. M. Zagi (2018).

Imevbore V. O & M. M. Zagi (2018): Environmental stewardship in Nigerian oil and gas industry–myth or reality. *Paper presented at Biennial Oil and Gas, Health, Safety and Environment Conference.*

Inyang P. B.E. (1973) Pollution of the most valuable environmental resources (land, water and air) in parts of the eastern states of Nigeria. Paper presented at the *Conference on Environmental Resource Management in Nigeria,* University of Ife, 9–14 July 1973 (Unpublished)

IUCN Niger Delta Panel (2018) Developing a Biodiversity Conservation Strategy for the Niger Delta: integrating biodiversity considerations into SPDC's land operations. SPDC 2018. Viii+36pp
https://doi.org/10.2305/IUCN. CH.2018.07.en

Jibiri N. N., Farai I. P. and S. K. Alausa (2006) Estimation of annual dose due to natural radioactive elements in ingestion of foodstuffs in tin mining area of Jos-Plateau, Nigeria. *Journal of Environmental Radioactivity*, 94(1), 31–40. DOI: 10.1016/j.jenvrad.2006.12.011 PMID: 17337103

Lar, Uriah. ((2018)). *Geology and Mineral Resources of Nigeria and their Uses.* Paper presented at a 3-day Capacity Building Workshop by Telpon Environment Ltd for the Nigeria Commodity Exchange (NCX) Staff, Abuja on "Product development for solid minerals in Nigeria." Tuesday 16th–Thursday 18th January 2018 at Elim Top Suites & Hotels, Rayfield, Jos. `https://www.researchgate.net/publication/3229631 GEOLOGY AND MINERAL`

Mortimore M. J. (1967) Land use and population pressure in Kano close-settled zone, Northern Nigeria. *Advancement of Science*, Vol. 23, pp. 677–686.

National Bureau of Statistics: *2009 Annual Collaborative Survey of Socio-Economic Activities in Nigeria.* Main Statistical Report Volume 1. August 2010.

NDES (2005) *Niger Delta Environmental Survey Phase I Report*, SPDC.

Odoh C. K., Kalu A. U. and Anyah F. (2017) Environmental impacts of mineral exploration and exploitation in Nigeria and their phytoremediation strategies for sustainable ecosystem. *Global Journal of Science Frontier Research: H Environment & Earth*, Volume 17 Issue 3 Version 1.0. Online ISSN: 2249-4626 & Print ISSN: 0975-5896

Odoh S. I. and Chilaka F. C. (2012) Climate change and conflict in Nigeria: a theoretical and empirical examination of the worsening incidence of conflict between Fulani herdsmen and farmers in Northern Nigeria. *Arabian Journal of Business and Management Review*, (OMAN Chapter) Vol. 2 No. 1, pp. 110–124.

Odu C. T.I. (1972) Microbiology of soils contaminated with petroleum hydrocarbons I: Extent of contamination and some soil and microbiological properties after contamination. *Journal of the Institute of Petroleum*, Vol. 58, pp. 201–208.

Odu C. T.I., 1977a Oil pollution and the environment. *Paper presented at SCOPE Workshop on Current and Potential Environmental Problems in Nigeria,* University of Ibadan, 25 March 1977.

Odu C. T.I., 1977b Microbiology of soils contaminated with petroleum hydrocarbons. II: Natural rehabilitation and reclamation of soils affected. *Journal of the Institute of Petroleum*, Vol. 63.

Odularu G. O. (2007) Crude Oil and Nigeria Economic Performance. `http://www.ogbus.ru/eng/authors/odularo/odularo 1.pdf`

Olade M. A. (2019) Solid mineral deposits and mining in Nigeria—a sector in transitional change *Achievers Journal of Scientific Research*, Volume 2, Issue 1, July 2019, p. 1–15. `http://www.achieversjournal.org/`

Olobaniyi S. B. and Efe S. J. (2007) comparative assessment of rainwater and groundwater quality in an oil producing are of Nigeria: environmental and health implications. *Journal of Environmental Health and Research*, Volume 6(2), 111–118.

Oyawoye M. O. (1970) The Basement Complex of Nigeria **In** Dessauvagie T. F.J. and Whiteman A. J. (eds.) *African Geology*, pp. 67–104. Ibadan: Ibadan University Press . Resources of Nigeria and Their Uses. Accessed July 4, 2020.

Tahal Consultants (Nig.) Ltd. (1974) *Ikeja Irrigation and Drainage Project Reconnaissance Survey*. Lagos: Lagos State Government.

Udo, R. K. (1970) *Geographical Regions of Nigeria.* London: Heinemann Educational Books.

UNDP (2012) *Niger Delta Biodiversity Project* UNDP Project Document. UNDP GEF PIMS no.:2047 / GEFSEC Project ID: 4090 172p. Niger Delta Biodiversity Prodoc.pdf

CHAPTER 9

Aderoju O. M., Okeke O. J., Nwadike B. K., Eta J., Onuoha H. U., Ihenacho N. M., and R. I. Eshiet (2014) Impacts of Urban Land use changes on flood events in Warri, Delta State Nigeria. *IOSR Journal of Environmental Science, Toxicology and Food Technology*. (IOSR-JESTFT) e-ISSN: 2319-2402, p-ISSN: 2319-2399. Volume 11, Issue 3, Ver. II (Mar 2017), pp. 01–07. `www.iosrjoournals.org` DOI: 10.9790/2502-1103020107

Akintola, F. O. (1974) *Parameters of Infiltration Equations of Urban Landuse Surfaces.* Unpublished PhD. Thesis, University of Ibadan.

Areola, Abiodun A. (2021) *Greening and the Social Ecology of Ibadan Metropolis, Nigeria* Unpublished PhD. Thesis, University of Ibadan.

Areola, O. (1976) Recreational Land use in Nigeria. *Nigerian Geographical Journal, 19(2), 227–236.*

Awosika L. F., French G. T.., Nicholls R. J. and C. E. Ibe (1992) The impacts of sea level rise on the coastline of Nigeria. In *Proceedings of IPCC Symposium on the Rising Challenges of the Sea. Margaritta, Venezuela 14–19, March 1992.*

Ayila A. E., Fabiyi O. O. and Y. A. Bello (2014) Statistical analysis of urban growth in Kano metropolis, Nigeria. *International Journal of Environmental Monitoring and Analysis*, Vol. 2 No. 1, pp.50–56. DOI: 10.11648/j.ijema.20140201.16

Bakare Wale (2019) Solid Waste Management in Nigeria *Bioenergy Consult*, April 4 2019. `https://ww.bioenergyconsult.com/solid-waste-nigeria/`

Bakoji Y. M., Elijah E., Philip A. H., Isa M. S. and U. J. Abba (2020) Urbanization and its impact on agricultural lands in growing cities in developing countries: a case study of Makurdi, Benue State, Nigeria *International Journal of Economy, Energy and Environment*, 5(4). 41–46. `htpp://www.sciencepublishinggroup.com/j/ijeee` `doi: 10.11648/j.ijeee.20200504.11` ISSN: 2575-5013 (Print): 2575-502 (Online)

Dan-Jumbo N. G., Metzger M. J. and A. P. Clark (2018) Urban land use dynamics in the Niger delta: the case of Greater Port Harcourt watershed *Urban Science*, 2, 108 pp. 1–24. doi:10.3390/urbansci2040108 www.mdpi.com/journal/urbansci

Dankani I. M. and Abubakar S. D. (2011) Uncontrolled urban growth around Dorayi area of Kano metropolis: the planning and infrastructural development implications. *Ife Research Publications in Geography*, Vol 10 Number 1, pp. 36–44. ISSN-0795-073X

Del Cerro Santamaria G. (2020) Building new knowledge about natures, ecologies and sustainability: a review article. *Urban Studies*, 00(0) 1–7. Urban studies review article.pdf. ORCID https://orcid.org/0000-0002-2791-3753

Elmqvist T., Bai X., Frantzeskaki N. *at al.* (eds) (2018) *Urban Planet: Knowledge toward Sustainable Cities.* New York: Cambridge University Press, 514 pp. ISBN 9781 107196933

Ernston H. and Sörlin S. (eds) (2019) *Grounding Urban Natures: Histories and Futures of Urban Ecologies.* Cambridge, MA: The MIT Press. 440pp.: ISBN 9780262537.

Fabiyi O. and Joseph O. (2018) Comparative analysis of spatial extent and population sizes of cities in Nigeria: implications for urban space administration. *Journal of Historical Archaeology and Anthropological Sciences*, 3(1): 87–95. DOI: 10.15406/jhaas.2018.03.00070 http://www.medcraveonline.com

Filani M. O. & S. Okafor (eds.) (2006) *Foundations for Urban Development in Africa: The Legacy of Akin Mabogunje.* The Cities Alliance 2006.
1818 H Street NW, Washington, DC 20433, USA. www.citiesalliance.org

Guttenberg A. Z. (1975) The urban system. **In** Levine N. D. (ed.) *Human Ecology.* North Scituate, Massachusetts: Duxbury Press, pp. 296–324.

Idowu A. A., Ayoola S., Opele A. I. and B. N. Ikenweiwe (2011) Impact of climate change in Nigeria. *Iranica Journal of Energy and Environment*, 2(2), 145–152.

Ike C. C., Ezeibe C. C., Anijiofor S. C. and N. N. Nik Daud, (2018) Solid waste management in Nigeria: problems, prospects and policies. *Journal of Solid Waste Technology and Management*, Volume 44, p. 163-172.
https://doi.org/10.5276/jswtm.2018.163

Imam, A., B. Mohammed, D. C. Wilson and R. Cheeseman, (2008). Solid waste management in Abuja, Nigeria. *Waste Management*, 28, 468–472.
http://dx.doi.org/10.1016/j.wasman.2007.01.006

Lasisi M., Popoola A., Adediji A., Adedeji O. and K. Babalola (2017) City expansion and agricultural land loss within the peri-urban area f Osun State, Nigeria. *Ghana Journal of Geography*, Volume 9(3), pp. 132–163.
https://www.researchgate.net/publication/323703694

Mabogunje A. L. (1968) *Urbanization in Nigeria*. London: University of London Press.

Maclaren International and Urwick, Currie and Partners Ltd. (1971) *Master Plans for Waste Disposal and Drainage, Ibadan.*, Vol. IV: Drainage and Solid Waste.

Mostafavi M. and Doherty G. (eds) (2017) *Ecological Urbanism. Revised Edition.* Zurich, Switzerland: Lars Muller Publishers. 656 pp. ISBN 9783037784679

Nigeria Census Board: Priority Tables Volume 1 Update, 2006 Census.

Odemerho F. (2014) Building climate change resilience through bottom-up adaptation to flood risk in Warri, Nigeria *Environment & Urbanization*, Vol 27(1): 139–160. DOI: 10.1177/0956247814558194 International Institute for Environment and Development (IIED). 139 `www.sagepublications.com`

Odemerho, F. O. (1976) *Perception of Environmental Hazards and Human Behaviour of Warri Residents*. BA Dissertation, Department of Geography, University of Ibadan.

Oguntoyinbo, J. S., Areola O., & M. Filani (eds) (1983) *A Geography of Nigerian Development* (2nd Edition). Ibadan: Heinemann Educational Books Plc.

Olanrewaju R., Ekiotuasinghan B. and G. Akpan (2017) Analysis of rainfall pattern and flood incidences in Warri metropolis, Nigeria. *Geography, Environment, Sustainability*, Vol. 10 No.4 pp.83–97. DOI: 10.24057/2071-9388-2017-10-4-83-97

Satherthwaite D., McGranahan, G. and C. Tacoli (2010) Urbanization and its implications for food and farming. *Philosophical Transactions of the Royal Society*, B: Biol Sci. 2010 Sep 27; 365(1554): 2809–2820. `doi: 10.1098/rstb.2010.0136`

Tokula A. E. and Ejaro S. P. (2017) The impact of urban expansion on agricultural land and crop output in Ankpa, Kogi State, Nigeria. *Landscape Architecture*, Vol. 1 pp. 1–13. `www.itspoa.com/journal/la`

CHAPTER 10

Adalemo I. A. (1978) Land management in Nigeria–the promise of the land use decree 1978. **In** Oguntoyinbo J. S., Filani M. O. and O. Areola (eds.) *Resources and Development in Africa III*, pp. 407–411.

Adebowale Nike (2019). Why Nigeria needs to pay more attention to environmental management–WHO. *Premium Times*, September 10 2019. `https://www.premiumtimesng.com/news/top-news/351580-why-nigeria-needs-to-pay-more-attention-to-environmental-management-who.html` [accessed October 27 2020]

Adegboye R. O. (1966) An analysis of land tenure structures in some selected areas in Nigeria. *Bulletin of Rural Economy and Society (Ibadan)*, 2, pp. 38–49.

Adejuwon J. O. (1968) *Savanna Patches in the Forest Areas of Western Nigeria; their Place in the Dynamics of Nigerian Vegetation.* Unpublished PhD Thesis. University of London.

Adejuwon J. O. (1971) A biogeographical survey of the dynamics of savanna vegetation in Nigeria. *Nigerian Geographical Journal*, 14, pp. 31–48.

Adekola P. O. and Ogundipe A. O. (2017) An assessment of the state of environmental management in Nigeria capital cities. *Journal of Geography, Environment and Earth Science International*, Vol. 12(2), pp. 1–13. Article No. JGEESI.34329 ISSN: 2454-7352 DOI: 10.9734/JGEEESI/2017/34329

Adeniyi P. O. (2013) *Spatial Disorganization and the Imperative of a National Geospatial Emergency for Development in Nigeria.* Lecture Presented at the College Lecture, College of Environmental Sciences, Bells University of Technology, Ota, Nigeria on April 30 2013.

Areola, Olusegun (2012) Geospatial Technologies for Resource Development in Nigeria. Guest Lecture *Annual Conference on Geospatial Technologies and Digital Cartography for National Security, Tourism and Disaster Management*, November 21–22, 2012. Geoinformation Society of Nigeria and Nigeria Cartographic Association in technical collaboration with the Regional Centre for Training in Aerospace Surveys (RECTAS), Ile-Ife.

Ayoade J. O. (1971) *Rainfall, Evapotranspiration and the Water Balance in Nigeria.* Unpublished PhD. Thesis, University of London.

Ayoade J. O. (1976) Evaporation and evapotranspiration in Nigeria. *Journal of Tropical Geography*, Vol. 43, pp. 9–19.

Bain A. D.N., Newton R. B. and A. C. Seward (1924) Nigerian Coalfield Section 1: Enugu Area. *Bulletin of the Geological Survey of Nigeria*, 6; 11.

Charter J. R. and Keay R. W.J. (1960) Assessment of Olokemeji fire control experiment (investigation 254) 28 years after institution. *Nigerian Forestry Bulletin*, No. 3.

Davies J. A. (1966) the assessment of evapotranspiration for Nigeria. *Geografiska Annaler*, Vol. 48a, pp. 139–156.

de Leeuw P. N. (1976) Fodder resources and livestock development in north-east Nigeria. *Savanna*, Vol. 5(1), pp. 61 74.

Dunston W. R. (1910) *Southern Nigeria: Report of the Mineral Survey 1905–6.* Miscellaneous Colonial Reports, 67.

Ebele N. E. and N. V. Emodi (2016) Climate Change and Its Impact in Nigerian Economy. *Journal of Scientific Research & Reports*, 10(6): 1–13 2016; Article no. JSRR.25162 ISSN: 2320-0227 SCIENCEDOMAIN international `www.sciencedomain.org`

Ejimudo K. B.O. and Nwador F. O. (2014) Environmental management and sustainable development in Nigeria's Niger delta. *Journal of Economics and Sustainable Development*, Vol. 5 No. 15 pp. 26–40. ISSN: 2222-1700 Paper ISSN: 2222-2855 (Online) http://www.iiste.org

Fair Observer (2020) Nigeria and Efforts at Environmental Management. https://www.fairobserver.com/region/africa/nigeria-and-efforts-at-environmental-management-63923/amp/ [accessed October 26 2020]

Falconer J. D. (1926) The Southern Plateau tinfields and the Sura volcanic line. *Bulletin of the Geological Survey of Nigeria*, 9.

Hagmann J., Chuma E., Murwira K., Connolly M. and P. Ficarelli (2002) Success factors in integrated natural resource management R & D–lessons from practice. *Conservation Ecology*, 5 (2): 29.
http://consecol.org/vol5/issu2/art29

Hopkins B. (1965) Observations on savanna burning in the Olokemeji forest reserve, Nigeria. *Journal of Applied Ecology*, 2 pp. 367–381.

Ikpe E. I., Sawa B. A and L. U. Ejeh (2017) Evidence of Climate Change and Adaptation Strategies among Grain Farmers in Sokoto State, Nigeria. *IOSR Journal of Environmental Science, Toxicology and Food Technology* (IOSR-JESTFT) e-ISSN: 2319-2402, p-ISSN: 2319-2399. Volume 11, Issue 3 Ver. II (Mar. 2017), pp 01–07. www.iosrjournals.org doi: 10.9790/2502-1103020107

Inakwu O. A. Odeh: The Legacy of Soil Mapping in Nigeria: from analogue to digital soil mapping. *Prepared for DES Seminar Series June 29 2012, Department of Environmental Science, Faculty of Agriculture and Environment, University of Sydney.*

Jacobson R. R.E. (1947) *The Younger Granite Complex of the Liruei Hills, Nigeria.* Unpublished PhD. Thesis, University of London.

King R. (1977) *Land Reform: A World Survey* London: Bell.

Nkwunonwo U. (2020) Geospatial technology for land resources management in Nigeria. In *Spatial Information Science for Natural Resource Management.* Chapter 4, pp. 62–87. IGI Global. doi:10.4018/978-1-7998-5027-4.ch004

Oguntoyinbo J. S. (1974) Land use and reflection coefficient (albedo) map for southern parts of Nigeria. *Agricultural Meteorology*, Vol. 13, pp.227–237.

Oluwasanmi H. A. (1966) *Agriculture and Nigerian Economic Development.* Ibadan: Oxford University Press.

Parkinson J. (1907) The structure of southern Nigeria. *Geographical Journal*, Vol. 24, pp. 56–64.

Raeburn C., Bain A. D.N. and W. Russ (1927) The tinfields of Zaria and Kano Provinces.

Raeburn C., Bain A. D.N. and W. Russ (1927) Tinstone in Calabar District. *Bulletin of the Geological Survey of Nigeria*, 11.

Richards PW. (1957) Ecological notes on West African vegetation I: The plant communities of the Idanre hills, Nigeria. *Journal of Ecology*, Vol. 45, pp. 563–577.

Ross R., 195 Ecological studies on the rain forest of southern Nigeria III: Secondary succession in the Shasha forest reserve. *Journal of Ecology*, Vol.w2 pp. 259–282.

Tuley P. and Alford M. T. (1975) *The Land Resources of Central Nigeria Phase I: Interim Report on the Landforms, Soils and Vegetation of the Jema'a Platform, Vol. 2: The Climate and Vegetation*. Surbiton, Surrey, England: Land Resource Division.

Udo R. K. (1977) *Land tenure, land reform and post-military Nigeria*. Unpublished Seminar Paper, Department of Geography, University of Ibadan.

Wilson R. C (1925) The geology of the eastern railway. *Bulletin of the Geological Survey of Nigeria*, 8.

Wilson R. C. and Bain A. D.N. (1922) The geology of the western railway, Section1: Iddo to Okuku. *Bulletin of the Geological Survey of Nigeria*, 2.

Yari, K. M. (2008) *The Sustainable Cities Nigeria Programme (1994–2006): Building platforms for environmentally sustainable urbanization*. SCP Documentation Series, Volume 7. UN-Habitat and UNEP. ISBN: 92-1-131743-6. ISBN: 978-92-1-132018-3 hs/1040/08

Index

B

D

E

F

G

H

I

J

K

L

M

N

O

P

Q

R

S

T

U

V

W

Y

Z